Lecture Notes in Mobility

Series Editor

Gereon Meyer, VDI/VDE Innovation und Technik GmbH, Berlin, Germany

More information about this series at http://www.springer.com/series/11573

Waressara Weerawat · Phumin Kirawanich ·
Anna Fraszczyk · Marin Marinov
Editors

Urban Rail Transit

Proceedings of the 6th Thailand Rail Academic Symposium

Springer

Editors
Waressara Weerawat
Faculty of Engineering
Mahidol University
Bangkok, Thailand

Phumin Kirawanich
Faculty of Engineering
Mahidol University
Bangkok, Thailand

Anna Fraszczyk
Legnica, Poland

Marin Marinov
Engineering Systems and Supply Chain Management (ESSCM)
Aston University
Birmingham, UK

ISSN 2196-5544 ISSN 2196-5552 (electronic)
Lecture Notes in Mobility
ISBN 978-981-15-5978-5 ISBN 978-981-15-5979-2 (eBook)
https://doi.org/10.1007/978-981-15-5979-2

This Springer imprint is published by the registered company Springer Nature Singapore Pte Ltd.
The registered company address is: 152 Beach Road, #21-01/04 Gateway East, Singapore 189721, Singapore

The Editors would like to dedicate this publication to all academia-industry partnerships in the field of urban rail transit.

Preface

The volume of proceedings of the 6th Thailand Rail Academic Symposium—TRAS2019—contains papers, selected on the basis of blind peer review, which were presented at the symposium from November 21 to 22, 2019. The 2019 symposium emphasized "urban rail transit" to increase understanding of the integration of rail technology in the urban context. It was not limited to the rail engineering only, but also to broaden the knowledge of the effects of rail development on the urban ecosystem, which is illustrated through passenger services within and around urban areas.

Railway development has a long history with Thailand's urbanization. Rail transport development in Thailand had started since Royal State Railway of Siam was formed in 1890, and then construction began throughout Thailand in order to be a backbone of the country's major transportation service. Not only was a rail transportation system provided, but also urbanization came along. This implied the start of the modern age of urban development in Thailand as well. The national rail network covered 4431 km, and many extensions and new rail services are still to come in the near future. This will be a renaissance of Thailand's rail transport and urbanization. Therefore, it requires several studies on rail engineering and urban development so that the country is ready for future development.

This book presents research articles that provide a knowledge of railway infrastructure development, which follows on from previous Thailand Rail Academic Symposiums to promote technical exchanges, experiences, and discussions on urban rail transit by concentrating on two main elements: innovation in rail education and industry, and rail transport policy and urbanization.

Urban rail transit includes all types of rail transport that serve local passengers in order to connect between places in both urban and suburban areas. These systems include rapid transit (i.e., subway and metro), tram, light rail, monorail, commuter rail, and cable car. Urban rail transit also plays a crucial role in the urban development around its service lines and networks. It has an enormous impact on the change of demographic structure, urban form, and employment. This year's symposium, therefore, addressed topics that were related to urban rail transit, which include urban rail service operational and system management, passenger behavior, urban socioeconomics, as well as transit-oriented development. Both domestic and international case studies are provided in the proceedings to overview a variety of research approaches as well as to follow up on specific problems that scholars have been working on recently.

The first six chapters focus on the soft side of urban rail transit, in which examples of transport policies, passenger behavior, key performance indicators, and socioeconomic impacts of a new system are based on case studies from Jakarta (Indonesia), Bangkok (Thailand), and the West Midlands (UK). Chapter "Rail Careers in ASEAN: Employers Search for Talents, Skills and Knowledge" looks at rail careers in ASEAN from a Thai perspective and highlights issues that need to be addressed in rail education and training cycles, especially in relation to soft skills. Chapters "The Development of a Railway Track Inspection System for Track Maintenance"–"Measuring the Changes of Subway Accessibility Through the Service Area Territories: A Case Study of Fukuoka Subway Network" look at a railway track inspection system as well as accessibility of urban rail stations from various perspectives, with examples from Thailand and Japan. The following chapters, from Chapters "Train Approach Information Platform and Service System" to "Design of Optimal Train Speed Profile for PMSM Railway Traction System Using Dynamic Programming with MTPA Control Method," focus their attention on metro operations, such as train approaching information or train rescheduling, with case studies from Thailand, Japan, and the UK.

The editors and the Organizing Committee express their gratitude to both local and international keynote speakers who were invited to share their knowledge and experience on railway infrastructure development in their areas of interest and case studies from Thailand and beyond. This aims to create international collaboration in railway research and drive the relevant innovation and technology in order to achieve sustainable rail system development. We would also like to acknowledge all TRAS participants for their research contributions to railway technical development. This book is a useful source for international academics and practitioners seeking pioneering research in the field.

We wish to express our gratitude toward the committee members of the 6th Thailand Rail Academic Symposium—TRAS2019.

Bangkok, Thailand
September 2020

Dr. Phumin Kirawanich
Dr. Waressara Weerawat
Dr. Anna Fraszczyk
Dr. Marin Marinov

Acknowledgements The editors would like to thank all the authors for their original contributions and good cooperation during the publishing process as well as all reviewers for their kind contributions. In addition, special thanks go to the MetroExchange team of academics and professionals from the UK and Thailand, who initiated this publication as well as to the funders of the MetroExchange project, Newton Fund (UK), and the Office of National Higher Education Science Research and Innovation Policy Council (Thailand).

Contents

Contributors

John Andrews Resilience Engineering Research Group, Faculty of Engineering, University of Nottingham, Nottingham, UK

Jatesada Borsub Department of Science Service, Innovative Robotics and Automation Group, Engineering Materials Division, Ratchatewi, Bangkok, Thailand

Ahmad Faisal Dahlan Faculty of Engineering, Mahidol University, Nakhon Pathom, Thailand

Anna Fraszczyk Faculty of Engineering, Mahidol University, Salaya, Nakhon Pathom, Thailand

Andrea Gualtieri Aston University, Birmingham, UK

Pawinee Iamtrakul Faculty of Architecture and Planning, Center of Excellence in Urban Mobility Research and Innovation, Thammasat University, Pathumthani, Thailand

Thanchanok Inmor Asian Institute of Technology, Klong Luang, Thailand

Takuro Inohae Department of Civil Engineering and Architecture, Saga University, Saga, Japan

Piyapong Jiwattanakulpaisarn ThaiRoads Foundation, Jomphol, Bangkok, Thailand

Kunnawee Kanitpong Asian Institute of Technology, Klong Luang, Thailand

Naphat Ketphat University of Leeds, Leeds, UK

Phumin Kirawanich Faculty of Engineering, Mahidol University, Salaya, Nakhon Pathom, Thailand

Klara Dwi Kristianingtyas Sophia University, Tokyo, Japan

Jirapan Liangrokapart Faculty of Engineering, Mahidol University, Nakornprathom, Thailand

Ronghui Liu Institute for Transport Studies, University of Leeds, Leeds, UK

Marin Marinov Aston University, Birmingham, UK

Ratthaphong Meesit Department of Civil Engineering, Faculty of Engineering, Burapha University, Mueang Chonburi, Thailand;
Resilience Engineering Research Group, Faculty of Engineering, University of Nottingham, Nottingham, UK

Nobuo Mishima Department of Civil Engineering and Architecture, Saga University, Saga, Japan

Masafumi Miyatake Sophia University, Tokyo, Japan

Lalita Narupiyakul Faculty of Engineering, Integrative Computational BioScience Center, Mahidol University, Phutthamonthon, Nakhon Pathom, Thailand

Apinya Padon Faculty of Architecture and Planning, Center of Excellence in Urban Mobility Research and Innovation, Thammasat University, Pathumthani, Thailand

Janene Piip JP Research and Consulting, Port Lincoln, SA, Australia

Sompat Praeknokkeaw Department of Science Service, Innovative Robotics and Automation Group, Engineering Materials Division, Ratchatewi, Bangkok, Thailand

Korntham Sathirakul Department of Science Service, Innovative Robotics and Automation Group, Engineering Materials Division, Ratchatewi, Bangkok, Thailand

Somsiri Siewwuttanagul Faculty of Engineering, Mahidol University, Salaya, Thailand;
Graduate School of Science and Engineering, Saga University, Saga, Japan

Somkane Srathongkhruen Faculty of Engineering, Mahidol University, Salaya, Nakhon Pathom, Thailand

Chaiyut Sumpavakup King Mongkut's University of Technology, North Bangkok, Bangkok, Thailand

Suradate Suratisak Department of Science Service, Innovative Robotics and Automation Group, Engineering Materials Division, Ratchatewi, Bangkok, Thailand

Somsak Vanit-Anunchai Suranaree University of Technology, Muang, Nakhon Ratchasima, Thailand

Waressara Weerawat Faculty of Engineering, Mahidol University, Phutthamonthon, Nakhon Pathom, Thailand;
Faculty of Engineering, Mahidol University, Salaya, Nakhon Pathom, Thailand

Anthony Whiteing Institute for Transport Studies, University of Leeds, Leeds, UK

Station Naming Strategies for a Metro System Expansion: A Case Study of Bangkok Metro Network

Anna Fraszczyk, Waressara Weerawat, and Phumin Kirawanich

Abstract In early 2019 Bangkok metro network included five lines operating on approx. 100 km of track in the city. The transport Master plan for the city sets an ambitious target of approx. 500 km of metro network by 2030. This will be achieved by expanding the existing and developing new metro lines in close collaboration between the governmental organisations and the private sector. However, in early stages of implementation of the expansion plan it was observed that metro lines operated by different companies use different strategies for station naming as well as station coding. This practice caused confusion as to which strategy to apply to the new lines. Therefore, this project aimed at looking into station naming strategies applied in other major cities in Asia and beyond in order to suggest practical solutions to all parties involved. In addition, an online survey was conducted in 2018, in which Bangkokians were asked to state their metro station names preferences and list problems they face with the existing names. Based on analysis of the literature review and survey data a set of tools was developed to facilitate a new strategy for station naming. These tools include: principles for metro station naming and a pyramid of station names categories. Also, potential name changes to four key metro interchanges are suggested. These findings are expected to be taken into account by the Office of Transport and Traffic Policy and Planning when facilitating expansion of the metro network in Bangkok in the future. Moreover, some of these tools could be used by other cities expanding or developing metro networks in Asia and beyond, but first they would have to be adjusted to a local context.

Keywords Metro · Station · Naming · Strategy · Policy

A. Fraszczyk · W. Weerawat (✉) · P. Kirawanich
Faculty of Engineering, Mahidol University, Salaya, Nakhon Pathom 73170, Thailand
e-mail: waressara.wee@mahidol.ac.th

W. Weerawat et al. (eds.), *Urban Rail Transit*, Lecture Notes in Mobility,
https://doi.org/10.1007/978-981-15-5979-2_1

1 Introduction

Currently there are over 160 metro systems in operation around the world and many new lines under development, especially in developing countries in Asia (UITP 2018). Each new system has at least one line and faces the task of naming its stations.

Historically, metro stations were usually named after streets, landmarks or districts, but there were also many exceptions and no clear rules to apply for name selection. For example, in the London Underground network Waterloo station was named after a battle (but the name is also used for a bridge and an area) and Angel station was named after a pub.

As there is no universal guidance on station naming available, each system applies their own station naming strategies. These rules are often based on policies regarding naming public places, such as streets or landmarks, which are influenced by a local culture and local language characteristics. However, these rules are rarely publicly available.

In order to promote transparent strategies for station naming this paper has two goals. Firstly, to investigate a selection of publicly available regulations regarding station naming applied in metro systems around the world in order to recommend common and clear guidelines for parties responsible for future station naming. Secondly, a case study of the Bangkok metro network is introduced. The existing metro station names are analysed and compared with a new set of data on people's preferences regarding station name characteristics. The analysis of results will provide quantitative evidence on the current station naming strategies versus metro users' preferences and help shaping the final recommendations addressed to metro stakeholders.

The paper is organised into nine sections, as follows. Section 2 looks into metro station naming policies, from principles to name changes to public places, and considers various decision makers involved in the process. Section 3 introduces a case study of the Bangkok metro network. Section 4 explains the methodology applied in the paper, the use of secondary data and collection of primary data via a questionnaire. Section 5 focuses on analysis of results and is divided into five sub-sections. Section 6 offers discussion, and highlights key findings presented in the paper. Section 7 presents overall conclusions of the paper and a list of specific recommendations addressed to decision makers. The final Sect. 8 lists limitations and avenues for further research in the area of metro station naming.

2 Metro Station Naming Policies

2.1 Principles

There is very little scientific literature available on station naming policies, with practically none regarding name categories, characteristics and translation strategies.

Table 1 Principles of station naming with examples

Principle	Explanation	London metro station example	Bangkok metro station example
Simple	Easy to remember, the shorter the better	Bank	Ari
Logical	Provide a mental link when trip planning; should be relevant to the area they reside	Hyde Park	Asok
Durable	Relevant as long as the station exists	Heathrow	Suvarnabhumi
Self-locating	Allow users to mentally locate themselves within the region; especially important for underground stations	Piccadilly Circus	Lumphini
Unique	Avoids confusion with any other name	Wimbledon	Bang Sue

Source Based on Metrolinx (2015a)

However, organisations involved in metro operations often develop their own tools that help them with the station naming process. Documents such as 'Station and Stop Naming Principles' developed by WMATA (2012) or a 'Naming Protocol' by Metrolinx (2017) provide a set of useful principles that help with managing growing transport networks and keep the station naming process under control. This is especially important when various stakeholders are involved in the process and where systems are not well integrated, e.g. different metro operators, lack of a common ticket. The key five principles proposed by Metrolinx (2015a) are for the names to be simple, logical, durable, self-locating and unique. Table 1 presents details of the principles when choosing station names with relevant examples of station names on London and Bangkok metro networks.

A simple and short station name is easier to remember (Metrolinx 2015a; WMATA 2012) and there are many examples of such names across different systems (e.g. Temple in London, 5th Avenue in New York City (NYC), Ari in Bangkok or Central in Singapore). A landmark, such as a university, a park or an attraction, is often seen as a logical choice for a station name, especially if it is located within a walking distance (WMATA 2012). Also, it is recommended that an institution's name could be added to a station name if it is the major reason metro passengers are using that stop. For example, St Paul's station in London is located nearby St Paul's Cathedral and '47–50 Streets Rockefeller Center' station in NYC is next to the Rockefeller Center building. However, it is recognised that priority should be given to established names embedded in local knowledge (Metrolinx 2017) and minor institutions or organisations' names should be avoided (WMATA 2012). Moreover, it is recommended for names to be consistently applied across all modes of transport (Metrolinx 2017), so that metro, rail and bus stations located nearby create transport hubs. Examples of this approach include Victoria station in London, which is a hub for metro, rail and coach services, or Ekkamai station in Bangkok, where metro and coach services meet.

In addition, Metrolinx (2017) specifies preferences for underground and surface stop names, where the first are recommended to prioritize street names and neighbourhoods while the latter include street intersections or landmarks. However, they also recognize that there will be exceptions to these naming protocols so it is advisable to stay flexible throughout the process.

2.2 *Station Name Change*

Usually, a metro station is given its name once. However, sometimes when a transport network in a city is dense, and different transport agencies are responsible for different parts of it, an inconsistency in station names and duplication might occur. If this is recognised at a planning stage, then it is relatively easy to amend it as the process involves mainly document-related actions. Therefore, a holistic view at a transport network is needed. If duplications or other issues with station names are identified at a later stage (e.g. construction or operational stage), it must be understood that the re-naming process involves more than just a change of the name plates. And so, the re-naming process considers the following aspects (WMATA 2012): station signage, vehicle signage, passenger information systems, announcements (on and off trains), metro network maps and neighbourhood maps at stations, customer information (printed and online), etc. In addition, the re-naming affects businesses and organisations, which use a metro station name as a geolocation for their customers (e.g. on marketing materials), spin off companies' names, etc.

To make the process of naming or re-naming a station clear Metrolinx (2015a) produced a simple and universal guide, which could easily be adapted by other metro networks. Table 2 lists seven steps recommended in naming or renaming a station.

Table 2 7 Steps for naming or renaming a station

Step number	Action
1	Consult the 'master list' of existing station names
2	Use the decision tree to guide possible names
3	Draft a shortlist of proposed names with supporting rationale
4	Solicit public input through surveys/stakeholder engagement processes, recognizing importance of balancing public input with technical requirements and regional perspective
5	Finalize names
6	Submit to relevant parties for use in public-facing communications
7	Public communications roll-out

Source Based on Metrolinx (2015a)

Step 1 in Table 2 involves consulting a 'Master List', which is a simple but effective way of identifying duplicates or any potential conflicts between existing and planned names. It is important to know where the List is stored, keep it up-to-date and give the relevant stakeholders access to it when required. Also, Step 2 mentions a 'Decision Tree' (Metrolinx 2015a), which is a tool that helps to arrive to the best possible name solution for a station eliminating not unsuitable names on the way.

2.3 *Metro Stations as Public Places*

In addition to specific metro-related principles for station naming, or in the absence of these, some authorities apply guidelines to naming public places, which can be easy adapted to the metro context. For example, the State Government of Victoria (2016), Australia published principles for naming places, where rules are listed and explained. Table 3 lists the principles in detail. Points 6 and 7 highlight the importance of indigenous heritage (e.g. Aboriginal in Australia) and could be generalised into other indigenous minority groups in other countries. Principle 9 clearly states that commercial and business names should be avoided. This is not always the case with metro systems and examples of such a use are presented in Sect. 2.5. Also, principle 11 which recommends an avoidance of directional names can be justified for public places, but in the context of metro stations this rule is actually applied by some of the metro systems. For example, 12% of London metro unique station names use directional terms and stations such as North Acton or West Acton that are well rooted in the community.

2.4 *Metro Station Naming Decision Makers*

In most of the cases governments or local authorities decide on metro station names at the metro planning and design stage. The original station names are usually not open for discussion with the general public and are decided and approved centrally. However, there are examples where metro operators seek the general public's opinion on metro station names. Such a public consultation initiative can have a form of a focus group (e.g. Metro 2015), an online consultation (Metrolinx 2015b) or an open competition (MetroTunnel 2017a, b). Table 4 shows four examples of metro systems where the public was actively engaged in the station naming process.

Table 3 12 Principles for naming public places

Number	Principle	Details
1	Ensure public safety	A name must not risk public safety (e.g. for emergency response) and cause confusion (e.g. for transport or mail services), considers local community and visitors
2	Recognise the public interest	Long-term benefits to the (future) community should be considered, short-term effects on private or corporate interest should be avoided
3	Link the name to place	Names should be relevant to the local area and used by a local community, if a person then he/she should have/had strong importance to the community
4	Avoid duplication	Duplication is not allowed within the same area
5	Names must not be discriminatory	Names must not discriminate
6	Recognition and use of indigenous languages in naming	Names rooted in the local context could be used, but are subject to agreement with traditional groups
7	Dual names	If community is bilingual or multicultural, this should be taken into account in choosing a name
8	Use of commemorative names	If named after an event, person or a place then additional rules would apply, which are usually listed separately
9	Avoid the use of commercial and business names	Places should not be named after commercial businesses, trade names, estate names or not-for-profit organisations
10	Language	Names from other languages need to be considered carefully, additional rules could apply (e.g. not exceed three words and/or 25 characters)
11	Avoid directional names	North, south, east and west names to be avoided
12	Assign extent to a road, feature or locality	The area must be clearly defined (e.g. coordinates)

Source Based on State Government of Victoria (2016)

2.5 *Naming Rights*

2.5.1 General Rules

Public transport is usually funded by a mix of a revenue from ticket sales and government subsidies (UITP 2013). Often, this is still not enough to cover the operational

Table 4 Examples of public engagement activities performed by metros in station naming process

Type of activity	Date	Details	Metro system
Focused outreach	March 2015	Two focus groups conducted, in English and Spanish	Los Angeles Metro, USA[a]
	2012	Focus groups regarding station name policies	Washington Metro, USA[b]
Open competition	October 2017	An open competition for naming 5 new stations in a Melbourne metro network expansion project. Also, engagement activities for primary and secondary schools	Melbourne Metro, Australia[c]
Online consultation	October 2015	Online public consultation on proposed new names for 11 stations, votes for alternative names on a 5-point Likert scale and comments	Eglinton Line, Toronto, Canada[d]

Source [a]Metro (2015), [b]WMATA (2012), [c]MetroTunnel (2017a), [d]Metrolinx (2015b)

costs of a system and alternative funding sources are explored. These alternative income streams include, e.g. commercial space renting (e.g. shops and offices at stations), advertising (e.g. outdoor, train, station, neighbourhood map) or naming rights (e.g. station names on sale) (MTA 2013; Delhi Metro Rail Corporation Limited 2016; RTA 2018; MRT 2018).

Metro owners can name their metro stations according to their own strategy, as long as it is in line with governmental rules, or offer a private company to name a station in exchange for a payment. Examples of the latter solution are visible in metro systems across the world, from the USA (NYC Subway), to Egypt (Cairo Metro) to India (Delhi Metro) to Malaysia (Kuala Lumpur monorail line). The scale of the phenomenon is rather small and usually includes few stations in a system only. In general, a station name could be either for sale or for rent for a specified period of time. The private company would then be able to choose the name of the station according to the naming rights conditions given by the station's owner and the company's promotional strategy (e.g. name of a company—Disney, name of a product—Mickey Mouse).

2.5.2 Naming Rights Conditions

A station can be an attractive asset to potential sponsors as, depending on its size and location, it gives them access to a number of potential customers on a daily basis. Station names therefore could go on sale or sponsors might request renaming

Table 5 Key advantages and disadvantages of station naming rights 'on sale'

Advantage	Disadvantage
Non-fare revenue	Disconnects the station from its urban landmarks
Interior (re)design	Creates confusion in customers
Large advertising space	Privatisation of public asset image

of a station. The majority of metro systems do not allow commercial station names on their networks, however some authorities do consider this option seriously but carefully (e.g. MRT 2018; RTA 2018).

Before metro station naming rights are put on sale/rent, detailed rules are usually agreed between the interested parties on conditions of such a deal. The conditions usually specify what names are and what are not allowed (e.g. offensive names not allowed, urban landmarks linked names only) as well as explain the re-naming approval process.

Although selling naming rights to private companies brings an additional income, metro users are not always happy. After conducting focus groups with their customers WMATA in Washington, US, stated that "the majority of customers strongly oppose commercial naming rights" supporting this opinion with an argument that users are connected to "historical value of the [capital] city" which should not be 'on sale' (WMATA 2012, p. 4). However, details of the research methodology and focus group size in their study is not disclosed, which makes it difficult to quantify these opinions. A different approach was applied in NYC by MTA, who produced a policy specifying a procedure for re-naming of their facilities at the request of a sponsor (MTA 2013). The 4-page policy document states its purpose, scope, definitions, policy statement and procedure. The procedure is divided into four sections: eligibility, valuations, corporate responsibilities (role of MTA's Director, the Board and the Chief Financial Officer) and termination. The key message regarding station naming included in the document is that an evident link between the station and the potential sponsor is expected. Interestingly, MTA recognises potential financial benefits of such a deal and declares to proceed with an independent evaluation of naming rights opportunities of such a deal if any re-naming application arrives. Moreover, the document encourages competition, therefore if a potential sponsor puts a re-naming request in, then the MTA Director has the right to contact other potential sponsors and ask if they are interested in submitting a competitive application.

Overall, commercial naming rights for metro stations have numerous pros and cons and the key issues are summarised in Table 5.

2.5.3 Examples

Very few examples of metro station naming rights sold to a private sector can be found in developed countries. On the other hand, other and more open approaches to naming rights have been applied by some developing countries. For example, in

Table 6 Examples of metro stations with naming rights sold to a private sector

Station name	Metro system	City/country	Details/duration
Air Asia—Bukit Bintang	KL monorail line	Kuala Lumpur, Malaysia	One of few commercial station names in 2019. Contracts between 3 and 10 years
HUDA City Centre	Yellow line/Delhi metro	Delhi, India	6+ stations, 10 year contracts since 2018
Tura El-Asmant	Line 1/Cairo metro	Cairo, Egypt	One of the two commercial station names in 2019
Atlantic Avenue-Barclays Center	NYC subway, 9 metro services/4 metro lines	NYC, USA	The only commercial station name. 20 year contract since 2012
Buxton Water (original Canada Water)	Jubilee line and London overground	London, UK	Name change for a sponsored event, 1 day

Kuala Lumpur, Malaysia there are 10 metro/rail lines, where some of them allow private names for their stations. As a consequence, stations such as: Air Asia—Bukit Bintang (8—KL Monorail Line), CGC—Glenmarie or BankRakyat—Bangsar (5—LRT Kelana Jaya Line) can be found on the metro system. In India, Delhi Metro sells naming rights to the private sector and some of their stations are named by private companies. Examples of commercial station names are listed in Table 6.

3 Case Study

Bangkok is facing a great expansion of its metro network with up to 500 km of metro line planned for operation after 2020 (Soodyodprasert 2019). The two BTS lines (light green and dark green), two MRT lines (blue and purple) and Airport Rail Link (commuter train) will be expanded with new metro lines (e.g. orange), new commuter train lines (e.g. red), monorail lines (e.g. yellow) and extensions of the existing BTS (e.g. green) and MRT lines (e.g. blue). Figure 1 shows the Bangkok metro network map according to the plans available in March 2019.

As seen on Fig. 1, the planned Bangkok metro network is quite dense and includes a total of 12 lines. In the early stages of the new lines planning process it was recognised that some duplications and inconsistencies in the way stations are named occurred across the network. A local authority requested an investigation into the name changes of four existing metro interchanges as well as recommendations regarding future station names. Therefore, a data collection tool was designed to gather new relevant data before any recommendations could be produced.

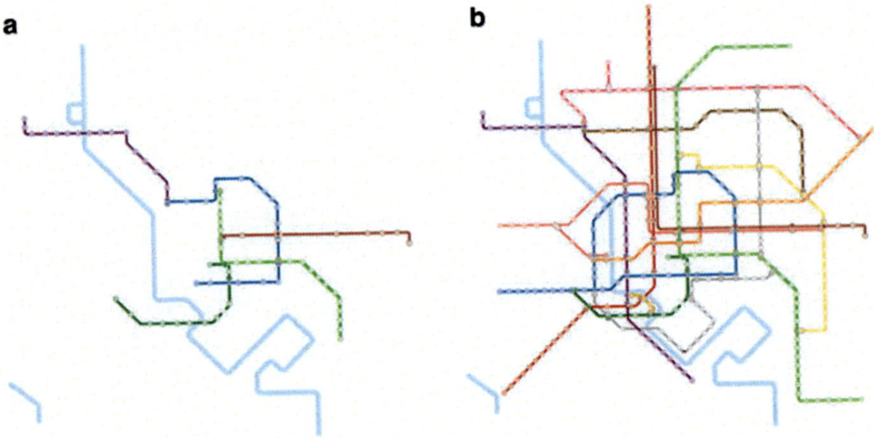

Fig. 1 Bangkok metro network schematic map: **a** with the 5 existing lines in March 2019; **b** with all planned lines and extensions after 2020. *Source* Based on Bangkok Mass Transitmap Version 3.0 March 2019 by CLARE

4 Methodology

4.1 Secondary Data Approach

In 2018 all existing station names in Bangkok metro network were studied and these included 76 stations across the three metro networks (BTS, MRT and ARL). The following three characteristics were considered:

1. name category (six options: street, area, landmark, organisation, building, person);
2. name length (in terms of syllables and characters);
3. name translation strategy (two options: Thai pronunciation, English meaning).

4.2 Primary Data Approach

Secondly, a questionnaire titled: "Station naming strategies for Bangkok metro system", designed in English and translated into Thai language, was put online. The survey was divided into three parts, where:

- Part 1 focused on the respondents' socio-economic data (age, gender, nationality, languages spoken, occupation, frequency of metro use, start/end metro use stops);
- Part 2 addressed questions related to desired characteristics of a station name and challenges with selected existing station names (priority categories when naming a metro station, length of a station name, language used, confusing

names, stations for re-naming, name change suggestions for specific 4 intersection stations, naming rights preference);
- Part 3 focused on travel route selection criteria and the use of mobile applications (important factors in route choice, specific route parameters checked on mobile apps).

This paper focuses of data collected in Part 2 of the survey and related to desired characteristics of a station name as well as preferences regarding (re)names of the four existing interchanges analysed in the context of some socio-economics data collected in Part 1 of the survey.

4.3 Data Collection

After the survey received an internal MU-CIRB ethical approval, it was launched on 25th June 2018 and closed on 10th October 2018. There were two separate language versions available: in Thai and in English. The final sample size included 399 responses, where 388 were given in Thai and 11 in English. After a data cleaning process all 399 records were included in analysis of data.

5 Analysis of Results

5.1 Respondents' Socio-economics

The sample included 399 respondents, where 137 were females, 254 were males and 8 did not state their gender. The male sub-sample was older with a majority of the males being age 25 or over. A great majority of respondents, 98% of the sample, were of Thai nationality and they completed the Thai version of the survey online. Overall, 71% of the sample spoke Thai (only), followed by 28% who could speak Thai and English (and few could speak some other languages, like e.g. Chinese, Japanese or German, too). Over half of the sample was represented by employees (57%) and a third by students (35%) with the remaining respondents being involved in private business or stated some other occupations (e.g. engineer, professor). Details of the sample's socio-economics are displayed in Table 7.

The respondents were divided into three groups, based on their frequency of metro use:

- **Frequent** users (n = 183): 46% of the sample, use metro at least few days a week;
- **Moderate** users (n = 122): 31% of the sample, use metro once or few times a month;
- **Rare** users (n = 94): 23% of the sample, use metro less than once a month or never.

Table 7 Socio-economics of the sample

Variable	Female (%)	Male (%)	Total
Sample size	n = 137	n = 254	n = 399[a]
Age[b]			
18–24	45	30	35
25–40	38	48	44
41–60	15	22	20
Over 60	2	0	1
Nationality			
Thai	99	98	98
Non-Thai	1	2	2
Language			
Thai	80	66	71
English	1	1	1
Thai + English +	19	33	28
Occupation[b]			
Student	43	29	35
Employee	52	61	57
Tourist	0	1	1
Private business	4	7	6
Other	1	2	2

[a]n = 8 respondents did not state their gender, their responses are included in the 'total' column
[b]Results are statistically significantly different between females and males

Details of respondents' frequency of metro use are displayed in Table 8.

5.2 *Name Categories*

The respondents were asked to select top three ways of naming metro stations, based on six categories as follows:

- Landmark;
- Building;
- Street;
- Person;
- Organisation;
- Area.

Table 8 Respondents' frequency of metro use

Respondents' frequency of metro use		Female n = 137 (%)	Male n = 254 (%)	Prefer not to say n = 8 (%)	Total	Overall (%)
Frequent n = 183	Every day	10	12	25	11	46
	Every weekday	21	16	13	17	
	Few days a week	12	21	0	18	
Moderate n = 122	Few times a month	23	23	25	24	31
	Once a month	5	8	12	7	
Rare n = 94	Less than once a month	27	18	25	21	23
	Never	2	2	0	2	

Next, the top three categories per respondent were identified and weights were applied to the results to calculate the popularity of each of the categories. The weights included:

- weight 3 for the 1st option (percentage value × 3);
- weight 2 for the 2nd option (percentage value × 2);
- weight 1 for the 3rd option (percentage value × 1).

Next, the percentages gained in each option were multiplied by suitable weights and results of final scores are displayed in Table 9. Overall, Table 9 includes views

Table 9 Twelve most popular top 3 options for naming metro stations

Option	Total votes (%)	Top 1	Value	Top 2	Value	Top 3	Value
1	16	Landmark	48	Building	32	Area	16
2	12	Landmark	36	Area	24	Building	12
3	8	Area	32	Landmark	16	Building	8
4	7	Landmark	21	Area	14	Organisation	7
5	6	Landmark	18	Area	12	Street	6
6	6	Area	18	Landmark	12	Street	6
7	5	Landmark	15	Street	10	Area	5
8	5	Landmark	15	Building	10	Street	5
9	4	Area	12	Landmark	8	Organisation	4
10	3	Area	9	Street	6	Landmark	3
11	3	Landmark	9	Building	6	Organisation	3
12	3	Landmark	9	Organisation	6	Area	3
Total 78%							

of the 78% of the total sample as the remaining 22% of the 'top three' combinations gained each less than 3% of popularity amongst the sample.

In the 12 options illustrated in Table 9 'Landmark', as the only parameter out of the six suggestions, is present in each of them. This is a clear suggestion that 'Landmark' is seen by respondents as an important and most popular option for station naming. Overall score for 'Landmark' is 210, which is the highest result out of the six options given. Secondly, 'Area' option is present in 10 out of 12 options, which makes it the second most popular option, also present in the first three highest suggestions. The value of 'Area' category is 145, which is the second highest option on the list. Thirdly, although 'Building' option for a station name is only mentioned 5 times in the 12 options, it is in the first three options with highest scores as well. The total value for 'Building' is 68, which is substantially lower that the top value, but still the third highest score out of the six categories.

Overall, a clear message is that the top three suggestions for station naming are related to:

1. Landmark (score 210),
2. Area (score 145),
3. Building (score 68).

Next, these results were compared with the existing station name categories and it was found that respondents' preferences are different from the existing strategy. Figure 2 compares the results and it can be seen that the existing stations are

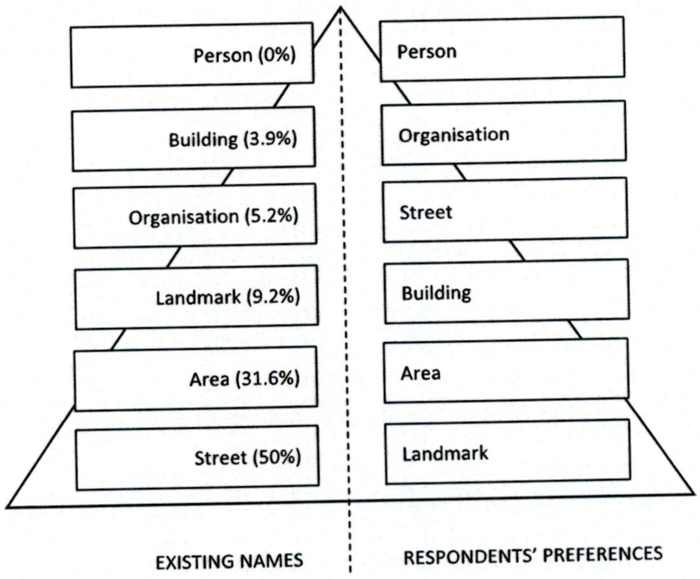

Fig. 2 Pyramid of station name categories for existing Bangkok metro stations and respondents' preferences

named mainly after streets (50%) and areas (32%) and the remaining categories are marginalised. However, the analysis of respondents' preferences show that they prefer landmark-based names, and then area and building names with street names on the 4th position only. Organisation and persons' based names are not popular and are not preferred by the respondents.

5.3 *Name Length*

The existing 76 station names were analysed in terms of three parameters: number of syllables, number of characters and translation strategy. The calculated results are compared against the respondents preferences for station names characteristics, as stated in the online survey.

5.3.1 Syllables

A majority of the respondents agreed that the number of syllables in a metro station name in Thai language should be shorter rather than longer. Figure 3 displays results divided between three groups of metro users, depending on their frequency of use. It can be seen that patterns of responses were similar in all three groups, with 2–3 syllables preferred by 58% of the total. Longer words, with between 4 and 5 syllables were preferred by 40% of the sample. No statistically significant differences were found between the three metro frequency use sub-groups.

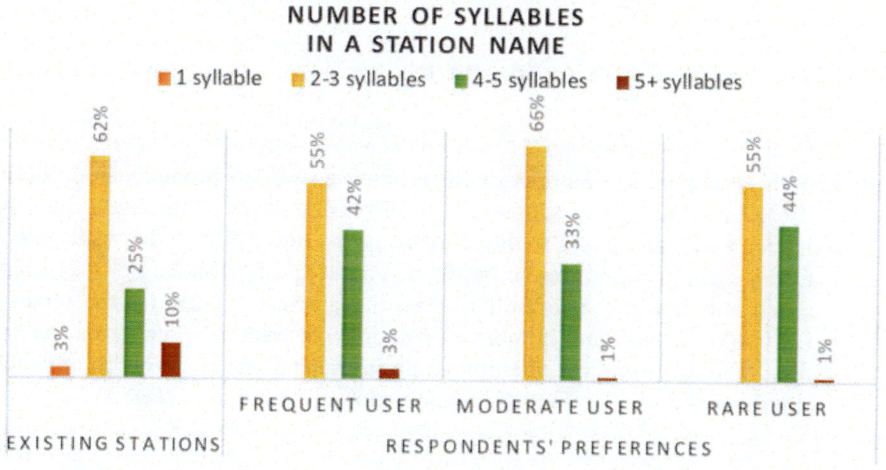

Fig. 3 Number of syllables in a metro station name: existing stations versus respondents' preferences

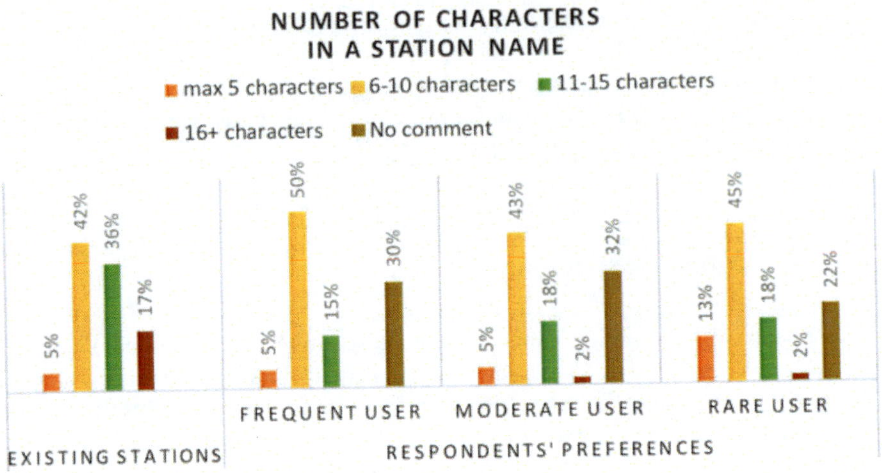

Fig. 4 Number of characters in a metro station name: existing stations versus respondents' preferences

5.3.2 Characters

A majority of the respondents prefers shorter rather than longer station names. Figure 4 displays results divided between three groups of metro users, depending on their frequency of use. Overall, 54% preferred names up to 10 characters long, with a focus on 5–10 characters in a name. The remaining responses split between: 11–15 characters (17%), 16+ characters (1%) or "No comment" (29%) answer.

5.4 Name Translation Strategies

Overall, when translating Thai station names into English the majority of respondents (64%) preferred a mixture of English meaning and Thai pronunciation. More specifically, this strategy was supported by 70% of 'Frequent' metro users and support for it decreased among less frequent metro users. Interestingly, the results were statistically significantly different between 'Frequent' and 'Moderate' users and 'Rare' users, although the latter still showed the greatest support for the mixed option. Only 10% of the total number of respondents primarily supported the option of using English meaning where possible in translations and a quarter of the sample supported the idea of English pronunciation of Thai names (Fig. 5).

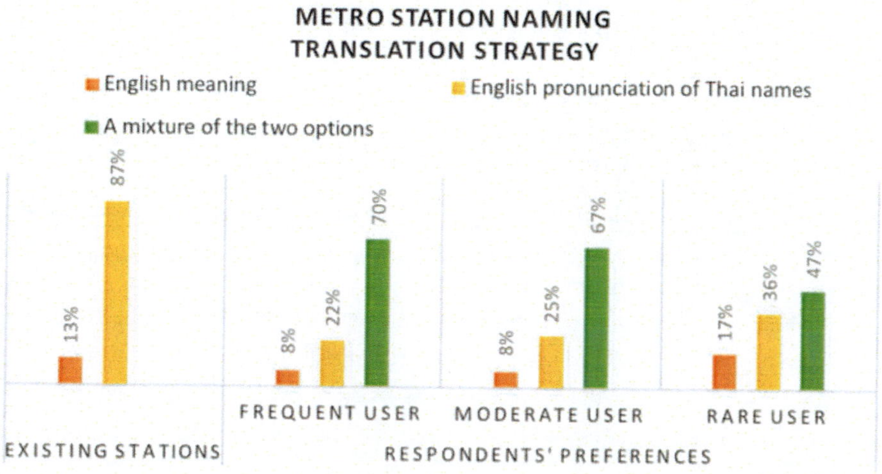

Fig. 5 Metro station names translation strategies: existing stations versus respondents' preferences

5.5 Bangkok Metro Network Interchanges

Four large existing interchanges between different metro lines in Bangkok were selected to test respondents preferences for station re-naming. At the time of data collection, each of the four interchanges had two separate stations with different names located close to each other. Also, in each of the four cases the crossing lines were operated by different companies. It was recognised that this set up could potentially cause confusion to metro users, especially those wishing to change lines.

The four interchanges investigated for name changes were:

- Asok (BTS Light Green) and Sukhumvit (BTS Blue);
- Mo Chit (BTS Light Green) and Chatuchak Park (MRT Blue);
- Makkasan (ARL) and Phetchaburi (MRT Blue Line);
- Sala Daeng (BTS Dark Green) and Si Lom (MRT Blue).

Figure 6 presents respondents preferences for interchange station re-names. Overall, a third of respondents in each question stated that "No change" was needed in the existing station names, where in each case the largest proportion of 'Rare' metro users supported this option. However, the remaining three quarters of respondents had various suggestions for stations re-renaming. The most popular suggestions were related to changes including line colour, and these were supported by between 37% (28% for Asok with line colour and 9% for Sukhumvit with line colour) and 40% (30% for Chatuchak Park with line colour and 10% for Mo Chit with line colour) of the respondents. Much less popular was the option of station names with a direction, which was supported by between 9 and 12% of respondents, depending on the interchange considered.

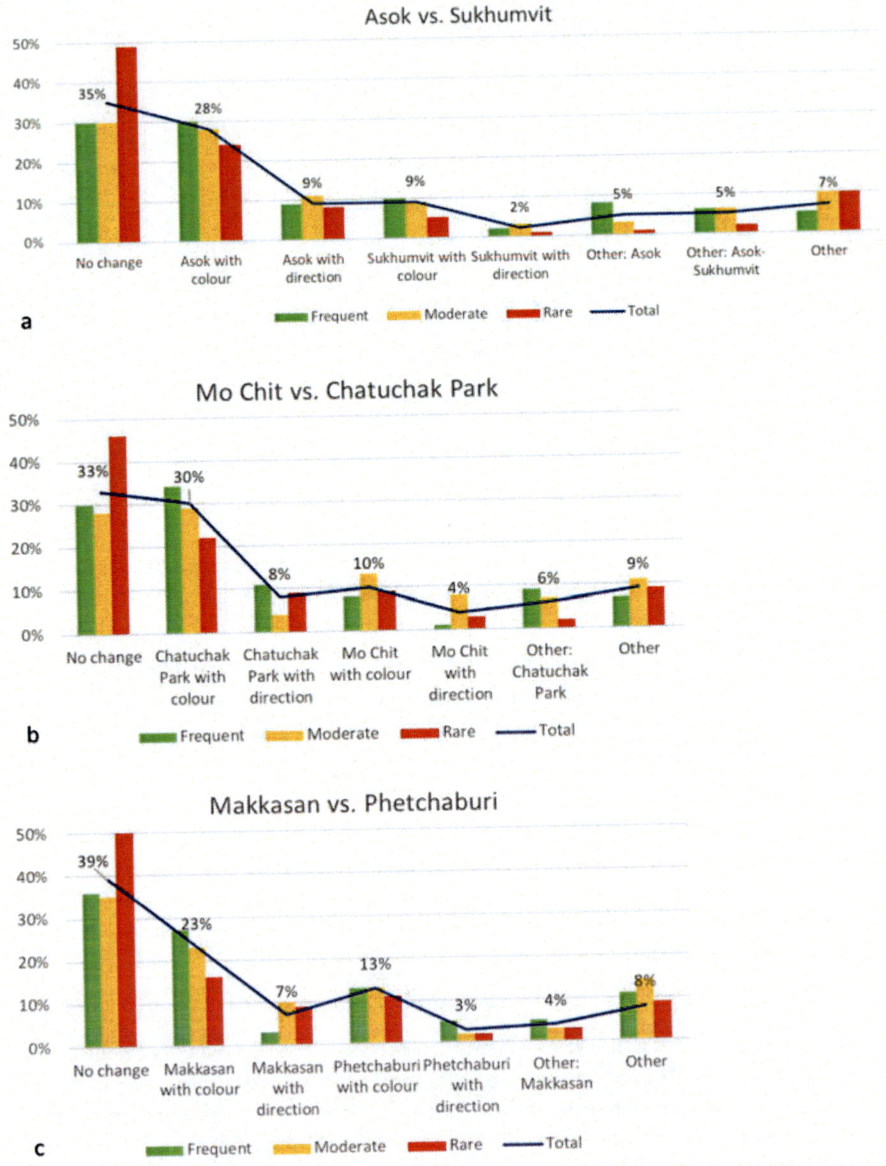

Fig. 6 Respondents preferences for interchange station re-names: **a** Asok/Sukhumvit; **b** Mo Chit/Chatuchak Park; **c** Makkasan/Phetchaburi; **d** Sala Daeng/Si Lom. Results are statistically significantly different between frequent/moderate/rare users for Asok/Sukhumvit and Mo Chit/Chatuchak Park

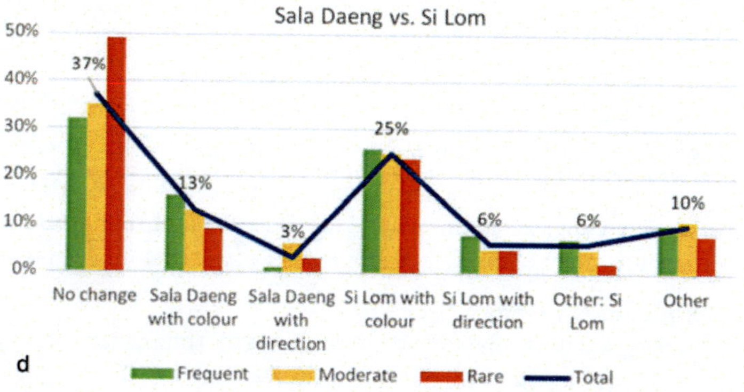

Fig. 6 (continued)

6 Conclusions

The paper presented results of an investigation related to station names in the Bangkok metro network. 399 respondents were asked to state their preferences regarding unique metro station names in terms of their length (number of syllables and characters) and translation strategies (English meaning vs. English pronunciation). Results revealed that majority of the sample preferred shorter station names (between 2–3 syllables and maximum 10 characters long) and a mixture of English meaning and pronunciation when applying translations from the original Thai names. Next, in terms of name categories preferences, respondents stated landmark, area and building names would be the top three best ways for station naming. Finally, analysis of respondents' preferences for re-naming of the four existing metro interchanges revealed that about a third of the sample preferred no changes at all. The remaining respondents supported various strategies from colour-coded (about 40% of the sample in different colour-coded combinations) to direction-coded (about 10% in different direction-coded combinations) names. Overall, the results were in line with the literature presented in Sect. 2, where names related to landmarks and areas are usually preferred and applied across different metro systems across the world. Also, the results showed that different frequency users have slightly different opinions regarding station names, but these are not statistically significantly different between frequent, moderate and rare metro users.

7 Recommendations

7.1 General Principles

Based on the results of the study and literature reviewed, a set of principles for metro station names is recommended for Bangkok metro network. The seven general principles for metro station names in Bangkok are that they should be:

1. **Simple** (easy to remember);
2. **Short** (priority for max. 2 words and max. 10 characters or 5 syllables);
3. **Durable** (relevant as long as the station exists);
4. **Geo-locating** (geographically linked with an area);
5. **Unique** (no duplicates or repetitions);
6. **Coherent** (one name for interchanges between different lines and other modes);
7. **Respectful** (not offensive or discriminatory).

7.2 Bangkok-Specific Station Name Categories

Analyses of the existing metro station names in Bangkok revealed that they are named mainly after streets (50%), areas (32%) and landmarks (9%). However, the respondents stated slightly different order of categories in their preferences, where: Landmark, Area and Building were the top three categories. It is recommended to use the six suggested options for station names, in the order of importance presented in Fig. 7, where 'Landmark' (the largest block) should be used as the first choice and 'Person' (the smallest block) as the last choice, possibly applied in unique circumstances only.

Fig. 7 A pyramid of station name categories with 'landmark', 'area' and 'building' as the base

8 Limitations and Next Steps

Firstly, results of this study are based on a Bangkok-based sample of respondents, who expressed their opinions and preferences related to Bangkok metro network only. It is recognised that preferences expressed by the sample are specific to the Thai context only and if a similar study was to be repeated elsewhere different results could be obtained. Therefore, findings presented in the paper, related to for example name length or name categories, apply to the Bangkok case only. However, it is also recognised that due to a very small amount of similar research, the paper adds a significant contribution to the area of research focused on metro station names. If similar studies regarding station names and involving local residents could be conducted in other geographical locations then it would be possible to compare results between different metro systems, but also factor other characteristics (such as culture, language or topography) into the findings.

Secondly, the way new data was collected for the study included online responses only. It is recognised that qualitative data collected via interviews and focus groups would allow better understanding of reasons behind people's responses and preferences that perhaps were not captured in a survey.

Acknowledgements The methodology and results presented in this paper are based on the outcomes of a project called 'Metro station naming strategies', which was part of a larger project titled 'A Study of Train Operations System Planning for Mass Rapid Transit Service Enhancement', funded by the Office of Transport and Traffic Policy and Planning completed in early 2019.

References

CLARE (2019) Bangkok mass transit map. Version 3.0, Mar 2019. Mahidol University

Delhi Metro Rail Corporation Limited (2016) Bid document for proposal for semi-naming rights at selected metro stations in Delhi (RFP-316M0001)

Metro (2015) Letter designations for fixed guideway lines. Presentation. Available at https://media.metro.net/board/Items/2015/04_april/20150408wesitem5.pdf. Accessed 20 Aug 2019

Metrolinx (2015a) Station and stop names. Eglinton Line. Presentation. Available at www.metrolinx.com. Accessed 8 Jan 2018

Metrolinx (2015b) Metrolinx engage report. Eglinton LRT station names, 11 Nov 2015. Available at www.metrolinxengage.com. Accessed 20 Sept 2018

Metrolinx (2017) Memorandum. Available at https://www.metrolinx.com. Accessed 4 Feb 2018

MetroTunnel (2017a) Station names. Available at https://metrotunnel.vic.gov.au/stations/station-names. Accessed 30 May 2018

MetroTunnel (2017b) Resources for teachers. Available at https://metrotunnel.vic.gov.au/library/resources-for-teachers. Accessed 07 June 2018

MRT (2018) Station naming rights. Available at https://www.mymrt.com.my/business/station-naming-rights/. Accessed 06 June 2018

MTA (2013) MTA real estate department policy and procedure. Re-naming of MTA facilities at the request of third-party sponsors

RTA (2018) Dubai metro naming rights home page. Available at https://www.rta.ae/links/NamingRights/index.html. Accessed 06 June 2018

Soodyodprasert S (2019) Rail infrastructure and investment in Thailand. Presentation, 4 Apr 2019. Office of Transport and Traffic Policy and Planning

State Government of Victoria (2016) Naming rules for places in Victoria. Statutory requirements for naming roads, features and localities 2016. Available at https://propertyandlandtitles.vic.gov.au. Accessed 30 May 2018

UITP (2013) Financing public transport. Press kit, 26–30 May 2013. UITP, Geneva. Available at https://www.uitp.org/sites/default/files/Financing%20public%20transport.pdf. Accessed 20 Aug 2019

UITP (2018) World metro figures 2018. Statistics Brief. Available at https://www.uitp.org/sites/default/files/cck-focus-papers-files/Statistics%20Brief%20-%20World%20metro%20figures%202018V4_WEB.pdf. Accessed 20 Aug 2019

WMATA (2012) Station naming policy. Available at https://www.wmata.com. Accessed 8 Jan 2018

An Assessment of Public and Private Transport Competitiveness in Jakarta with the Focus on MRT

Ahmad Faisal Dahlan and Anna Fraszczyk

Abstract The provision of Jakarta MRT is expected to reduce congestion in Jakarta by shifting motorists to the MRT. Neverthless, shifting drivers to the MRT is much more complex since the process of decision-making involves various technical and social aspects. A comprehensive study of decision making was undertaken to see drivers' response towards Jakarta MRT development. Hence, the research aim is to examine the competitive level of Jakarta MRT in shifting motorists to the MRT. The data collected via a questionnaire from residents of Jakarta is divided into two groups: Private Vehicle Ownership (PVO) and Private Vehicle Non-Ownership (PVNO). The results indicated that the MRT becomes a preferred transport mode for both groups, where the shortest travel time becomes the greatest reason. Even though the MRT has waiting and walking time as parts of effort, and private vehicles appear to be superior in this occasion that requires less effort, the respondents still consider the shortest travel time as the most important parameter. However, 45% of PVO's respondents still prefer choosing a motorcycle or a car to MRT. Shifting motorists to use the MRT requires a lot of attentions in terms of reliability, cost, and security. The other factors such as connectivity, headway, comfort, and cleanliness are also valued as important for the MRT development in the future.

Keywords Jakarta MRT · Competitiveness · Public perception · Metro · Survey

1 Introduction

Jakarta is the capital city of the Republic of Indonesia and considered as the biggest city in southeast Asia (Susilo et al. 2007). Its developing economy attracts many people, who live in rural and suburban areas, to move to the capital. Consequently, Jakarta is inhabited by many people from various regions of the country and reflects

A. F. Dahlan (✉)
Faculty of Engineering, Mahidol University, Nakhon Pathom, Thailand
e-mail: fais.dahl@gmail.com

A. Fraszczyk
Mahidol University, Nakhon Pathom, Thailand

W. Weerawat et al. (eds.), *Urban Rail Transit*, Lecture Notes in Mobility,
https://doi.org/10.1007/978-981-15-5979-2_2

Table 1 Comparison of Jakarta's population with private and public transports

	Year		
	2010	2014	2015
Population	9,640,406	10,075,310	10,177,924
Private vehicles	7,340,783	9,902,917	–
Public transports	89,978	112,724	–

Source BPS

its diversity (Susilo et al. 2007). Bureau of the statistic in Indonesia (BPS) shows that the population in Jakarta tends to rise, and the city was already classified as megacity with over 10 million people in 2015, as shown in Table 1 (Badan Pusat Statistik 2017). Those numbers excluded people who live in the surrounding areas of Jakarta, such as Bogor, Depok, Tangerang, and Bekasi, but commute to work or school in Jakarta on a daily basis.

Rapid growth of Jakarta residents triggers a huge motorization. In 2014, population of Jakarta was 10,075,310 people and the number of private vehicles and public transport was 9,902,917 and 112,724 units, respectively (Badan Pusat Statistik 2017; Dinas Perhubungan Provinsi DKI Jakarta 2018). Hence it is fair to assume that some residents in Jakarta have more than one private vehicle. The main reason why people are still using the private vehicle is because of unreliable public transport (Mukti and Prambudia 2018). Further, people from the surrounding areas who commute to/from Jakarta, contribute to the traffic and generate 7 million trips/day (JABODETABEK Urban Transportation Policy Integration Project in the Republic of Indonesia 2012). These conditions are the reasons why congestion in Jakarta is getting worse.

It is estimated that the traffic congestion causes economic losses up to 2.5 trillion rupiahs/day in Jakarta (JABODETABEK Urban Transportation Policy Integration Project in the Republic of Indonesia 2012). Therefore, Jakarta MRT, the first metro system in Indonesia, was initiated by the government of Indonesia as a solution to a severe congestion problem. The integrated MRT with other transport modes i.e. BRT and LRT, was expected to shift drivers from their private vehicles. The MRT project originally consisted of two construction phases. Phase 1 started in October 2013 and began the operation on 24 March 2019, while Phase 2 would be started after the completion of Phase 1.

Neverthless, shifting drivers to the MRT is much more complex since the process of decision-making involves various technical and social aspects. A comprehensive study of decision making was undertaken to see drivers' response towards Jakarta MRT development. Hence, the research aim is to examine the competitive level of Jakarta MRT in shifting motorists to the MRT. The study used a comparison model between the MRT with private vehicles in order to see the MRT's competitiveness level. The results appear to be beneficial for the operator of Jakarta MRT and improvement of the service in the future.

The paper is organised as follows. First, introduction which explains problem statement, research aim and includes a contextual of literature reivew are presented in Sects. 1 and 2. Then, a set of methodology to conduct the research is established is

Sect. 3. Subsequently, results and discussion of key findings are presented in Sects. 4 and 5. Section 6 includes conclusion and recommendations and limitations and next steps are provided in Sect. 5.

2 Literature Review

2.1 Competitiveness

In many transport research, the term of competitiveness is often used to compare one transport mode to another. For example, Danapour et al. (2018) compared the planned high speed rail mode (HSR) with current air transport (AT) in Tehran-Isfahan route according to ticket price, travel time, hospitality, and convenience. The binomial logit model was employed to estimate the probability of choosing each mode. The results indicated that shorter travel time, cheaper cost, the availability of meal, and wider aisles could increase the likelihood of HSR being chosen. Gundelfinger-Casar and Cotto-Millan (2017) conducted similar research about comparing the existing HSR and AT in Spain. The ten-year estimation demand was proposed in the study, which revealed that HSR's demand might increase due to shorter travelled distance and travel time. In some cases, HSR was seen as a competitor for AT. However, the study from Albalate et al. (2015) attempted to balance competition and cooperation between HSR and AT service in Europe. It was recognized that there was a direct competition between the two modes, but on the other hand HSR could also be a feeder system for AT. Behrens and Pels (2012) also compared HSR and AT, a study case in London-Paris passenger market. The results indicated, similar to previous studies, that the competition between the two modes would decline over the years since HSR was in more demand than AT, according to travel time. Another research about competitiveness was conducted by Lee (2018), which compared LRT, BRT, and Bi-Modal Tram (BT) by considering the type of city, either metropolitan or small and medium. The result indicated that BT was considered to be the most suitable for small and medium cities, while LRT was desirable for both city categories. Yen et al. (2018) also conducted a competitive research, which compared bus way and heavy rail in Queensland, Australia. They concluded that bus way is more competitive than heavy rail because of frequency of service and accessibility.

2.2 The Parameters of a Travel

Travel mode choice is the decision making process that always involves a mix of technical and social aspects including parameters that influence people's perception. In the past, people only considered travel time and cost when commuting (Papaioannou and Politis 2019). Several studies also confirmed that cost and time as the two key parameters for choosing a transport mode. Munira et al. (2013) conducted a research about public perception of van users in Bangkok, Thailand. The result indicated that users emphasized more on poor ticketing system and long travel time compared to Bus, Taxi, MRT, or BTS. Munawar (2007) assessed public transport in Yogyakarta, Indonesia and highlighted that (long) travel time could influence people's decisions for not using public transport. De Vos et al. (2019), who conducted a research about travel patterns in Ghent, Belgium also mentioned that people who used a car due to the short distance and travel time tend to be more satisfied with both commute and leisure trips. Another research by Herwangi et al. (2015), who sought a correlation between transport disadvantage and motorcycle usage of low-income people in Yogyakarta, Indonesia indicated that one of the reasons for low-income people using a motorcycle was short travel distance, which also meant short travel time. Redman et al. (2013), who conducted a literature review study about attributes of public transport that attract car users also found that short time and cheap cost were still effective to shift car users to public transport. Further, social ascpects especially effort are also considered for travel mode choice. Zhang et al. (2017) concluded that the amount of effort required in term of distance traveled can also influence the fact that people prefer choosing a private mode. Ko et al. (2019) also confirmed that travel time influenced a preferred transport mode since the shorter travel time, the less effort required. Further, the nonquantifiable parameters in travel choice i.e. green transport choices also contribute to a person travel choice (Porter et al. 2013).

Transport emission policies were driven by governments to shift people's paradigm of travel choice. Many reports (Porter et al. 2013; Deakin et al. 2010; Gross et al. 2009; Sloman et al. 2010) explain that some people are aware of emissions as a factor in travel choice based on regulation of governments, but still this issue is often not a priority. Even though governments encourage people to be aware of harmful emissions, emission regulations are not fully effective. The awareness of emissions for green transport has emerged and slowly become a trend, although it does not become a primary factor yet.

These four parameters, which are cost, time, effort and emmisions, can be quantified in units for different modes of transport objectively. This allows for a fair comparison of modes in terms of these parameters. However, as mentioned earlier, in addition to technical aspects there are also other social aspectd influence people travel choices. Hence, it is required to expand the parameters in order to reveal what other factors could influence transport mode choice.

2.3 The Expectation Factors

The expectation factors cover for both technical and social aspects. Currently, modern lifestyle expand people consideration in choosing a transport mode, which also take into account various social aspects (Papaioannou and Politis 2019). The social aspects i.e. effort, comfort, safety, cleanliness, are considered to be included to identify choice mode. Birago et al. (2017) attempted to measure service quality of Metro in Accra,

Ghana. The result indicated that though Metro has cheap ticket price, users were still less satisfied with the service since reliability, comfort, and accessibility were often neglected. Dell'Olio et al. (2011) emphasized that cleanliness and comfort should be considered for public transport improvement. Similar with the latter, Tyrinopoulos and Antoniou (2008) also found that frequency and cleanliness contribute greatly to a satisfaction level of public transport. Another interesting research was conducted by Jain et al. (2014) and Susilo et al. (2010), which specified that safety was the most important parameter for shifting people from private to public transport. Purba et al. (2015) strengthened the fact that cost and time were no longer the only parameters for choosing a transport mode, where closer bus stops to home and destination was the most important parameter. Beirao and Cabral (2007) conducted a qualitative study about travellers' attitudes towards private car and public transport. Even though public transports have a cheaper travel cost than a car, lack of social aspects i.e. reliability, comfort, and flexibility, influenced people to avoid public transport. Javid et al. (2013) compared satisfaction level of different modes in Lahore, Pakistan. In terms of importance level factors', interestingly comfort, safety, punctuality, routes and headway should be addressed for public transport improvement. Joewono et al. (2015) focused on a public perception research about paratransit in three cities: Jakarta, Bandung, and Yogyakarta, Indonesia. The respondents from the cities claimed that security was the most important aspect that should be addressed for its improvement.

Overall, variety of research focused on additional paremeters influencing people's travel choices and their results show that these factors play an important role in mode shift. Therefore, not only time, cost, effort and emissions, but also additional factors, such as cleaneliness or reliability, as well as many others, should be taken into account when studying competetiveness of different modes of transport.

3 Methodology

The study attempts to assess the MRT service competitiveness by comparing the service offered by MRT with private vehicles and TransJakarta (BRT) in Jakarta. After reviewing several literatures, the comparison utilizes the key four parameters: time, cost, effort, and emissions, which can be quantified in units for different transport modes. The points of comparison start from Lebak Bulus station (A) to Bundaran Hotel Indonesia station (B) of Jakarta MRT. Further, six factors: cleanliness, comfort, security, connectivity, reliability, and headway, which are obtained from literatures, are also employed to gain more insight about MRT service competitiveness in the future. The factors will be focused on the MRT improvement according to public perception. The expectation factors are delivered on a 5-point Likert scale (from 1—very negative to 5—very positive). The study was planned to be conducted before the MRT was officially launched on 24 March 2019. The online questionnaire was used to distribute the survey to Jakarta and surrounding areas.

Table 2 Estimated cost and travel time from A to B

Transport modes	Cost (IDR)	Travel time (min)
Private		
Car	30,000	40
Motorcycle	10,000	40
Public		
MRT	14,000	30
BRT	3500	45

Source Travel time is based on Google Maps estimations out of peak hours and for MRT ticket price is based on Jakarta MRT information

3.1 The Cost

In the analysis, the cost of the MRT and BRT is fixed, while private vehicles' cost is in the form of fuel purchases at a gas station, but other costs (e.g. purchase, maintenance, and spare parts cost) are not included. It is estimated that private car needs 3 L of fuel and a private motorcycle needs 1 L of fuel for an A-B journey. Table 2 shows an estimated cost for a journey from A to B for various transport modes in Jakarta.

3.2 The Travel Time

The travel time is calculated in non-jammed conditions in Jakarta and the morning hours between 10 a.m. and 12 p.m. are selected. Google Maps tool is employed to estimate the travel time from A to B. Table 2 shows an estimated travel time from A to B. The travel time covers the time required to reach a destination point. Since private vehicles and BRT have so many route options, then the quickest route is chosen.

3.3 Effort

The effort factor determines how much effort is required to travel from A to B. The effort required for each transport mode is very different, some vehicles (e.g. private vehicles) require minimal effort while others (e.g. BRT) require a great effort. The effort parameter should be quantifiable according to numbers and the two-indicator is used to assess effort parameter: waiting time and walking time. Previous studies (Munira et al. 2013; Birago et al. 2017; Dell'Olio et al. 2011; Tyrinopoulos and Antoniou 2008; Jain et al. 2014; Javid et al. 2013; Prasertsubpakij and Nitivattananon 2012) also involved the waiting time according to service improvements in the public transport systems. The shorter of the waiting, the smaller the effort needed by passengers to conduct a journey. The operator gives 10 min for waiting

Table 3 Effort score

Transport modes	Waiting time	Walking time	Scale of waiting time	Scale of walking time	Average score
Private					
Car	–	–	1	1	1.00
Motorcycle	–	–	1	1	1.00
Public					
MRT	10	5	3	2	2.50
BRT	10	20	3	5	4.00

for the train in non-peak hours. Then, it is estimated that waiting time for BRT is also 10 min. Walking time is included since the term "walking" refers to effort parameter. The MRT users require walking time of about 5 min, from buying the ticket until standing in the platform for the train. Subsequently, BRT users require 20 min of walking, where the first 10 min are used to walk from Lebak Bulus station to Lebak Bulus terminal, then the other 10 min are used to walk from Menteng terminal to Bundaran Hotel Indonesia station. The assessment of effort parameter uses a Likert scale from 1 to 5, where 1 is very easy and 5 is very hard. All indicators will be converted into scale, then average score can be calculated by using Eq. (1), where scale of waiting time is summed by scale of walking time, then divided by two and the result is shown in Table 3.

$$\text{Average Score} = \frac{\text{scale of waiting time} + \text{scale of walking time}}{2} \tag{1}$$

3.4 *The Emissions*

The emission calculation framework from Department for Environment, Food and Rural Affairs of United Kingdom (DEFRA) will be adapted in order to estimate the emissions (Conversion Factors 2017). The study used distance travelled and emission factor data for estimating the emissions. The distance travelled data is obtained from Google Maps, while DEFRA provides the emission factors. If a certain transport mode has more than one possible route to reach the destination place, then the shortest route will be chosen.

The formula to calculate the total emissions of the vehicle is presented in Eq. (2). Table 4 summarizes the calculation of total emissions for all transport modes. The total emissions will be changed to the standard emissions in order to facilitate the respondents in understanding them. The standard emission will be divided into three ranges: low (0.00–1.50 kg CO_2 eq); medium (1.51–3.00 kg CO_2 eq); and high (more than 3.00 kg CO_2 eq).

Table 4 Standard emissions

Transport modes	Distance traveled (km)	Emission factor (kg CO_2 eq)	Total emissions (kg CO_2 eq)	Standard emissions
Private				
Car	14.50	0.19490	2.8260500	Medium
Motorcycle	14.50	0.10323	1.4968350	Low
Public				
MRT	15.70	0.01225	0.1923250	Low
BRT	17.50	0.10259	1.7953250	Medium

Table 5 A comparison of MRT with other transport modes in terms of the four parameters

Transport modes	Cost (IDR)	Travel time (min)	Effort			Emissions
			Waiting time (min)	Walking time (min)	Scale	
Private						
Car	30,000	40	–	–	Very easy	Medium
Motorcycle	10,000	40	–	–	Very easy	Low
Public						
MRT	15,000	30	10	5	Moderate	Low
BRT	3500	45	10	20	Hard	Medium

$$\text{Total emission} = \text{the distance traveled} \times \text{the emission factor} \tag{2}$$

Table 5 explains a comparison of MRT with private vehicles in terms of four parameters in order to reveal people's travel preference in Jakarta. This data will be presented to the respondents with a request to select one suitable transport mode to travel from A to B.

3.5 *Questionnaire Design*

The questionnaire is divided into three sections: socio-demographics, travel preference, and expectations. The section of socio-demographics collects data about gender, income, occupation, and commute transport mode. This section might support travel preference data by providing detail information about respondents. The travel preference section provides a comparison study between the MRT with the other modes in terms of the four parameters. Table 6 will be delivered to respondents to obtain their travel preference. Respondents will be asked for a suitable mode and their intention of using it for an A-B journey. The expectation section provides six factors: cleanliness, comfort, security, connectivity, reliability, and headway, which

Table 6 Results of chi-square of homogeneity test

Variables	χ^2	*Df*	*p*
Gender	17.092	1	0.000
Occupation	7.062	2	0.029
Income	6.635	4	0.156
Daily commute transport	122.491	5	0.000
Travel preference	28.520	3	0.000

aim to improve the MRT service in the future. This section is delivered on a 5-point Likert scale (from 1—very negative to 5—very positive).

4 Results

4.1 Socio-Demographics

The data was collected from 457 respondents in Jakarta, Depok, Tangerang, and Bekasi by using an online questionnaire. The data was separated into two groups: Private Vehicle Owner (PVO) and Private Vehicle Non-owner (PVNO) in order to see the scale of demand for the MRT according to motorists' view. Interestingly, the female population within the sample has slightly more private vehicles than the male population in PVO, while in PVNO the female population is much more dominant than the male population, as shown in Fig. 1. At a 5% level of significance, from the data, there is sufficient evidence to conclude that the distributions of gender for PVO and PVNO are not the same, χ^2 (1, $N = 457$) $= 17.092$, $p < 0.05$. The employee

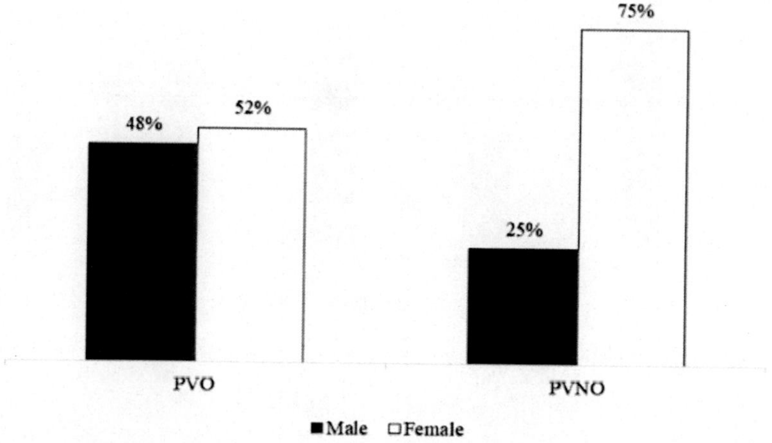

Fig. 1 Respondents' private vehicle status

population greater in PVO than in PVNO, while the student population greater in PVNO than in PVO, χ^2 (1, $N = 457$) = 7.062, $p < 0.05$.

Since the number of students in PVNO is higher than in PVO, it is reasonable that the number of respondents without income in PVNO is higher than in PVO, as shown in Fig. 3. The PVOs' respondents prefer using a motorcycle and a car for commute, while ojek online and walking are preferred modes for the PVNOs' respondents, as shown in Fig. 4, χ^2 (5, $N = 457$) = 122.491, $p < 0.05$ (Figs. 2, 3 and 4).

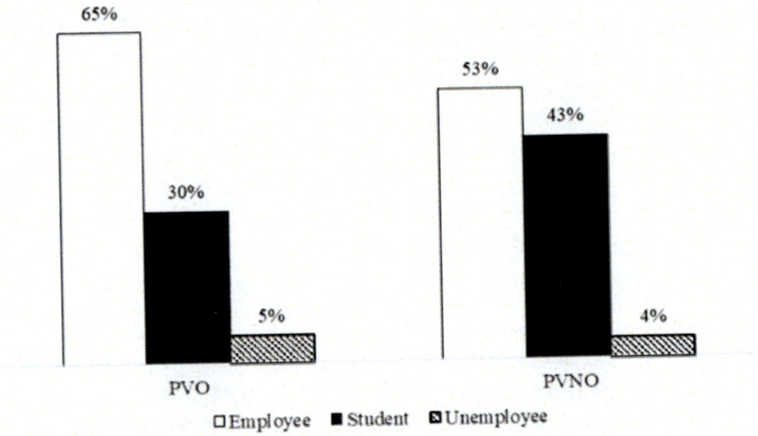

Fig. 2 Respondents' occupation

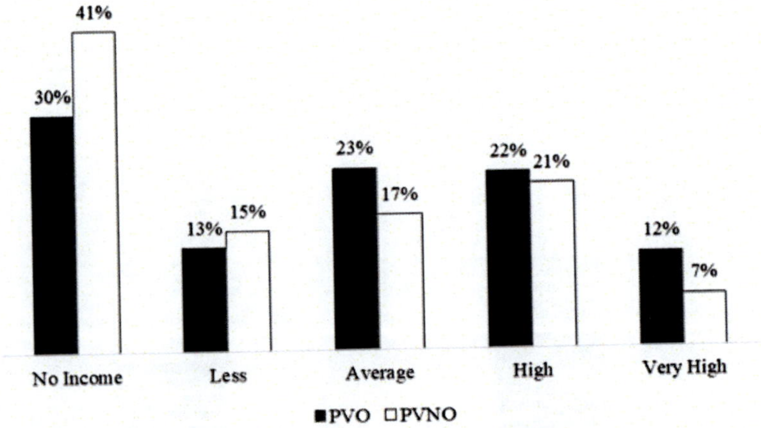

Fig. 3 Respondents' income

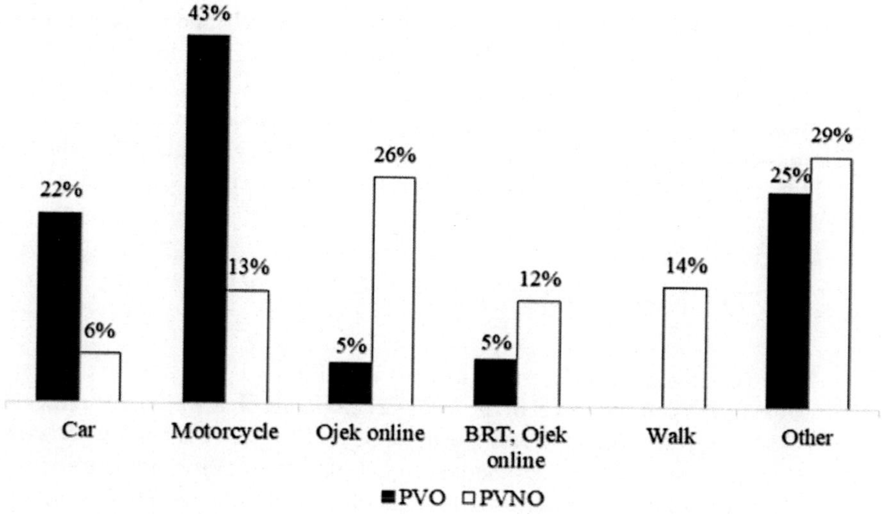

Fig. 4 Respondents' daily commute transport

4.2 *Travel Preference*

For the travel preference from A to B, surprisingly MRT becomes a preferred transport mode for both groups, where the shortest travel time is the greatest reason, as shown in Fig. 5. Even though the MRT has waiting and walking time as parts of effort, and private vehicles appear to be superior in this occasion that requires less

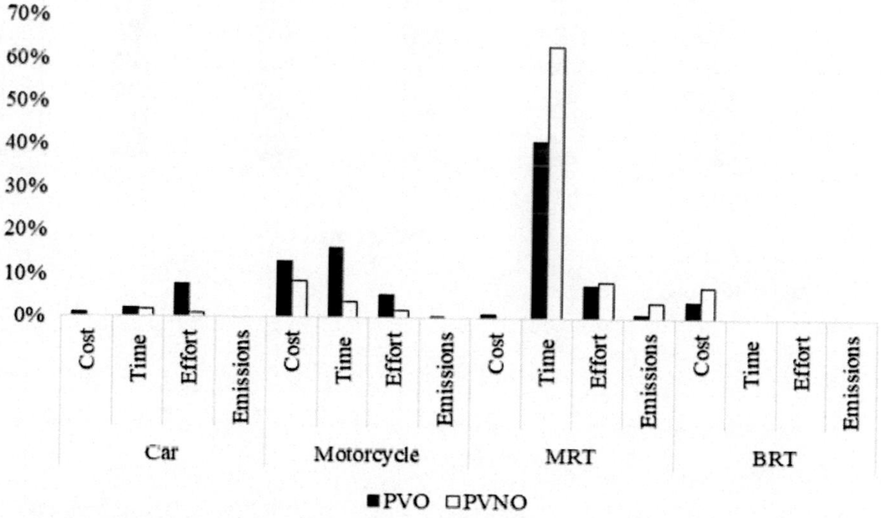

Fig. 5 Respondents' travel preference

effort, the respondents still consider the shortest travel time as the most important parameter. The respondents tend to neglect the amount of effort required to use the MRT. However, some respondents with private vehicles still prefer choosing a motorcycle or a car to MRT, while a great majority of respondents without private vehicles chooses MRT, χ^2 $(3, N = 457) = 28.530, p < 0.05$. In general, the respondents from both groups still choose MRT for their suitable transport mode for an A-B journey.

4.3 The Expectation Factors

The expectation factor's graph, as shown in Fig. 6, displays how the respondents assess the importance level of the MRT development on a 5-point Likert scale (where 1—very negative and 5—very positive). Overall, the six factors were assessed on the high score, where the respondents highlighted reliability and security as the most important and the second most important factor, respectively. There is no significant difference between the two groups according to independent t-test, as shown in Table 7.

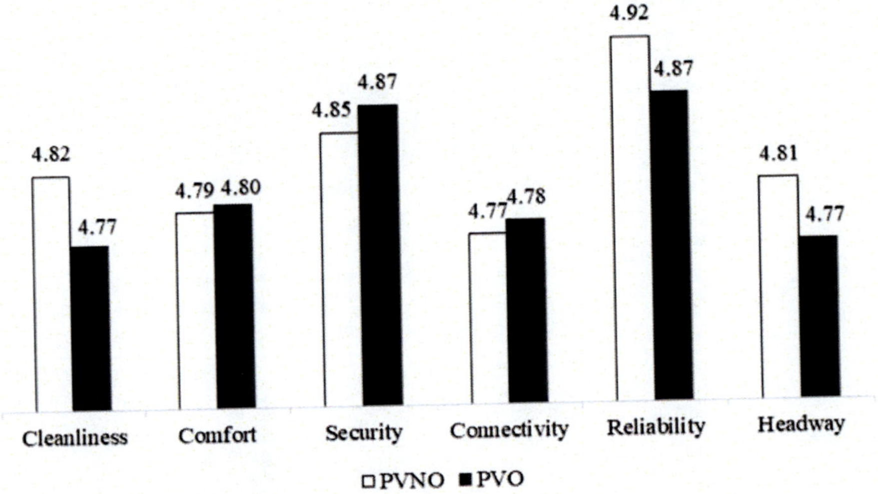

Fig. 6 Respondents' expectation factors

Table 7 Results of t-test for the expectation questions by classification

	Classification						95% CI for mean difference	*t*	*df*	*p*
	PVO			PVNO						
	M	SD	N	M	SD	N				
Cleanliness	4.76	0.713	351	4.82	0.493	106	−0.197, 0.094	−0.694	455	0.138
Comfort	4.79	0.562	351	4.79	0.511	106	−0.114, 0.125	−0.086	455	1.00
Security	4.86	0.483	351	4.84	0.473	106	−0.085, 0.124	0.373	455	0.521
Connectivity	4.78	0.554	351	4.77	0.503	106	−0.108, 0.128	0.164	455	0.939
Reliability	4.87	0.448	351	4.91	0.368	106	−0.134, 0.053	−0.846	455	0.104
Headway	4.76	0.587	351	4.81	0.554	106	−0.171, 0.813	−0.699	455	0.187

5 Discussion

5.1 Moving Motorists to the MRT

It is surprising that a great majority of the respondents from both groups chooses the MRT for a suitable transport mode for an A-B journey. Though the quickest travel time become a selling-point for the MRT, it attracts 51% of PVOs' respondents only and 45% still insists to choose private vehicles rather than the MRT. Shifting respondents who own a motorcycle to the MRT will be much more difficult than respondents with a car. Majority of the PVOs' respondents who choose a car based their decision on travel effort, while the respondents who choose a motorcycle based their decision on travel time and cost. A motorcycle has longer travel time than the MRT, but some motorists insist that a motorcycle is still quicker than the MRT. If the operator improves the system, where users only spend a little effort to use the MRT, car owners might be interested to shift to the MRT. However, this improvement may not be enough to attract motorcycle owners, who might require improvement on time and cost as well.

On the other hand, the respondents from both groups assessed the reliability as the most important factor that should be addressed for the MRT development. It might be possible that reliability and headway will become another MRTs' selling-points, which could attract car and motorcycle owners. The other factors should also be considered by the operator since shifting motorists to the MRT will face difficulties if the operator fail to capture passengers' voices.

In term of cost, it is expected that at the beginning, the demand for the MRT might be high since the MRT is the first such a system in Indonesia. However, there is a possibility that the demand will drop with time when financial issues take over from the respondents' curiosity about the new system. As seen in the literature section, cost is considered as one of the key factors when determining a transport mode selection, so the MRT ticket price is expected to play a significant role in attraction and retention of its customers. This view is also reinforced by Javid et al. (2013) stating that the ticket price of public transport should be (periodically) examined to see the effect of ticket price on the demand over time periods. Further, shifting motorists to the MRT also requires a good connectivity with other public transports (e.g. BRT) in order to create seamless journeys. Currently, majority of the respondents integrates BRT with ojek online. In the next years, the MRT might enable the integration of public transport with ojek online and BRT. This will prevent missing links so that potentially motorists should be much more interested to use the MRT.

5.2 Safety Issue

The image of public transports in Jakarta has been receiving criticism over years related to and described as overcrowded, uncomfortable and unsafe, as well as unclean and unreliable option (Susilo et al. 2010). Those problems keep worsening the congestion issue in Jakarta since the transit system is unable to provide an optimal service. On the other hand, the desire to use public transport, especially Jakarta MRT, turns out to be very high. The six expectation factors received very high scores from respondents, all being over 4.77 on a 5-point Likert scale (where 1—very unimportant and 5—very important). On the top of that, security might be viewed as important as reliability. The literature confirms that crimes have always been a problem of public transports in Indonesia (Munawar 2007; Susilo et al. 2010; Purba et al. 2015). A great majority of the respondents choose the MRT, which offers a safer travel option and guarantees the quickest travel time along the A-B corridor. However, the operator should consider the major findings of the study, such as importance of the security factor, which is connected to the safety issue in the public transport.

6 Conclusions and Recommendations

The huge number of private vehicles that operate on Jakarta's road causes severe traffic congestion. Jakarta MRT is expected to shift motorist from private to public transport. However, shifting people from private to public is much more complex since it involves social aspects. The aim of the study was to assess the competitive level of Jakarta MRT compared to private vehicles and the existing public transport. The comparison involved four parameters: cost, time, effort, and emissions. Then, the six expectation factors were also included in the study in order to capture passengers' voices about the MRT development in the next years.

Unsurprisingly, a great majority of PVNOs' respondents tend to choose the MRT since it has the quickest travel time for an A-B journey. Yet, nearly half of PVOs' respondents still prefer choosing private vehicles to the MRT. They claim that private vehicles are still superior to the MRT according to time and cost. Shifting motorists to use the MRT requires a lot of attentions in terms of reliability, cost, and security. The other factors, such as connectivity, headway, comfort, and cleanliness, are also seen as important for the MRT development in the future.

Taken together, the findings presented in the study support strong recommendations for the MRT operator, government, and the people of Jakarta and other surrounding areas to succeed in implementing a sustainable urban transport system. This recommendations are proposed taking into account the results of the study and the current conditions of the Jakarta MRT after the launch. There are three suggestions for the further development of the MRT:

1. The local government and the MRT operator should seriously consider the ticket price if they do not want to lose the market share and keep their prices competitive. Once the ticket price has been set, an evaluation should be undertaken, i.e. after one semester or one quarter, to analyze how high the demand in certain periods is, especially for workers and students, in using the MRT service. Users should also actively voice their opinions about ticketing through Jakarta MRT's public

relation channels e.g. website or social media. Lessons should be learnt and updated ticket prices and/or discounts applied, if needed.

2. In the future, how the MRT operator runs the business will greatly affect travel demand for Jakarta MRT and its competitiveness against other mode of transport in the city. Business process improvement emerges as greatly suitable to tailor the business service in line with the passengers' requirements. It allows the business process to be far leaner and more efficient regarding e.g. reliability on the system, the service time in the peak hours along the system, etc.
3. The MRT operator is obliged to pay attention to the factors that make all passengers comfortable and safe while using the MRT service. For example, a wagon equipped with surveillance cameras would be able to increase a public trust in security, prevent crimes and allow people to feel safe when using Jakarta MRT.

7 Limitations and Next Steps

Despite best efforts to run the study according to the plan, there were some limitations identified. Firstly, as the idea for the research was to investigate travel options, opinions and preferences of Jakarta residents in the context of the new MRT service before its launch on 24th March 2019, the time to collect new data was very limited, especially when internal procedures (e.g. university ethical approvals) were taken into account. The actual data collection took place from 1st February to 7th March 2019, so it was completed just weeks before the official launch of Jakarta MRT. This short timeframe for data collection, prior to the official launch, could have some impact on people's perception of the service. However, it is not clear whether this impact would be positive or negative. Secondly, a great majority of the respondents were young-adults, and the participation of adults or elderly in the survey was much lower. This most likely was caused by the fact that adults, especially the elderly, could have had some limitations in accessing the Internet. In the field, they were often busy with work, in a hurry or felt uncomfortable when approached to complete the survey. These conditions made their participation in the study limited.

Few options for further research work have been identified. Firstly, a greater research focus on ticketing and process business improvement could produce interesting findings that would account for increasing demand and improving the business service. Users' attitudes towards MRT's pricing system as well as their expectations regarding other payment options (e.g. smart cards) could help the operator to innovate and meet their customers' needs and wants.

Secondly, a research on satisfaction and perceived value of time would also assist the MRT operator in improving the quality of services and its overall business. There is no doubt that MRT has a travel time advantage for customers travelling between A and B, but it might be worth investing into an ongoing monitoring of the remaining three parameters studied (cost, effort and CO_2 emissions), compare them against different modes of transport and use the results to actively promote sustainable travel choices in Jakarta and highlight the MRT's competitiveness in fighting congestion and pollution.

Finally, further research on the effectiveness of Jakarta MRT in supporting the intermodal transport system in Jakarta would help stakeholders, including the local government, to understand the complexity of transport network interrelations, from ticket prices to interchanges to sustainability options and more, from their customers' perspectives. This also could assess the extent to which Jakarta MRT could be smoothly integrated with other transport modes, so that customers overall experience could be enriched.

Acknowledgements The authors would like to thank National Science Technology and Innovation Policy Office as a sponsor for The 6th Thailand Rail Academic Symposium (TRAS) 2019. We also thank Mahidol University and Thai Railway Engineering Association as co-organizers of the event. Moreover, the data collection procedure, including an online questionnaire, was approved by the Institutional Review Board, which is the internal Mahidol University ethical committee, with Certificate of Approval (COA) No. MU-CIRB 2019/013.2201. All respondents who completed the questionnaire were volunteers and all the data collected was anonymised, no sensitive details (name, address, etc.) were recorded.

References

Albalate D, Bel G, Fageda X (2015) Competition and cooperation between high-speed rail and air transportation services in Europe. J Transp Geogr 42:166–174

Badan Pusat Statistik (2017) Population and population growth rate by regency/city in DKI Jakarta Province, 2010, 2014, and 2015

Behrens C, Pels E (2012) Intermodal competition in the London-Paris passenger market: high-speed rail and air transport. J Urban Econ 71(3):278–288

Beirão G, Cabral JS (2007) Understanding attitudes towards public transport and private car: a qualitative study. Transp Policy 14(6):478–489

Birago D, Mensah SO, Sharma S (2017) Level of service delivery of public transport and mode choice in Accra, Ghana. Transp Res Part F Traffic Psychol Behav 46:284–300

Conversion factors 2017—condensed set for most users (2017)

Danapour M et al (2018) Competition between high-speed rail and air transport in Iran: the case of Tehran-Isfahan. Case Stud Transp Policy 6(4):456–461

De Vos J, Ettema D, Witlox F (2019) Effects of changing travel patterns on travel satisfaction: a focus on recently relocated residents. Travel Behav Soc 16:42–49

Deakin E, Frick K, Shively K (2010) Markets for dynamic ridesharing? Case of Berkeley, California. Transp Res Rec J Transp Res Board 2187:131–137

Dell'Olio L, Ibeas A, Cecin P (2011) The quality of service desired by public transport users. Transp Policy 18(1):217–227

Dinas Perhubungan Provinsi DKI Jakarta (2018) Data Perbandingan Jumlah Kendaraan Pribadi Dan Angkutan Umum DKI Jakarta

Gross R et al (2009) What policies are effective at reducing carbon emissions from surface passenger transport? A review of interventions to encourage behavioural and technological change

Gundelfinger-Casar J, Coto-Millán P (2017) Intermodal competition between high-speed rail and air transport in Spain. Util Policy 47:12–17

Herwangi Y et al (2015) Making the connection between transport disadvantage and motorcycle usage of low-income people in Yogyakarta urbanized area, Indonesia. J East Asia Soc Transp Stud 11:90–109
JABODETABEK Urban Transportation Policy Integration Project in the Republic of Indonesia (2012) Ministry of Economic Affairs Republic of Indonesia, Jakarta
Jain S et al (2014) Identifying public preferences using multi-criteria decision making for assessing the shift of urban commuters from private to public transport: a case study of Delhi. Transp Res Part F Traffic Psychol Behav 24:60–70
Javid MA et al (2013) Comparison of commuters' satisfaction and preferences with public transport: a case of wagon service in Lahore. Jordan J Civ Eng 7(4):426–472
Joewono TB, San Santoso D, Susilo YO (2015) Paratransit transport in Indonesia: characteristics and user perceptions. J East Asia Soc Transp Stud 11:1346–1361
Ko J, Lee S, Byun M (2019) Exploring factors associated with commute mode choice: an application of city-level general social survey data. Transp Policy 75:36–46
Lee D-J (2018) A multi-criteria approach for prioritizing advanced public transport modes (APTM) considering urban types in Korea. Transp Res Part A Policy Pract 111:148–161
Mukti I, Prambudia Y (2018) Challenges in governing the digital transportation ecosystem in Jakarta: a research direction in smart city frameworks. Challenges 9(1):14
Munawar A (2007) Public transport reform in Indonesia, a case study in the city of Yogyakarta. J Eng Technol 22
Munira S, Sarm S, San Santoso D (2013) Perception of public van users in Bangkok. J East Asia Soc Transp Stud 10:1501–1515
Papaioannou P, Politis I (2019) Applying behavior change theory to predict travel behavior of university commuters. In: The practice of spatial analysis. Springer, pp 219–251
Porter CD et al (2013) Effects of travel reduction and efficient driving on transportation: energy use and greenhouse gas emissions
Prasertsubpakij D, Nitivattananon V (2012) Evaluating accessibility to Bangkok metro systems using multi-dimensional criteria across user groups. IATSS Res 36(1):56–65
Purba A et al (2015) Evaluating TransJogja service quality based on user perceptions and expectations. J East Asia Soc Transp Stud 11:1173–1187
Redman L et al (2013) Quality attributes of public transport that attract car users: a research review. Transp Policy 25:119–127
Sloman L et al (2010) The effects of smarter choice programmes in the sustainable travel towns: summary report
Susilo YO et al (2007) A reflection of motorization and public transport in Jakarta metropolitan area. IATSS Res 31(1):59–68
Susilo YO, Joewono TB, Santosa W (2010) An exploration of public transport users' attitudes and preferences towards various policies in Indonesia. J East Asia Soc Transp Stud 8:1230–1244
Tyrinopoulos Y, Antoniou C (2008) Public transit user satisfaction: variability and policy implications. Transp Policy 15(4):260–272
Yen BT, Mulley C, Tseng W-C (2018) Inter-modal competition in an urbanised area: heavy rail and busways. Res Transp Econ 69:77–85
Zhang R, Yao E, Liu Z (2017) School travel mode choice in Beijing, China. J Transp Geogr 62:98–110

Safety Culture in a Railway Maintenance Environment: A Case Study of Bangkok Metro Network

Somkane Srathongkhruen and Anna Fraszczyk

Abstract A safety culture is an organisational culture, which places a high level of importance on safety beliefs, values and attitudes and these are shared by the majority of people within the company or workplace. This paper discusses the complexity of the safety culture approach in the railway maintenance environment. The aim of the paper is to better understand the rail safety culture from a developing country's perspective in order to advise how theory and practice could be integrated to improve it. The research explores factors that influence the relationship between safety culture and safety behaviour and examines differences in perceptions of safety culture within three metro operators in Bangkok, capital city of Thailand. Results of an online survey completed by the total of 97 representatives of the three companies show some differences in their sub-samples' characteristics, in terms of age and years of service, as well as in the way safety culture is perceived by employees involved in metro maintenance supervisory tasks. Overall, the study shows that training and supervision culture within companies scores highest while management commitment issues score lowest, on average. However, the scores are still high with all being between 70 and 90% positive range. Also, across four out of five different safety culture issues discussed, order of responses from highest to lowest is similar for the three companies, where one company is always at the top and the other always at the bottom. This shows that responses are consistent and highlight overall perception of safety culture within each company. Specific outcomes of the study and recommendations are widely applicable to any safety critical environment, but especially railways in developing countries.

Keywords Railway · Metro · Safety culture · Maintenance · Railway safety management

---

S. Srathongkhruen (✉) · A. Fraszczyk
Faculty of Engineering, Mahidol University, Salaya 73170, Nakhon Pathom, Thailand
e-mail: somkane1817@gmail.com

W. Weerawat et al. (eds.), *Urban Rail Transit*, Lecture Notes in Mobility,
https://doi.org/10.1007/978-981-15-5979-2_3

1 Introduction

Safety culture is a complex issue and many organisations came up with their own definitions of it. Safety culture, as defined by the Health and Safety Executive (HSE) (1993) is: "the product of individual and group values that determine the commitment to an organisation's health and safety management", in addition to technical safety issues (Smith 2016). The issue of safety culture in an organisation is very important because it involves people's safety, from managers to workers and to customers, and it often requires an investment of money and of time. In a weak safety culture environment, an incident might happen and, as a consequence, this might involve additional costs (Stevens 2014).

There is a large volume of transport safety culture research in developed countries, where dedicated organisations set standards and regulations for specific modes of transport and environments. For example, at a national level there is the Railway Standards and Safety Board (RSSB) in the UK, which is a not-for-profit organisation supporting its members (key rail industry players) in improving their safety and performance by developing and managing railway standards (ORR 2019). These standards refer to a catalogue of issues, from management and training to infrastructure and operations (RSSB 2019). At an international level, there is the International Union of Railways (UIC), which includes various industry safety working groups focused on issues such as: human factors, occupational health or level crossings (UIC 2019a). UIC is also involved in setting up international railway guidelines and standards across the railway system. Both of these organisations, and their many national and international equivalents, are involved in various safety culture and human factor research projects and publications: the majority of these are conducted in developed countries. For example, Gibson (n.d) looked at the role of human factors in supporting railway system safety in the UK and distinguished three layers of human factors related to: individuals (e.g. experience), job/workplace (e.g. communication) and organisation (e.g. supervision). Blais and Duda (2014) worked on a development of safety culture tools for the railway industry in Canada and recommended the use of a survey supported with interviews and workshops to measure safety culture within an organisation. Lee and Chung (2017) studied safety culture awareness in the context of improvements needed in methods and procedures of future accident investigations in South Korea. They employed a survey completed by 147 railroad staff and found that staff-management relationships and monitoring should be improved. Cheng (2011) focused on correlations between safety culture and safety performance in Taiwan. He sampled 536 train drivers and found that safety culture was positively and significantly related to safety performance of the Taiwan Railway Administration.

Overall, very little research dedicated to rail safety culture issues has been found in developing countries and these are the places where many new railway projects, especially new metro systems are under development (UITP 2018). Few exceptions, related to both general concept of safety culture and to railways specifically, include research by Khan (2013) or He et al. (2012). Khan (2013) from Pakistan investigated safety culture in developing countries and highlighted the complexity of the problem, where key issues are related to perception, data management, legislation, corruption and exploitation. He et al. (2012) from China looked very briefly at the safety culture framework and reminded that the relationship between safety culture and safety performance is both qualitative and quantitative.

Thailand is an example of a developing country which is rapidly expanding its railway infrastructure. The new railway projects vary, from Bangkok metro network expansion to high speed rail in other parts of the country. However, not much research attention is paid to the issues of workforce, which is often new to the rail sector and its WHS procedures, and safety culture as a whole. Therefore, a case study of safety culture in the railway maintenance environment of the metro networks in Bangkok, capital city of Thailand will be investigated in the paper.

The paper is organised in seven sections. Section 2 looks at the overview of safety culture issue and key factors associated with it. Section 3 presents a case study of the Bangkok metro network. Section 4 introduced methodology applied in the paper. Section 5 presents analysis of results and is organised in seven sub-sections, where each looks at safety culture from a slightly different perspective. Section 6 offers conclusions of the study and recommendations addressed to organisations involved in safety critical environments. Finally, Sect. 7 lists limitations of the study and suggests further research options.

2 Overview of Safety Culture Literature

2.1 *Safety Culture Framework*

Safety culture is a complex area, which can be influenced by various factors and actors involved in it. A positive safety culture is seen as a commitment made by managers and individuals involved in an organisation to always act safely (ERA 2017). Overall, three aspects of safety culture could be distinguished (HSE 2005):

•Psychological (‘safety climate’);

•Behavioral (‘organizational’);

•Situational (‘corporate’).

These aspects can be referred to an organisation and measured subjectively through the use of for example safety climate questionnaires, which aim to uncover the workforce’s attitudes and perceptions at a given point in time. This approach is summarized in Fig. 1, where links between psychological and behavioural aspects and also between behavioural and situational aspects are made. This means that the way people feel at work will affect their actions. These actions should be linked to standards and procedures, but due to ‘safety climate’ at work people might not be following them strictly.

Safety Management System (SMS) is a set of arrangements within an organisation, including documents and actions, which aim to ensure that the organisation achieves

Fig. 1 Adapted from HSE, 2005, p. 3–4

Fig. 2 Adapted from Drews, 2017, p. 4

its business objectives in a safe way (ERA 2019).The interaction of safety culture between the requirements of the SMS on what people say, what people have and what they do is referred in the ERA standard presented in Fig. 2. In a similar manner to Fig. 1, it shows that what people believe in will influence their actions in safety of talking and doing, and this will have an impact on safety outcomes. For example, in a rail context, if a manager believes in the importance of Health and Safety (H&S) training on a new equipment, he/she will ensure that all their staff is trained on this topic. As a consequence, staff will be aware of safety issues related to the topic and should 'talk' and 'walk' according to the rules.

2.2 *Key Safety Culture Factors*

Literature distinguished various factors that contribute to safety culture within an organisation. Key factors and their details are presented below.

2.2.1 Management

Management in an organisation has an important role to play in executing safety and making sure all procedures are respected and communicated to workers (Zizzo 2011). Management strategy includes issues such as commitment, ability, leadership, participatory style, or flexible culture co-ordination (Farrington-Darby et al. 2005). However, it is important to ensure that the SMS is operated by experienced and trained staff, who do not compromise on safety issues (Smith 2016).

2.2.2 Rules and Procedures

Safety rules and procedures in an organisation should be clear and practical (Farrington-Darby et al. 2005). Organisations are responsible for encouraging positive safety behaviours and providing appropriate operational procedures to their staff (Drews 2005; Smith 2016), so that it is clear who is responsible for what within an organisation.

2.2.3 Reporting System

The reporting system is another important factor that contributes to safety culture. A reporting system should include the following elements: reporting near misses, encouraging 'open door' policy and 'no blame' culture (Farrington-Darby et al. 2005; Drews 2005). In addition, risk and incident analysis should be regularly conducted and confidentiality should be guaranteed. Also, encouraging feedback from staff on H&S issues is a good practice. Investigation and analysis of any incidents should focus on system performance and factors that contributed to an event and avoid an individual responsibility approach (Drews 2005).

2.2.4 Communication

It is recommended that visibility of leadership style and commitment should be clearly communicated across an organisation (Farrington-Darby et al. 2005). Organisations should remember that communication is a human factor issue and it applies to all levels within a company (e.g. within teams, worker-supervisor). Therefore, methods of communication should be regularly assessed to ensure they are suitable and effective in the way they promote H&S messages to staff (Smith 2016) and beyond. Transport Canada (2010) encourages two-way (management-employee) communication via tools, such as for example: safety meetings and forums, mentoring, performance reviews, but also newsletters, brochures and training.

2.2.5 Training

Training is crucial in safety critical environments and an appropriate training offer should be available to staff before they start their specialised duties. For example, RSSB produced a number of documents on staff training in (RSSB 2019): on-train emergency procedures, ETCS principles for driver training or controller positions. The importance of training is also highlighted in the context of individual and behavioural factors (Farrington-Darby et al. 2005), staff (continuous) development or the continuous improvement of organisations (Drews 2005).

2.2.6 Relationships

The supervisor-subordinate relationship can affect the safety culture within an organisation, therefore it is important for supervisors to lead by example, promote 'open door' policy and encourage correct behaviour from staff (Farrington-Darby et al. 2005). Organisations should create an environment where in cooperation with staff they can continuously work on improving safety culture principles and practices and encourage positive safety behaviours (Drews 2005).

2.2.7 Good Practice Sharing

Sharing of good practice in safety culture is important so that organisations can learn from each other and avoid similar mistakes. RSSB currently has eight good practice guides available, where five of them relate to train operating companies (RSSB 2019). Examples of good practice documents include topics on technical issues, such as for example the use of de-icing agents on stations, and staff issues, such as for example fatigue management or individual risk factors. UIC also promotes sharing of good practice in safety within their working groups (UIC 2017) or supporting events promoting safety (e.g. International Level Crossing Awareness Day) (UIC 2019b).

2.2.8 Benchmarking

It is recommended to regularly review the maturity of the safety culture within an organisation and benchmark it against others in the railway industry and in other safety critical environments (Drews 2005). Transport Canada (2010) encourages self-evaluation of safety culture and this includes benchmarking as well as lessons learnt, so that continuous improvement can be achieved at all levels within an organisation.

2.2.9 Incentive Programme

In addition to all the above factors, an incentive programme could encourage workers to show their best safety behaviour and obey safety rules at work (Zizzo 2011). Such programmes have been implemented by various railway companies, for example Network Rail in the UK (Network Rail 2015). However, it is important to make sure that the incentive programme adapted is positive, non-discriminatory and encourages (not discourages) people from reporting incidents and accidents at work (Zizzo 2011; Johnson 2015). Balancing the for and against of a reward system should be considered by organisations when looking for tools promoting safety culture.

3 Case Study: Bangkok Metro Network

Bangkok metro network is the only metro network in Thailand. It currently includes three different systems and three operators running each of them. BTS was the first metro line in Bangkok opened in 1999, which today includes two lines (BTS Sukhumvit and BTS Silom) and 44 stations in total (BTS 2019). It is an elevated metro system, also called 'sky train', and all stations are open stations. BTS employs approx. 2000 people, where 300 of them work in a maintenance section.

MRT is another metro system in Bangkok, which currently includes one underground (Blue Line) and one elevated (Purple Line) line (MRTA 2019). Both lines are operated by BEM, who looks after 39 stations in total. BEM employs approx. 2500 people, including 500 maintenance workers.

Airport Rail Link (ARL) is a metro line operated by State Railways of Thailand (SRTET), which stretches from Phaya Thai station in the city centre to Suvarnabhumi station at the airport located in the Eastern part of the city (SRTET 2019). The line includes eight stations and there are plans to extend it to another airport (Don Mueang) located in the north of Bangkok. SRTET employs approx. 500 people, including 200 maintenance staff.

In addition to these three systems and five lines currently in operation, which total about 100 km of track, Bangkok is expanding the metro network with plans under way to reach 500 km of track (MRTA 2019; BTS 2019). New lines are currently under development, which include extensions of existing lines and construction of new, including two monorail lines. These infrastructure developments will require additional staff, which will have to be recruited to operate and maintain new lines and stations. This will increase the pool of people employed in the Bangkok metro network and who will require appropriate H&S training to ensure they positively contribute to their companies' safety cultures.

4 Methodology

Methodology for the study was designed in three steps: questionnaire design, data collection and data analysis and each is described in details below.

4.1 Questionnaire Design

A questionnaire is a common way of collecting data on safety culture. Despite some criticism that this method allows showcasing of a single perspective of a specific group of people only (He et al. 2012), it does provide some (quantitative) insight into a company's safety culture. Examples of organisations employing safety culture surveys can be found in the UK (Nexus 2015), Australia (NSW, n.d.) or Canada (Blais, n.d.). Therefore, a questionnaire was created for this study to collect data on safety culture. The questionnaire was available online in English and Thai versions and included seven parts. The selection of questions for the survey was inspired by the two safety culture surveys by Nexus (2015) and NSW (n.d.).

It is a common practice to ask respondents about their socio-economics so that links, if any, between their responses and characteristics could be identified. A similar approach of socio-economic questions was used in Nexus and NSW surveys. Therefore, in the introduction to the survey respondents were asked to state their:

- company name (ARL, BEM, BTS, sub-contractor);
- age (less than 25, between 26 and 34, between 35 and 49, between 50 and 64, 65 or over);
- gender (male, female);
- years of service (less than 1 year, 1–5 years, 6–10 years, 11–15 years, over 15 years);
- and position (supervisor or worker).

Next, the first five parts of the questionnaire were designed with a 5-point Likert answer scale, where:

- 1—never;
- 2—not often;
- 3—sometimes;
- 4—nearly always;
- 5—always.

Part 1 focused on training and supervision and included five questions related to: induction safety training, quality of supervision, company's encouragement, awareness of safety issues and training and access to information about safety training schedule.

Part 2 included five questions on safe work procedures linked to: reality of procedures, risk assessment, volume of H&S procedures, involvement in reviewing procedures and self-assessment of procedure following.

Part 3 asked five questions about consultation culture regarding: H&S requests from supervisor, involvement in safety matters, supervisor's approach to feedback and their involvement in H&S and provision of safety talks to staff.

Part 4 focused on reporting safety and asked five questions on: H&S reporting procedures, reporting (or not) of near misses and accidents and updates in safety procedures after an incident.

Part 5 included five questions on commitment of management, more specifically on: understanding of safety issues by managers, rapid response to H&S concerns, supervisor's involvement in H&S talks, management ignorance of H&S issues and company's involvement in H&S.

Part 6 provided three pictures of rail maintenance workers on a site and asked respondents to assess any issues they can see with H&S in each case.

Part 7 was the final and offered additional space for other comments.

No ethical issues regarding the questionnaire were raised and it was approved at the design stage by the university internal review board.

4.2 Data Collection

The data collection was designed to investigate H&S behaviour and opinions of signalling maintenance workers representing three metro operation companies in Bangkok. The participants sample was targeting two groups: Safety Management and Maintenance Workers. The Safety Management included those who design safety plans and safety rules and the employees were those involved in safety critical maintenance as part of their normal duties. More specifically, management workers work on track electrification, track renewals, switches, signalling and other tracks equipment.

Data collection would target three metro operating companies in Bangkok: ARL, BEM and BTS. Estimated size of each company's maintenance team and the planned sample size is presented in Table 1. Overall, a sample of 20 people from each company

Table 1 Estimated population and sample size in the study

Company	Estimated population size[a]	Sample size	Estimated percentage of the population represented by the sample
ARL	Approx. 200	20	10
BEM	Approx. 500	20	4
BTS	Approx. 300	20	6.7

[a]This data is not available publicly and it was estimated based on the authors' knowledge and informal discussions with contacts in each company

was planned. Respondents would be volunteers recruited via personal contacts and professional reference to suitable individuals.

It is recognised that the sample size, of 20 individuals per company, represents a small proportion of each company's population of maintenance supervisors and workers only. However, based on the authors' experience, it is estimated that this sample is realistic to achieve results in the timeframe dedicated to the project and using an online questionnaire as a data collection tool.

4.3 Results Display

As mentioned in Sect. 4.1, the first five parts of the questionnaire use a 5-point Likert scale for answer options. Values of responses to each question will be added and converted into percentages. For example, if all answers to Q1 in Part 1 had a value of '5′ for the 1st company, value of '4' for the 2nd company and value of '3' for the 3rd company, their conversion into percentages would be: 100%, 75% and 50%, respectively. Next, results to all questions in the first five sections will be converted into percentages, displayed on a graph and compared between the three companies studied. However, exceptions include five questions, where respondents rate negative statements (e.g. Q19: "Sometimes accidents are not reported") and ratings of 4 ('nearly always') or 5 ('always') actually describe a negative safety culture. The negative statements are included in Part 2: Q6, Q8; Part 3: Q14; Part 4: Q19; and Part 5: Q24.

5 Results

5.1 Sample Size

Three metro companies in Bangkok were approached for data collection and it was expected to collect 20 responses per company. Overall, the actual number of respondents who participated in the survey was higher than expected and included:

- ARL: 38 respondents;
- BEM: 30 respondents;
- BTS: 29 respondents;
- Sub-contractor: 2 respondents.

Table 2 displays respondents' numbers and their characteristics divided into three company's representatives and the sub-contractors. Interestingly, 100% of respondents are classified as supervisors, which means that it will not be possible to compare views of supervisors and workers, as originally planned. Also, ARL provided nearly double the number of respondents than originally expected and they were all included

Table 2 Respondents' characteristics [count]

Questions	ARL	BEM	BTS	Sub-contractor	Proportion of the total[a] sample (%)
Supervisor	38	30	29	2	
Worker	–	–	–	–	
Years of service					
Less than 1 year	–	–	–	–	–
1–5 years	11	13	6	–	32
6–10 years	21	10	19	–	53
11–15 years	6	6	3	–	15
Over 15 years	–	–	–	–	–
Age					
Less than 25 years	–	–	1	–	1
25–34 years	12	11	19	–	43
35–49 years	19	13	9	–	42
50–64 years	7	6	–	–	14
Over 64 years	–	–	–	–	–
Gender					
Male	30	24	27	2	84
Female	8	6	2	–	16

[a]The total sample excludes the two sub-contractors

in the analysis. The two sub-contractors responses were excluded from further analysis due to the small sub-sample size.

Overall, just over half of the sample has been working for their companies for the period between 6 and 10 years, where 32% are less experienced (1–5 years) and 15% are more experienced (between 11 and 15 years). None of the respondents has been employed by their company for more than 15 years, which means that they started their employment after the companies launched operation of the Bangkok's metro lines (the first BTS line was opened in 1999). In terms of age, the youngest workforce member amongst the respondents is within the BTS sub-sample, where 67% are 34 years old or younger in comparison to 32% in ARL and 37% in BEM in the same age category. Both ARL and BEM have a fifth of their sub-samples age 50 or over, and this age category is not represented within the BTS sub-sample at all.

An overwhelming majority of respondents are male (84%) with the largest proportion of females in ARL (8 out of 38 equals 21%) and BEM (6 out of 30 equals 20%) and the smallest in BTS (2 out of 29 equals 7%).

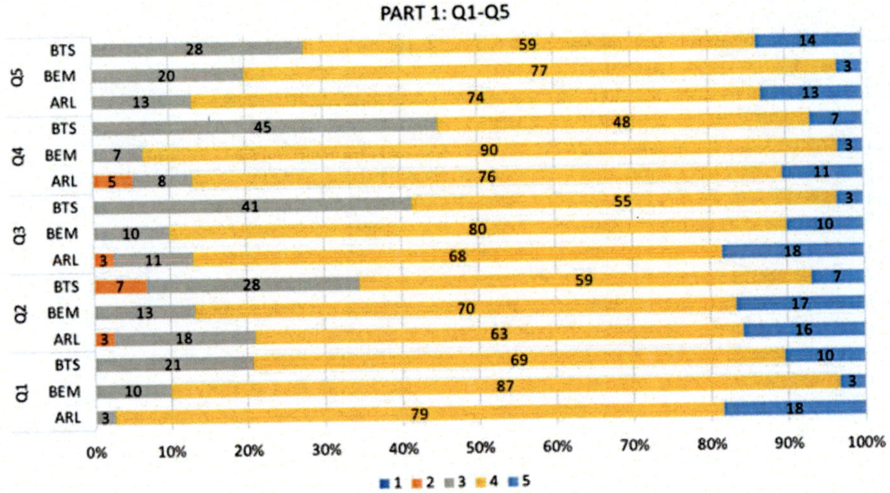

Fig. 3 Respondents' answers to questions 1–5 regarding training and supervision split between the three sub-samples

5.2 *Training and Supervision*

The five training and supervision questions were answered on a 5-point Likert scale, then numbers within each sub-sample were converted into percentages. Results displayed in Fig. 3 show responses to questions 1–5 split between the three companies. Overall, majority of respondents within each sub-sample scored 4 or 5 in each question. However, the largest proportion of employees who scored 3 or less was always from BTS, with the lowest scores for Q3 ("The company encourages suggestions on how to improve health and safety") on suggestions how to improve safety (41%) and Q4 ("I am always made aware of safety issues and safety training") on H&S training awareness (45%). On the contrary, the highest proportion of respondents who scored 5 was always from ARL, who formed between 11% (Q4) and 18% (Q1: "I got induction safety training when I started" and Q2: "Supervisors rarely check that people here are working safely"). On average across the five questions, 80% or more respondents from ARL and BEM ranked the five questions positively (score 4 or 5).

5.3 *Safe Work Procedure*

On average, responses to the five questions in Part 2 of the survey were slightly less positive than in the previous section, but still about 70% of respondents ranked them at 4 or 5. However, two out of five questions were actually using negative statements

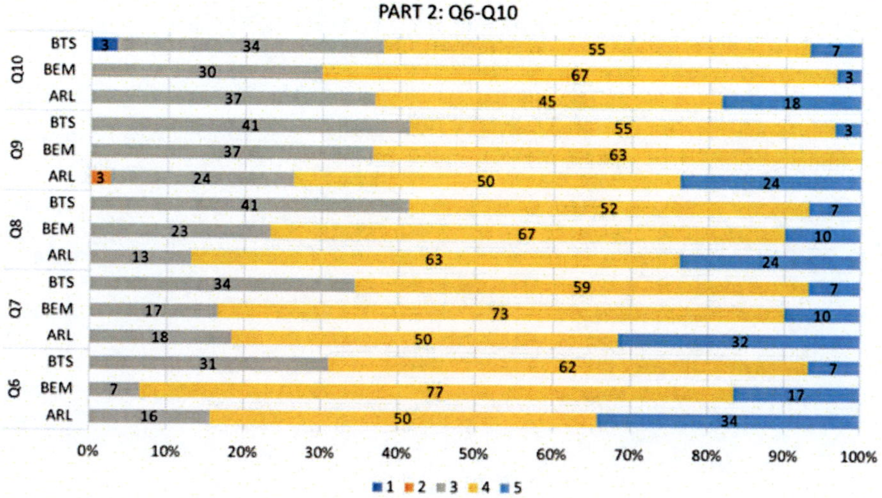

Fig. 4 Responses to safe work procedure questions

(Q6 and Q8), so higher scores translate into negative responses. Details are shown on Fig. 4.

BTS respondents' answers to Part 2 questions include lower scores (up to 41% for Q8 and Q9) than in the other two sub-groups. The highest positive score of 94% was achieved by BEM sample for Q6 ("Some health and safety procedures do not reflect how the job is actually done") and this is the only example in this part where high scores of 4 and 5 reach over 90%. However, this actually means that respondents agree with the statement that some H&S procedures in their workplaces are not adequate to their work environments, which overall brings a negative message. Similarly, responses to Q8 ("There are too many health and safety procedures given the real risks of my job") scored between 59% (BTS) and 87% (ARL) of 4 and 5 values, which means that respondents agree with the statement that the volume of H&S procedures applicable to their workplace is too high. Overall, the lowest scores across the three sub-samples were given to Q10 ("I always respect and follow any safety work procedures from my company"), with values between 62% (BTS) and 70% (BEM), meaning that the remaining 30% + respondents (between 30 and 38%) respect the H&S procedures 'sometimes' (value 3) only.

5.4 *Consultation*

Part 3 of the survey explored issues of H&S consultations between employees and their supervisors or managers. Within ARL sample a quarter (in Q15) or a third (in Q11 and Q13) declared that they always receive appropriate H&S support from their supervisors. However, Q14 highlighted a common view across all respondents

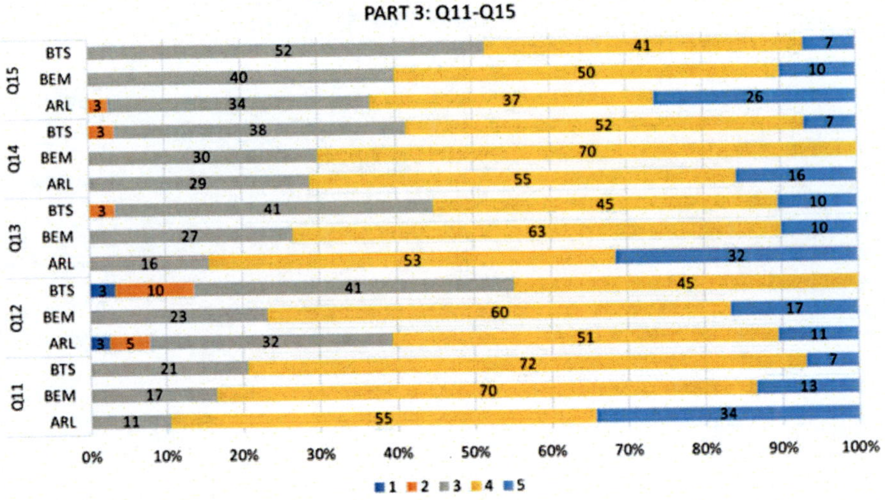

Fig. 5 Responses to consultation questions

that their supervisors do not do enough to secure H&S in their workplaces. BTS sample had less positive views about these issues with about a half of respondents being only 'sometimes (value 3) satisfied with their supervisors' approach to H&S consultations. Details are displayed in Fig. 5.

5.5 *Reporting Safety*

A great majority of respondents declared that they know H&S procedures and they 'always' (score 5) or 'nearly always' (score 4) use them, if necessary. However, responses to Q17 regarding 'near misses' reveal that they are not always reported with 45% of BTS respondents, 24% of ARL and 17% of BEM agreeing to this. Also, responses regarding procedures updates after incidents reveal that this happens only 'sometimes (score 3) and not 'always' (score 5) as would be required by H&S procedures. More details are shown on Fig. 6.

5.6 *Management Commitment*

Between 80 and 90% of respondents agree that managers must 'always' (score 5) or 'nearly always' (score 4) understand H&S issues and their consequences. The majority confirms that their supervisors talk to them about H&S, but still between 23 and 31% states that this happens only 'sometimes' (score 3). A worrying majority declares that their management 'sometimes' or 'nearly always' turns a 'blind eye'

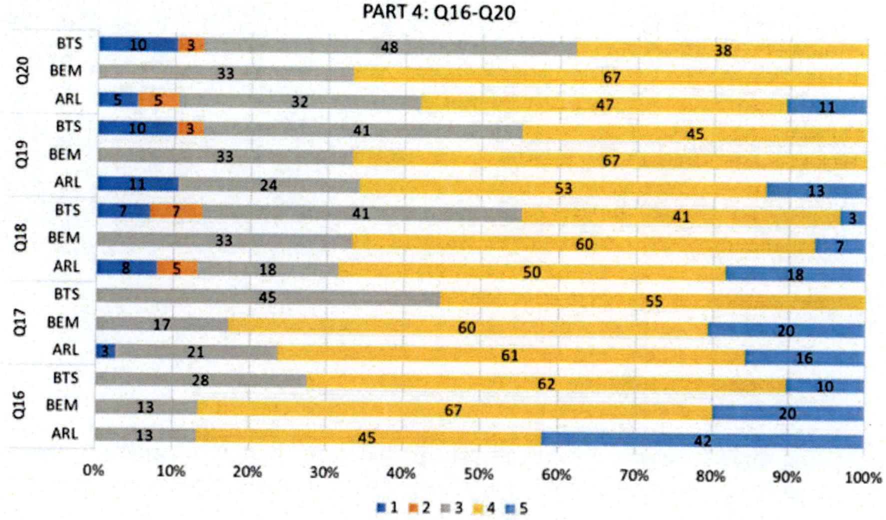

Fig. 6 Responses to reporting safety questions

on breaking H&S procedures at work. About 70% of the sample believes that their companies really care about H&S of their workers, but the remaining 30% says that this happens 'sometimes' only. More details are displayed on Fig. 7.

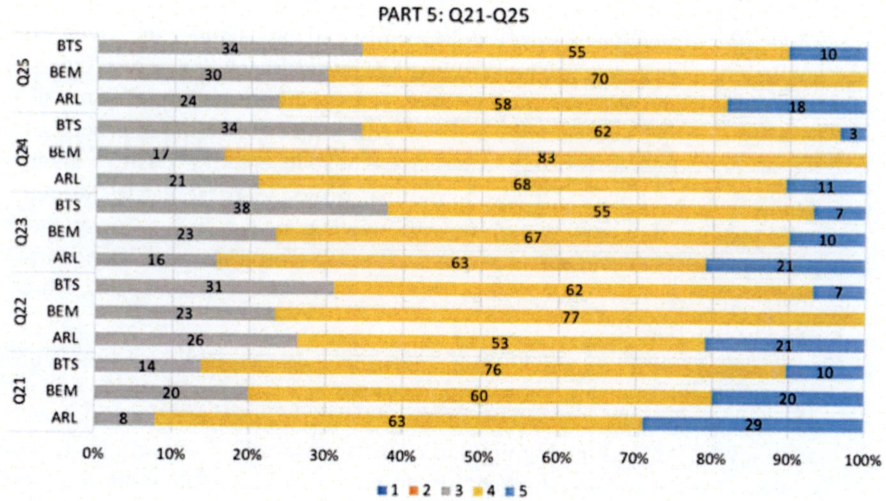

Fig. 7 Responses to management commitment questions

Fig. 8 Three case study scenarios displayed in the questionnaire

5.7 *Case Study*

The three scenarios in Fig. 8 were included in the questionnaire to assess H&S knowledge of respondents. They were asked to think independently and identify any H&S issues with case studies displayed. The pictures displayed various H&S rule violations regarding for example supervision or safety equipment. All respondents answered according to WHS rules which shows that they have a good safety awareness and can identify hazards displayed in the cases correctly.

6 Conclusions and Recommendations

This paper investigated safety culture in a railway maintenance environment by employing a questionnaire to study work safety culture as perceived by respondents from three Bangkok metro companies. The results highlighted some differences between the three sub-samples in the way they perceive training, procedures, supervision, consultation, reporting or management in creating positive safety culture within their workplaces. It was found that respondents from ARL are on average more positive about safety culture within their company than the respondents from BEM or BTS. However, all responses on average scored between 70 and 80% (around 4 on a 5-point Likert scale) out of 100% (5 of a 5-point Likert scale), which is seen as a positive result.

It is recommended for the three employers to follow-up on these findings by organising internal interviews and focus groups as well as reviews of rules and procedures to ensure that safety culture is well monitored and communicated to staff at all levels. Any issues highlighted in the survey and of concern to the employers could and should be investigated further internally. These follow-up activities should then inform the development of tailored action plans to further improve WHS.

The companies could also consider introducing some form of an incentive programme where employers could be rewarded for their positive safety culture behaviour. Examples of such programmes are present in other railway companies,

Table 3 Factor in unsafe acts and the recommendation to the three companies

Factors	Recommendations to employers
Training	• Provide H&S training to all managers, supervisors, workers, contractors and subcontractors • Ensure that training is provided in the language(s) and at a literacy level that all workers can understand
Management leadership	• Communicate the policy to all workers and, at appropriate times and places, to relevant parties • Reinforce management commitment by considering safety and health in all business decisions, including contractor and vendor selection, purchasing, and facility design and modification
Reporting system	• Discuss management issues at staff meetings and involve workers in the process • Encourage workers to comment on the H&S of maintenance activities they are involved in • Collect regular feedback from employees with suitable suggestions to improve the reporting process
Communications	• Improve communication between the host employer and contractors and staffing agencies to determine responsibilities and duties of each party involved and to ensure protection of all on-site workers before work begins • Identify issues that may arise during on-site work and include procedures to be used by the host employer and contractors and/or staffing agencies for resolving any conflicts before work starts

for example in Network Rail in the UK, but as literature suggests these need to be implemented carefully with suitable measures encouraging fair and safe reporting.

Specific recommendations are linked to key four factors of: training, management leadership, reporting system and communications and are listed in Table 3.

7 Limitations and Next Steps

A number of limitations of the study can be identified. Firstly, it is recognised that this study focused on quantitative data only with a questionnaire as a key data collection tool. If additional data collection methods, such as interviews, focus groups or observations, were used, the final outcome of the study could have been more complex. Therefore, a follow up on qualitative data regarding safety culture could be the next step with the study.

Secondly, the sample included 97 metro maintenance workers employed on Bangkok metro networks who completed the online survey voluntarily. However, it is not exactly clear what proportion of the maintenance workers they represent and how representative the sample is in terms of socio-economics and overall safety culture attitudes and perceptions across the population of interest. A larger sample size would allow conclusions to be drawn on safety culture represented by the companies' populations rather than a small sample of their workers.

Thirdly, the survey attracted responses from supervisors only who formed 100% of the sample. Although it was originally planned to compare views of supervisors and workers, due to the composition of the sample this was not possible. If views of both groups were collected, then results could have highlighted differences between the two perspectives when executing H&S procedures.

Finally, results presented in the paper are based on simple statistical analysis including average values converted into percentages. If more sophisticated analyses were employed it could have been possible to search for some correlations between respondents' socio-economics and their views on specific elements of safety culture investigated in the study as well as links between answers to different questions (e.g. training vs. reporting).

Acknowledgements The authors would also like to extend their gratitude to the ARL, BEM and BTS companies, their safety and maintenance departments as well as the participants for their help in this project.

References

Blais D (n.d.) Developing a tool to measure and improve the safety culture of Canada's Railway Industry. Presentation. Available at: https://www.safetyculturesymposium.ca/uploads/2/6/7/3/26734471/presentation_safety_culture_halifax.pdf. Accessed 14 Aug 2019

Blais D, Duda R (2014) Development of a safety culture measurement and improvement tool for the Canadian Railway Industry. International Railway Safety Council, 12–17 Oct 2017, Berlin, Germany

BTS (2019) Company's profile. Available at: https://www.bts.co.th/eng/index.html. Accessed 16 Aug 2019

Cheng YH (2011) Safety culture, Safety behavior and safety performance in Railway companies. World Congress on Railway Research (WCRR), Lille, France

Drews K (2005) Safety culture. HU training June 2017, pp 1–10

ERA (2017) Promoting a positive railway safety culture. Leaflet. Available at: https://www.era.europa.eu/sites/default/files/library/docs/leaflets/promoting_a_positive_safety_culture_en.pdf. Accessed 14 Aug 2019

ERA (2019) Safety management system. Available at: https://www.era.europa.eu/activities/safety-management-system_en. Accessed 14 Aug 2019

Farrington-Darby T, Pickup L, Wilson J (2005) Safety culture in railway maintenance. Safety Sci 39–60

Gibson H (n.d.) RSSB and the role of human factors in supporting railway system safety. Available at: https://www.ipa.go.jp/files/000042513.pdf. Accessed 14 Aug 2019

He A, Xu S, Fu G (2012) Study on the basic problems of safety culture. Procedia Eng 43(2012):245–249

Health and Safety Executive (1993) Safety culture. Health and Safety Executive

Human Engineering (2005) A review of safety culture and safety climate literature for the development of the safety culture inspection toolkit. Health Safety Executive 3–42.

Johnson D (2015) Safety incentives & OSHA. In: Industrial Safety & Hygiene News. Available at: https://www.ishn.com/articles/102417-safety-incentives-osha. Accessed 15 Aug 2019

Khan MI (2013) Developing a safety culture in developing countries. In: International Conference on Safety, Construction Engineering and Project Management (ICSCEPM 2013), 19–21 Aug, Islamabad, Pakistan

Lee S, Chung S (2017) Correlation study on safety culture awareness and accident investigation recognized by railroad staff. Int J Appl Eng Res 12(21), 10617–10624. ISSN 0973-4562 (2017)

MRTA (2019) MRT project. Available at: https://www.mrta.co.th. Accessed 16 Aug 2019

Network Rail (2015) Annual Incentive Plan 2015–16. Available at: https://www.networkrail.co.uk/wp-content/uploads/2016/11/Annual-Incentive-Plan-2015-2016.pdf. Accessed 15 Aug 2019

Nexus (2015) Nexus safety climate survey

NSW (n.d.) New South Wales safety culture survey

ORR (2019) RSSB. Available at: https://orr.gov.uk/about-orr/who-we-work-with/safety-bodies/rssb. Accessed 14 Aug 2019

RSSB (2019) Standards catalogue. Available at: https://catalogues.rssb.co.uk. Accessed 14 Aug 2019

Smith, P (2016) Safety case for the introduction of new technology into an existing railway system, 21. Imperial College London Department of Civil and Environmental Engineering Centre for Transport studies

SRTET (2019) https://www.srtet.co.th. Accessed 16 Aug 2019

Stevens, J. (2014) Why safety culture is more important than productivity. Available at: https://www.ehstoday.com/blog/why-safety-culture-more-important-productivity. Accessed 20 Apr 2018

Transport Canada (2010) Achieving an effective safety culture. Available at: https://www.tc.gc.ca/eng/railsafety/rsc-615.htm. Accessed 15 Aug 2019

UIC (2019a) Safety. Available at: https://uic.org/safety/. Accessed 14 Aug 2019s

UIC (2017) Safety at UIC. Brochure. Available at: https://uic.org/IMG/pdf/brochure_safetynew.pdf. Accessed 15 Aug 2019

UIC (2019b) International level crossing awareness day. Available at: https://uic.org/landing-pages/ilcad. Accessed 15 Aug 2019

UITP (2018) World metro figures. Statistics in brief. Report

Zizzo S (2011) 8 steps to a strong safety culture. Available at: https://www.ishn.com/articles/91474-8-steps-to-a-strong-safety-culture. Accessed 21 Apr 2018

Passenger Behavioral Response to Off-Peak Fare Reduction in Airport Rail Link, Bangkok, Thailand

Thanchanok Inmor, Kunnawee Kanitpong, and Piyapong Jiwattanakulpaisarn

Abstract This study has the main objective to analyze the effect of fare reduction on passenger travel behavior. A questionnaire has been used as a tool to analyze the travel characteristics and factors that motivate passengers to shift travelling time. Also, the amount of time that travelers are able and willing to change in Bangkok has been investigated and analyze from the questionnaire. The population are the Airport Rail Link (ARL) passenger in Thailand. The sampling size is 740 passengers using Taro Yamane formula with the 97.5% confident level. The distribution of questionnaires is both onsite and online surveys. The Binary Logit Model was using to analyze. The data has been analyzed by the STATA program. The results showed that the passenger who are older, higher income, higher level of education degree are not likely to change their travel times compare with the lower groups. For the factors that motivate them to shift time, the findings suggested that the rate of discount, length of time shift, and the travelling distance have a significant effect at 0.01. The Binary Logit Model analysis found that the shift time of 1 h with the satisfaction rate of discount between 8.40–19.20 baht in the morning and in the evening with 6.60–12.60 baht are the times that people are able and willing to change for a given discount. The study results can be used with other policies to give discount rates to passengers who are willing to shift their travel times. This will help at least to solve the overcrowding problem during the rush hours at ARL stations.

Keywords Off-peak discount · Travel time displacement · Crowding · Peak demand management · Peak spreading

T. Inmor (✉) · K. Kanitpong
Asian Institute of Technology, P.O. Box 4, Klong Luang 12120, Thailand
e-mail: thanchanokinmor@gmail.com

P. Jiwattanakulpaisarn
ThaiRoads Foundation, Ladprao Road, Jomphol, Bangkok 307-30810900, Thailand

W. Weerawat et al. (eds.), *Urban Rail Transit*, Lecture Notes in Mobility,
https://doi.org/10.1007/978-981-15-5979-2_4

1 Introduction

Suvarnabhumi Airport Rail Link (ARL) and City Air Terminal has been opened since 2010. The passengers who wish to travel from downtown to the airport can use ARL with good service at a speed of 160 km per hour on an elevated platform parallel to the eastern railway. The main purpose of ARL is to provide more convenient, faster, and a more flexible transportation service with 8 stations along the route between downtown Bangkok to Suvarnabhumi airport (Phayathai Station, Ratchaprarop Station, Makkasan Station, Ramkhamhaeng Station, Hua Mark Station, Ban Thab Chang Station, Lad Krabang Station and Suvarnabhumi Station). ARL has a total distance of 28 km and passenger capacity of 14,000–50,000 passengers per day per direction. Two train systems, express line and city line, are in service nowadays. The express line, or Suvarnabhumi Airport Express starts from Makkasan Station directly to Suvarnabhumi Airport Station. It only stops at the two terminals taking about 15 min. The city line provides service between Phayathai Station, which is located in Bangkok downtown as an interchanging point to other mass transit systems and Suvarnabhumi Airport Terminal. The city line stops at six stations including City Air Terminal (Airport Rail Link 2017).

Since 2011, the ridership of ARL has tremendously increased with incremental expansion of the line (Fig. 2). In 2015, over 19 million of passengers were reported, and a total of 80 million passengers were reported to use ARL.

Recently, ARL has faced congestion problems especially in rush hours in both the morning and evening periods (Fig. 1). Congestion affects passengers in such ways being uncomfortable with significant delays. Many cities are employing time-based Public Transport fare pricing strategies for spreading peak hour travel demand in face of intense resource constraints for expanding Public Transport capacity. This is not a recent trend as it is has been used under a different name as time-of-day differential fares in the US in the 70s (Cervero 1986). This strategy is characterized by activities such as management of price using peak surcharges and off-peak discounts. Peak surcharges have been used extensively in the transport sector especially in the private

Fig. 1 Passenger congestion at Airport Rail Link Station

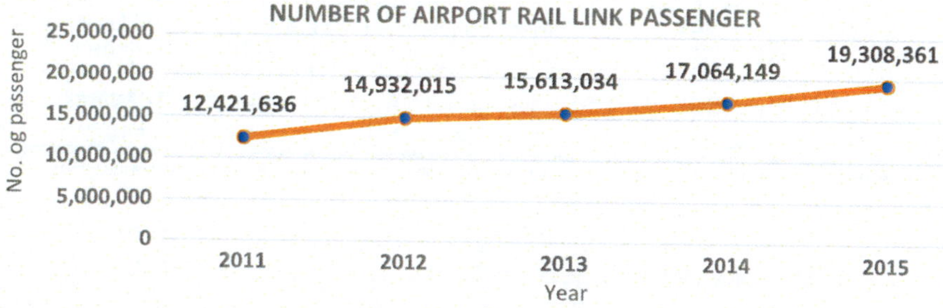

Fig. 2 Number of Airport Rail Link passenger

domain during rush hour in the form congestion charge. In recent years, there has been use of fare-free public transportation policies in Melbourne and Singapore. In theory, this can be considered as the largest discount that could be offered to commuters before peak travel hours to encourage them to travel less during peak morning period. It should be noted that such discounts are very rare (Currie 2010).

The specific objectives of this study are:

(1) To analyze the characteristics of people willing to shift time of travel.
(2) To study the factors that motivate people to shift time of travel.
(3) To investigate the amount of time that people are able and willing to change for a given discount.

2 Background

Based on previous studies, there are many factors that could influence passengers to shift the time of travel to off-peak periods. Many studies consider the fare reduction and surcharge as a factor contributing to passenger behavior.

Nature (2009) found that increasing the fee by 20% during rush hours could motivate passengers to change their time to an earlier or later time by 13% and a 20% off-peak fare discount in integration with a 20% high peak surcharge could motivate passengers to change travel times by 20%. RailCorp (2008) found that when comparing peak fare surcharges and off-peak discounts, passenger's willingness to travel during off-peak hour increased from 43 to 53%. The survey found that fare reduction in off-peak hour in the morning time could motivate passengers to travel outside of peak period by 40% (Consolidated Communications 2006). Whelan and Johnson (2004) conclude that increasing fare differentials between peak and off-peak hour can reduce overcrowding on the train more substantial.

Furthermore, there are factors which also affect the decisions of passenger to shift time of travel, such as time flexibility and socioeconomic conditions. Faber (2007) found that 45% of passengers do not have the flexibility to change travel time. However, passengers around 55% of passengers have the flexibility to change travel times including those who could shift this time by around 30 min. In addition,

Fig. 3 ARL station for survey location

high income people are more flexible in changing travel times than those with low incomes, and passengers with long distances are likely to change travel times while short-distance travelers need the fare benefits to motivate them. There was a strong relation between morning and afternoon peak travel. If commuters changed their morning travel time, then changes were likely in the evening. The results also found that, the 56% of passengers had more flexibility to travel earlier rather than later, this was considered to reflect the need to arrive at work by a certain time.

3 Methodology

This study uses the Stated Preference (SP) survey. The questionnaire will ask passengers if they willing to shift time of travel from peak hour or not if provided a fare discount in the off-peak period. Passengers are provided a set of fare reduction options in any scenario. The respondents must consider between fare discount and amount of time of travel that they have to change. And each situation has alternatives for respondents to choose. SP survey can be used to explain or forecast the assumed situations that will happen in the future.

3.1 Specification of Study Area

The questionnaire survey by onsite survey conducted in all of 8 Airport Rail Link stations. The name of stations is shown in Fig. 3.

3.2 Observation Periods

The questionnaires handed to passengers who use the Airport Rail Link during the peak period (Morning 07.00–09.00 and Evening 17.00–20.00) on the weekday (Monday–Friday).

3.3 *Parts of Questionnaire*

This questionnaire consists of 3 main parts. The First part consists of demography is information of passengers. The second part consists of ARL usage information, and travel information, and the third part consists of alternatives for determining about fare discounts and attitudes of fare discounting strategy.

- Demographic information
  This part captures the socio-demographic factors of the individual include:
  - Gender
  - Age
  - Education
  - Status
  - Occupation
  - Income
- Travel information
  This part captures the travel information of the person completing the questionnaire including:
  - Number of Accommodation
  - Access Mode
  - Travel purpose
  - Frequency
  - Card type
  - Origin
  - Destination
  - Distance
  - Fare
  - Time for Using ARL
- Decision making due to fare discounting
  This part captures the decision of travelers towards changing time when offered a fare discount. This survey included the following question:
  - If the fare is discounted from the normal price (random in a range of 10 to 40%) in off peak hour, will you be willing to shift your arrival time at the ARL station to the time which is not in the period of considered time (which are 07.00–08.30, 07.30–09.00, 17.00–19.00 and 18.00–20.00)?

There are four scenarios in the questionnaire which depend on the arrival time at the origin station as shown in Fig. 4. For the fare discount, the percentage of discount will be separated in to 2 cases. Firstly, the passengers who buy tickets from the machines or travel by using adult cards were asked to determine the minimum discount that make passengers change the arrival time by randomness between 10 and 40%. On the other hand, passengers who travel using student cards were asked about

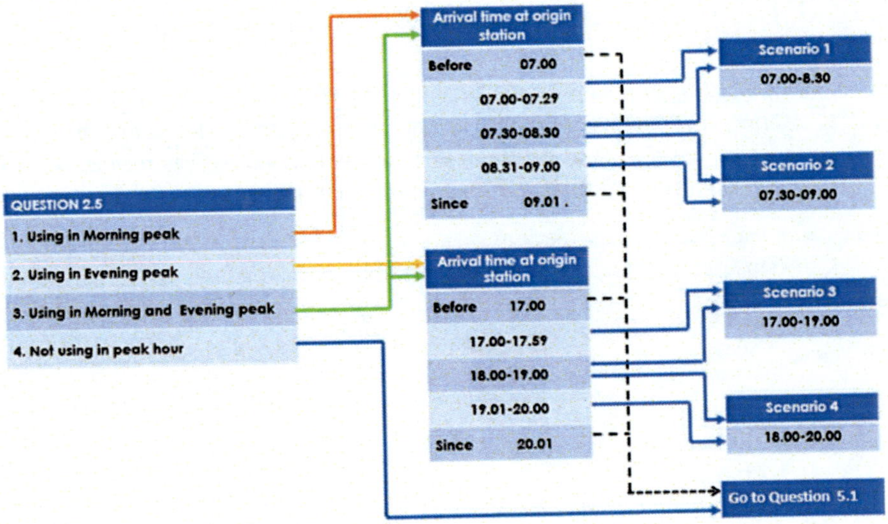

Fig. 4 Plan to determine passenger to the scenario set of questionnaires

random discount from 30 to 40%. However, the senior card holders who already get 50% discount will not be asked because this study does not consider a discount of more than 40% as illustrated in Fig. 5.

There are 4 conditions in each time period that are 10, 20, 30 and 40%. The responders have to decide between 2 alternatives, which are shift or not shift the arrival time of travel.

4 Results

This study determines characteristics of passengers who use ARL for traveling in peak hours who are willing to shift times of travel when motivates with fare discounts. This part analyzed the effect of fare reduction on passenger travel behavior. For analyzing the characteristics of travelers willing to shift time, there are two alternatives of travel options to decide which are willing to change travel times and not willing to change travel times. This study analyzes using Binary Logit Model (BNL).

$$Y = e^{\beta_0 + \beta_1 x_1 + \beta_2 x_2 + \cdots \beta_n x_n} \quad (1)$$

"Y" is a dependent variable which represents the decision of Airport Rail Link passengers when they get fare discounts to shift times of travel. The responder has two choices which are shift time of travel to early or later scenario times or not shift times of travel. The independent variables shown in "X_1, X_2, …, X_n" include; Age, Gender, Education etc. These variables would affect the dependent variable "Y".

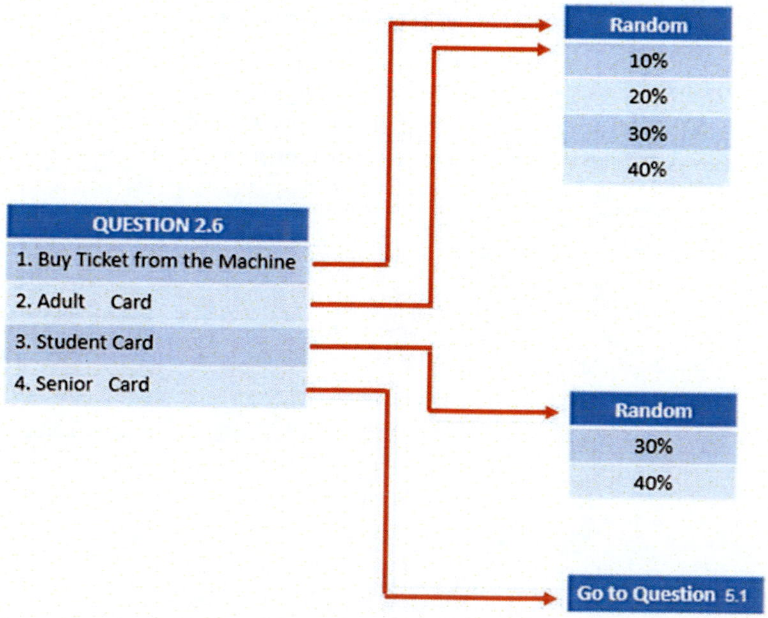

Fig. 5 Plan to determine test about fare discount

4.1 Results and analysis of all scenario 1–4

The Binary Logistic Regression Model was used to analyze the characteristics of passengers willing to shift time of travel from peak period. From the analysis checked Correlation variables are using Pearson correlation value, The value of correlation variables accepted at 0.8 in High co-linearity probability of Correlation Coefficient.

The test of multicollinearity indicates the existence of strong correlation among age of passengers lower than 24 year correlates with the age between 24 to 40 years, office worker and age lower than 24 years, age lower than 24 years and income lower than 15,000 baht with the correlation coefficient of −0.82, −0.83 and 0.83. The highest correlation also was found between office worker with income lower than 15,000 baht, study trip and work trip, work trip and office worker with coefficient −0.89, −0.81 and 0.85 respectively. There was a correlation coefficient of 0.82 and −0.83 between income lower than 15,000 with purpose study and purpose work. Another was found between purpose study and purpose work with the coefficient of −0.89. Therefore, one of each variables for all correlations will be excluded from the model, which are Age_L24, Occu_3, IncL15K and Purp_s. The definitions of dependent and independent variables for binary logistic regression model of scenario 1-4 as shown in Table 1.

As shown in Table 2 the model was analyzed by Binary logistic Regression to analyze the characteristics of passengers' responses to shifting time of travel to off peak hours. The results were presented by four different models. After analyzing

Table 1 Definitions of dependent and independent variables for binary logistic regression model of all scenarios, 1–4

Variable Type	Variable	Definition	Average
Dependent	Timeshift	1 = Shift travel time, 0 Not shift travel time	0.26
Independent	Male	1 = Male, 0 Otherwise	0.38
	Age_L24	1 = Age lower than 24 years, 0 Otherwise	0.35
	Age_2440	1 = Age between 24 and 40 years, 0 Otherwise	0.56
	Age_M40	1 = Age more than 40 years, 0 Otherwise	0.09
	Edu_LB	1 = Education lower than bachelor's degree, 0 Otherwise	0.26
	Edu_B	1 = Education with bachelor's degree, 0 Otherwise	0.58
	Edu_MB	1 = Education higher than bachelor's degree, 0 Otherwise	0.15
	Single	1 = Single, 0 Otherwise	0.76
	Occu_1	1 = Student, 0 Otherwise	0.19
	Occu_2	1 = Scholar, 0 Otherwise	0.14
	Occu_3	1 = Office worker, 0 Otherwise	0.64
	Occu_4	1 = Freelance, 0 Otherwise	0.03
	Inc_L15K	1 = Income lower than 15,000 baht, 0 Otherwise	0.34
	Inc_15to25	1 = Income between 15,000 and 25,000 baht, 0 Otherwise	0.23
	Inc_M25K	1 = Income higher than 25,000 baht, 0 Otherwise	0.43
	Accom	1 = Travel alone, 0 Otherwise	0.89
	Transfer	1 = Transfer mode with public transportation, 0 Otherwise	0.58
	Purp_S	1 = Traveling for study's purpose, 0 Otherwise	0.28
	Purp_W	1 = Traveling for work's purpose, 0 Otherwise	0.67
	Purp_P	1 = Traveling for personal' s purpose, 0 Otherwise	0.05
	TimeFix	1 = Fixed time of travel, 0 Otherwise	0.79
	Freq5days	1 = Travel 5 days or higher per week, 0 Travel lower than 5 days per week	0.86
	Card	1 = Travel using passenger card, 0 Buy ticket from the machine	0.65
	Sta_M3	1 = Travel more than 3 stations, 0 Otherwise	0.45
	Dist	Distance for traveling	13.84
	Opinion	1 = Agreed with the price reduction, 0 Otherwise	0.60
	Delay	1 = Travel Delay, 0 Otherwise	0.50
	Tchange	Changing Time (Minute)	53.71
	Discount_baht	Discount (Baht)	6.81

Table 2 Regression analysis results of factors affecting decisions by fare discount of all scenario 1–4

Independent variable	Model 1 Coef	Model 2 Coef	Model 3 Coef	Model 4 Coef	Model 5 Coef
Male	−0.13100***	−0.11860***	−0.12900***	−0.00974	
Age_L24					
Age_2440					
Age_M40	−0.15900*	−0.17570**	−0.16100 *	−0.17063**	
Edu_LB					
Edu_B	−0.47450***	−0.48700***	−0.47100***	−0.68952***	
Edu_MB	−0.94930***	−1.0258***	−0.95500***	−1.34450***	
Single	−0.23030***	−0.23480***	−0.23100 ***		
Occu_1	−0.65920***	−0.69410***	−0.66400***	−0.94747***	−0.34315***
Occu_2	−0.48230***	−0.49390***	−0.48600***	−0.44667***	−0.44371***
Occu_3					
Occu_4					
Inc_L15K					
Inc_15to25					
Inc_M25K	−0.16480***	−0.16730***	−0.16600***	−0.21417***	−0.46162***
Accom	−0.11070	−0.12580*	−0.11100		
Transfer	0.07525*	0.08142*	0.07500*	0.17886***	0.19590***
Purp_S					
Purp_W					
Purp_P	−0.28810**	−0.26140**	−0.28200**	0.02425	−0.04472
TimeFix	−0.34870***	−0.33470***	−0.34700***	−0.22812***	−0.28043***
Freq5days	−0.04020	−0.02110	−0.04100		
Card	−0.73540***	−0.74680***	−0.73400***		
Sta_M3	−0.37710***		−0.33600***	−0.36957***	−0.46236***
Dist	0.00426	−0.01420***			
Opinion	0.55042***	0.54360***	0.55100***		
Delay	−0.05390	−0.05440	−0.54000		
Tchange	−0.02080***	−0.02070***	−0.02100***	−0.02010***	−0.01981***
Discount_b ~ t	0.11286***	0.10736***	0.11424***	0.10781***	0.10518***
Summary of statistics					
Observations	1684	1684	1684	1684	1684

(continued)

Table 2 (continued)

Independent variable	Model 1 Coef	Model 2 Coef	Model 3 Coef	Model 4 Coef	Model 5 Coef
Log likelihood	−6636.7507	−6651.4374	−6637.1232	−6880.2166	−6973.8633

***Indicates significance at the 1% level
**Indicates significance at the 5% level
*Indicates significance at the 10% level

these 4 models, the result showed that males tend not to shift time of travel when compared with females. The model also found that passengers of age more than forty years were not likely to shift times of travel compared with passengers of age between twenty-four and forty. Education of passengers was also significant to the model. The results showed, passengers who have education degrees of bachelor and higher than bachelor's degree will tend not to shift times of travel from their current times compared with passengers who have education lower than bachelor's degree. Occupation of passengers who travel in peak hours was also significant. The results of all the models show that students and scholars tend not to change the time to use ARL early or later from the current time when compared with the freelance worker group. The result about income of passengers shows. Those who have income rates more than 25,000 baht are not likely to shift times for using ARL when compared with income rates at 15,000–25,000 baht.

For the travel information, most variables are significant with all four models. First, discount and time changing variables are the most important of all variables. From the result, when the time that passengers have to change from the normal time using ARL, increases, passengers are likely to not shift time of travel from the current time. For discount variables, the results show that if the discount was higher, passengers will tend to change travel time to off-peak hours. That means more discounts can motivate passengers to shift times to use ARL to off-peak hours. The results also found that passengers who travel more than 3 stations were not likely to change time of travel compared to passengers who have to travel less than 4 stations. Purpose of travel, passengers who have a time fixed for traveling and passengers who use cards for travel, was also significant. The results show that passengers who have a personal purpose of travel do not tend to shift travel time when compared with those who go to work, and this may be because passengers who have to go for personal purposes in peak hours have to travel at this time.

Passengers who have no flexibility of travel time tend to not change times of travel and passengers who use ARL cards for travel tend to not shift travel times when compared with passengers who buy tickets from the automated machines. Another variable that is significant from the model is the opinions of passengers with the question "Do you think fare discount in off-peak hour can motivate people to change travel time to off-peak hour or not?" The result shows that passengers who think fare discount can reduce congestion in peak hours tend to change travel time compared with passengers who think fare discounts cannot solve congestion problems in peak hours.

4.2 Results and Analysis of Scenario 1 (07.00–08.30)

A high Pearson correlation of ARL passengers responding to fare reductions was found between age lower than twenty four and age twenty four to forty with the correlation coefficient of −0.81, age lower than twenty four and office work occupation with −0.84 correlation coefficient, age lower than 24 and income lower than 15,000 baht with correlation coefficient at 0.82, it and also found correlation at 0.81 between age lower than twenty four and purpose of going to study. Office workers correlated with income lower than 15,000 baht, study purpose and work purpose with −0.88, −0.83 correlation coefficient value. Income lower than 15,000 baht also correlated with study purpose and work purpose at correlation coefficient values 0.83 and −0.82. Another correlation coefficient at −0.91 was found between study purpose and work purpose. Therefore, one variable that correlated with some variables was cut from the model for analysis which are Age_L24, Occu_3, Inc_L15K, Purp_S. The definitions of dependent and independent variables for Binary Logistic Regression Model of scenario 1 as shown in Table 3.

As shown in Table 4, a Binary logistic Regression Model was used to analyze the socio-economic characteristics and travel information affecting behavior of passengers willing to shift time of travel to off-peak hour. The results of analysis from scenario 1 are both negative and positive direction of significant variables which are described below.

Age of passenger: From the results, it was found that passengers of age more than forty years tend not to change travel time compared with passengers of age between twenty-four and forty years old. From the result it can be said that passengers who are older tend not to shift time of travel compared with younger age passengers.

Education level of passengers was also found to be significant. The result show that respondents with a bachelor's and higher than bachelor's degree are more likely not to change time of travel from 07.00–08.30 compared with who have education degree lower than bachelor's degree. If focused on coefficient value, the coefficient value of education more than bachelor's degree variable, is higher than bachelor's degree variable. That means passengers who have education higher than bachelors tend not to shift times of travel more than passengers with bachelor's degree when compared with education lower than bachelor's degree.

Status of passenger: A single variable was found that has a negative effect on the dependent variable. It means single passengers tend not to shift travel times if compared with female passengers.

Two group occupations of passengers, which are student and collegian, show negative direction of significant variables. Student and collegian have a tendency to not shift time of travel from the current time compared with the freelance working group.

Income of passengers which is more than 25,000 baht shows negative effect on the dependent variable when compared with passenger's rate income between 15,000 and 25,000. From the result, it can be said that passengers with income more than

Table 3 Definitions of dependent and independent variables for binary logistic regression model of scenario 1 (07.00–08.30)

Variable Type	Variable	Definition	Average
Dependent	Timeshift	1 = Shift travel time, 0 Not shift travel time	0.27
Independent	Male	1 = Male, 0 Otherwise	0.38
	Age_L24	1 = Age lower than 24 years, 0 Otherwise	0.31
	Age_2440	1 = Age between 24 and 40 years, 0 Otherwise	0.60
	Age_M40	1 = Age more than 40 years, 0 Otherwise	0.09
	Edu_LB	1 = Education lower than bachelor's degree, 0 Otherwise	0.25
	Edu_B	1 = Education with bachelor's degree, 0 Otherwise	0.59
	Edu_MB	1 = Education higher than bachelor's degree, 0 Otherwise	0.16
	Single	1 = Single, 0 Otherwise	0.74
	Occu_1	1 = Student, 0 Otherwise	0.18
	Occu_2	1 = Scholar, 0 Otherwise	0.12
	Occu_3	1 = Office worker, 0 Otherwise	0.68
	Occu_4	1 = Freelance, 0 Otherwise	0.02
	Inc_L15K	1 = Income lower than 15,000 baht, 0 Otherwise	0.30
	Inc_15to25	1 = Income between 15,000 and 25,000 baht, 0 Otherwise	0.26
	Inc_M25K	1 = Income higher than 25,000 baht, 0 Otherwise	0.44
	Accom	1 = Travel alone, 0 Otherwise	0.89
	Transfer	1 = Transfer mode with public transportation, 0 Otherwise	0.56
	Purp_S	1 = Traveling for study's purpose, 0 Otherwise	0.26
	Purp_W	1 = Traveling for work's purpose, 0 Otherwise	0.71
	Purp_P	1 = Traveling for personal' s purpose, 0 Otherwise	0.04
	TimeFix	1 = Fixed time of travel, 0 Otherwise	0.83
	Freq5days	1 = Travel 5 days or higher per week, 0 Travel lower than 5 days per week	0.91
	Card	1 = Travel by using passenger card, 0 Buy ticket from the machine	0.69
	Sta_M3	1 = Travel more than 3 stations, 0 Otherwise	0.44
	Dist	Distance for traveling	13.88
	Opinion	1 = Agreed with the price reduction, 0 Otherwise	0.58
	Delay	1 = Travel Delay, 0 Otherwise	0.50
	Tchange	Changing Time (Minute)	45.21
	Discount_baht	Discount (Baht)	6.75

Table 4 Regression analysis results of factors affecting decisions by fare discount of scenario 1 (07.00–08.30)

Independent variable	Model 1 Coef	Model 2 Coef	Model 3 Coef	Model 4 Coef
Male	−0.09614	−0.08474	0.09149	
Age_L24				
Age_2440				
Age_M40	−0.33832***	−0.37424**	−0.35902**	
Edu_LB				
Edu_B	−0.96413***	−0.97111***	−1.05471***	
Edu_MB	−1.31088***	−1.39813***	−1.54831***	
Single	−0.19662*	−0.19676*		
Occu_1	−1.40760***	−1.44828***	−1.44813***	−0.49627***
Occu_2	−0.63423***	−0.59388***	−0.63335***	−0.66623***
Occu_3				
Occu_4				
Inc_L15K				
Inc_15to25				
Inc_M25K	−0.06960	−0.06218	−0.11840	−0.35997***
Accom	0.00033	−0.01678		
Transfer	0.03824	0.04858	0.14464*	0.16126*
Purp_S				
Purp_W				
Purp_P	−0.70284**	−0.61000**	−0.04233	−0.18459
TimeFix	−0.35493**	−0.31449*	−0.14088	−0.24773*
Freq5days	−0.47862***	−0.48919***		
Card	−0.97319***	−1.00578***		
Sta_M3	−0.57425***		−0.61931***	−0.72216***
Dist	−0.00325	−0.03162***		
Opinion	0.71423***	0.70423***		
Delay	−2.06112***	−2.0569***		
Tchange	−0.02117***	−0.02095***	−0.03327***	−0.03238***
Discount_b ~ t	0.14708***	0.13777***	0.12388***	0.11925***
Summary of statistics				
Observations	481	481	481	481
Log likelihood	−1566.2087	−1574.3272	−1826.3679	−1859.4497

***Indicates significance at the 1% level
**Indicates significance at the 5% level
*Indicates significance at the 10% level

25,000 bath per month tend not to shift travel times when compared with those who have income between 15,000 and 25,000.

Transfer mode for traveling of passenger also found significance to the model in positive effect. From the result, it means passengers who have to transfer modes for traveling with ARL tend to change time of travel compared with passengers who travel only by using ARL per trip.

Purpose for traveling was also found to be significant in the model. The result found that passengers traveling on private business during peak times are unlikely to change their time of travel from rush hours compared to those who have the purpose of travel to get to work.

Fixed time for traveling. Passengers who have fixed time for traveling tend to not change time of travel from the current time compared with passengers who do not have time fixed for travelling.

Frequency of using ARL was also found to be significant after analysis. The result showed that passengers who travel by ARL five to seven days per week tend to not change travel time compared with those using ARL for travel less than five days per week.

The result found that passengers using the ARL service with a ticket card are more likely to change their travel time than those who purchase tickets from automated ticketing machines.

The respondents who travel more than three stations tend not to change travel of times compared with those who travel less than four stations.

Distance for traveling: the result found that when distance of traveling increased, passengers were unlikely to shift times of travel from the current time that they use every day.

Opinion of passengers who believed that discounting fares would reduce congestion during rush hours were tending to switch travel times compared with who did not agree that fare reductions can solve congestion problems in peak hours.

The most important variables are discount and time changing. From the 4, price reductions are an important variable in motivating passengers to change their times of travel from rush hour. The result indicates that when fare reductions increase, the number of passengers likely to change their times of travel will also increase. But contrary to travel times, the results show that, if required to change the times of travel from current time using ARL increases, the tendency for the passenger to change the travel time will decrease.

The delay independent variable is also significant to the model. This variable can explain about the direction of the shifting time of travel from 07.00 to 08.30. From the result ARL passengers who travel in the period 7.00–8.00 AM. tend not to shift travel time to later than 08.30 AM.

Value of Time (VOT)

The estimated coefficients from the developed models have been used to estimate VOT measures and the overall performance of the ordered logit, and the generalized linear mixed model has been found to be superior to the binary logit model.

Time value of passengers who travel at 07.00–08.30 AM: using the data from the model of 481 samples, the model can calculate the value of time based on the formula:

$$\text{Value of Time (VOT)} = \text{Time Coefficient /Cost Coefficient} \quad (2)$$

And reference data will be equal to (−0.02117)/(0.147082) = −0.14 baht per minute.

The time value from the above model shows that the passengers in morning peak (07.00–08.30 AM.) are willing to shift time of travel 1 min if ARL fare discounts 0.14 baht, or with discounts of 8.40 baht are willing to shift time of travel 60 min.

4.3 Results and Analysis of Scenario 2 (07.30–09.00)

The high Pearson correlation for passengers who travel in the 7.30–09.00 peak period responds to shifting travel time earlies or later from this period when receiving fare discount was found between scholar and worker at −0.81, scholar and income lower than 15,000 at 0.81 of coefficient value. Office worker also correlated with income lower than 15,000 baht and purpose for work at −0.80 and 0.82 of correlation coefficient value. Another was found between purpose for study and purpose for work with −0.83. Therefore, one of each variable for all correlations will be cut off from the model which are Occu_2, Occu_3 and Purpose. The definitions of dependent and independent variables for binary logistic regression model of scenario 2 as shown in Table 5.

As shown in Table 6, The Binary logistic Regression Model was used to analyze the socio-economic factors and travel information influence on behavior of passengers willing to shift time of travel to off peak hours. The results of analysis from scenario 2 of significant variables are explained below.

The most importance variables are discount and time changing. From the table, Price reductions are an important variable in motivating passengers to change their time of travel from rush hour. The result indicates that when reducing fare increases, the number of passengers likely to change their times of travel will also increase, but contrary to travel times. The results show that, if the time required to change the time of travel from the current time using ARL increases, the tendency for the passenger to change the time of travel will decrease.

Delay independent variable is also significant to the model. This variable can explain about the direction of the shifting time of travel from 07.30–09.00 AM. From the table, ARL passengers tend not to shift travel times to later than 09.00 AM.

Value of Time (VOT)

The estimated coefficients from the developed models that have been used to estimate VOT measures and the overall performance of the ordered logit and the generalized linear mixed model have been found to be superior to the binary logit model.

Table 5 Definitions of dependent and independent variables for binary logistic regression model of scenario 2 (07.30–09.00)

Variable Type	Variable	Definition	Average
Dependent	Timeshift	1 = Shift travel time, 0 Not shift travel time	0.28
Independent	Male	1 = Male, 0 Otherwise	0.38
	Age_L24	1 = Age lower than 24 years, 0 Otherwise	0.18
	Age_2440	1 = Age between 24 and 40 years, 0 Otherwise	0.69
	Age_M40	1 = Age more than 40 years, 0 Otherwise	0.12
	Edu_LB	1 = Education lower than bachelor's degree, 0 Otherwise	0.14
	Edu_B	1 = Education with bachelor's degree, 0 Otherwise	0.67
	Edu_MB	1 = Education higher than bachelor's degree, 0 Otherwise	0.19
	Single	1 = Single, 0 Otherwise	0.69
	Occu_1	1 = Student, 0 Otherwise	0.02
	Occu_2	1 = Scholar, 0 Otherwise	0.16
	Occu_3	1 = Office worker, 0 Otherwise	0.78
	Occu_4	1 = Freelance, 0 Otherwise	0.05
	Inc_L15K	1 = Income lower than 15,000 baht, 0 Otherwise	0.20
	Inc_15to25	1 = Income between 15,000 and 25,000 baht, 0 Otherwise	0.26
	Inc_M25K	1 = Income higher than 25,000 baht, 0 Otherwise	0.54
	Accom	1 = Travel alone, 0 Otherwise	0.91
	Transfer	1 = Transfer mode with public transportation, 0 Otherwise	0.59
	Purp_S	1 = Traveling for study's purpose, 0 Otherwise	0.13
	Purp_W	1 = Traveling for work's purpose, 0 Otherwise	0.82
	Purp_P	1 = Traveling for personal' s purpose, 0 Otherwise	0.05
	TimeFix	1 = Fixed time of travel, 0 Otherwise	0.74
	Freq5days	1 = Travel 5 days or higher per week, 0 Travel lower than 5 days per week	0.88
	Card	1 = Travel by using passenger card, 0 Buy ticket from the machine	0.62
	Sta_M3	1 = Travel more than 3 stations, 0 Otherwise	0.44
	Dist	Distance for traveling	13.85
	Opinion	1 = Agreed with the price reduction, 0 Otherwise	0.52
	Delay	1 = Travel Delay, 0 Otherwise	0.50
	Tchange	Changing Time (Minute)	45.18
	Discount_baht	Discount (Baht)	6.78

Table 6 Regression analysis results of factors affecting decisions by fare discount of scenario 2 (07.30–09.00)

Independent variable	Model 1 Coef	Model 2 Coef	Model 3 Coef	Model 4 Coef
Male	0.062896	0.118408	0.3296833***	
Age_L24				
Age_2440	0.72011***	0.68483***	0.67684***	
Age_M40	0.61639*	0.55826*	0.39768	
Edu_LB				
Edu_B	−0.71238***	−0.73855***	−0.87976***	
Edu_MB	−1.36079***	−1.50705***	−1.68523***	
Single	−0.09589	−0.10864		
Occu_1				
Occu_2				
Occu_3				
Occu_4	−0.30618	−0.31859	0.07297	0.66223*
Inc_L15K				
Inc_15to25	0.74169***	0.85887***	0.60556**	0.58238***
Inc_M25K	0.27310	0.37799	0.12802	−0.01688
Accom	−0.39255*	-0.50378**		
Transfer	−0.02686	−0.02638	0.04091	0.12443
Purp_S				
Purp_W				
Purp_P	0.06142	0.03271	0.44287	0.31336
TimeFix	−0.76945***	−0.77523***	−0.32971*	−0.13742
Freq5days	−0.58406***	−0.54655**		
Card	−0.80962***	−0.79706***		
Sta_M3	−0.65092***		−0.57095***	−0.69583***
Dist	0.00024	−0.03354***		
Opinion	0.71866***	0.71378***		
Delay	−1.66088***	−1.65131***		
Tchange	−0.02697***	−0.02672***	−0.03414***	−0.03320***
Discount_b ~ t	0.08535***	0.07586***	0.07223***	0.06998***
Summary of statistics				
Observations	245	245	245	245
Log likelihood	−815.28652	−820.37986	−928.16193	−957.38835

***Indicates significance at the 1% level
**Indicates significance at the 5% level
*Indicates significance at the 10% level

Time value of passenger traveling from 07.30–09.00 AM: using the data from the model of 245 samples, the model can calculate the value of time, based on the formula:

$$\text{Value of Time(VOT)} = \text{Time Coefficient /Cost Coefficient}$$

Reference data will be equal to (−0.02697)/(0.085352) = −0.32 baht per minute.

The time value from the above model shows that the passengers in Morning peak (07.30–09.00AM.) are willing to shift time of travel 1 min if ARL fare discounts 0.32 baht, or a discount of 19.20 baht are willing to shift time of travel 1 h.

4.4 *Results and Analysis of Scenario 3 (00.19–00.17)*

The high Pearson correlation, was found between passenger's age lower than 24 years and passenger's age 24 to 40 years, worker and passenger's age lower than 24 years, income lower than 15,000 baht and age lower than 24 years, purpose travel for study and passenger's age lower than 24 years, travel purpose for work and age lower than 24, education lower than bachelor and student, worker and income lower than 15,000 baht, purpose traveling for study and worker. Purpose traveling for working and worker, purpose traveling for study and income lower than 15,000, purpose traveling for work and income lower than 15,000 baht, and the last correlation found between purpose traveling for study and purpose traveling for working. Some variable has to cut out before analysis which are age lower than 24 years, education lower than bachelor, worker, income lower than 15,000 baht and purpose traveling for study. The definitions of dependent and independent variables for binary logistic regression model of scenario 3 as shown in Table 7.

As shown in Table 8, The Binary Logistic Regression Model was used to analyze the characteristics of passengers willing to shift travel time from 17.00–19.00 to earlies or later when receiving fare discount. The results were presented by four different models. The result found that (passenger's gender, passenger's education, passenger's occupation and passenger's income) are significant to the model.

Gender of passenger: The Result found that males tend not to change travel time compared with females.

Education of passenger: passengers with education level higher than bachelor's degree were not likely to shift time of travel compared with passenger with bachelor's degree.

Occupation: From the result, it has a negative effect to the dependent variable which, are student and scholar. It means students and scholars tend to not shift the time of travel compared with freelance workers.

Table 7 Definitions of dependent and independent variables for binary logistic regression model of scenario 3 (19.00–17.00)

Variable Type	Variable	Definition	Average
Dependent	Timeshift	1 = Shift travel time, 0 Not shift travel time	0.26
Independent	Male	1 = Male, 0 Otherwise	0.39
	Age_L24	1 = Age lower than 24 years, 0 Otherwise	0.43
	Age_2440	1 = Age between 24 and 40 years, 0 Otherwise	0.49
	Age_M40	1 = Age more than 40 years, 0 Otherwise	0.08
	Edu_LB	1 = Education lower than bachelor's degree, 0 Otherwise	0.33
	Edu_B	1 = Education with bachelor's degree, 0 Otherwise	0.54
	Edu_MB	1 = Education higher than bachelor's degree, 0 Otherwise	0.14
	Single	1 = Single, 0 Otherwise	0.78
	Occu_1	1 = Student, 0 Otherwise	0.27
	Occu_2	1 = Scholar, 0 Otherwise	0.15
	Occu_3	1 = Office worker, 0 Otherwise	0.55
	Occu_4	1 = Freelance, 0 Otherwise	0.03
	Inc_L15K	1 = Income lower than 15,000 baht, 0 Otherwise	0.42
	Inc_15to25	1 = Income between 15,000 and 25,000 baht, 0 Otherwise	0.20
	Inc_M25K	1 = Income higher than 25,000 baht, 0 Otherwise	0.38
	Accom	1 = Travel alone, 0 Otherwise	0.87
	Transfer	1 = Transfer mode with public transportation, 0 Otherwise	0.57
	Purp_S	1 = Traveling for study's purpose, 0 Otherwise	0.35
	Purp_W	1 = Traveling for work's purpose, 0 Otherwise	0.59
	Purp_P	1 = Traveling for personal's purpose, 0 Otherwise	0.06
	TimeFix	1 = Fixed time of travel, 0 Otherwise	0.79
	Freq5days	1 = Travel 5 days or higher per week, 0 Travel lower than 5 days per week	0.83
	Card	1 = Travel by using passenger card, 0 Buy ticket from the machine	0.65
	Sta_M3	1 = Travel more than 3 stations, 0 Otherwise	0.46
	Dist	Distance for traveling	13.67
	Opinion	1 = Agreed with the price reduction, 0 Otherwise	0.64
	Delay	1 = Travel Delay, 0 Otherwise	0.50
	Tchange	Changing time (min)	60.15
	Discount_baht	Discount (Baht)	6.79

Table 8 Regression analysis results of factors affecting decisions by fare discount of scenario 3 (17.00–19.00)

Independent variable	Model 1 Coef	Model 2 Coef	Model 3 Coef	Model 4 Coef
Male	−0.17192**	−0.16884**	−0.10311	
Age_L24				
Age_2440				
Age_M40	0.14595	0.13357	0.16381	
Edu_LB				
Edu_B				
Edu_MB	−0.43380***	−0.48639***	−0.54503***	
Single	−0.11166	−0.10964		
Occu_1	−0.54039**	−0.54988**	−0.59684**	−0.57872**
Occu_2	−0.44514***	−0.45785***	−0.42058**	−0.41766**
Occu_3				
Occu_4				
Inc_L15K				
Inc_15to25				
Inc_M25K	−0.32158***	−0.33098***	−0.35519***	−0.47973***
Accom	−0.20547*	−0.20816*		
Transfer	0.08923	0.09480	0.21496***	0.25659***
Purp_S				
Purp_W	−0.25705	−0.24555	−0.28771	−0.22736
Purp_P				
TimeFix	−0.01601	−0.01082	0.02056	0.00291
Freq5days	0.17529	0.19626		
Card	−0.72917***	−0.74237***		
Sta_M3	−0.28682**		−0.33738***	0.08036
Dist	−0.00438	−0.01817***		
Opinion	0.57480***	0.57429***		
Delay	1.17979***	1.17729***		
Tchange	−0.01677***	−0.01670***	−0.01497***	−0.01494***
Discount_b ~ t	0.14106***	0.13693***	0.12544***	0.04120***
Summary of statistics				
Observations	548	548	548	548
Log likelihood	−2098.2236	−2100.83	−2283.3773	−2283.329

***Indicates significance at the 1% level
**Indicates significance at the 5% level
*Indicates significance at the 10% level

Income: passenger income is also found a negative effect in passenger income more than 25,000 baht tend to not change travel time compared with passengers with income at 15,000–25,000 baht.

Number of travelers for traveling: the model shows passengers who travel alone tend not to shift travel time from the current time using ARL in morning peak hours as shows in negative effect to the model.

Transfer mode with ARL for traveling: this dependent variable has a positive effect to the model. From the result, passengers who have mode transfer for traveling were likely to change time of travel when compared with passengers using only airport rail link without mode transfer for traveling.

In the travel information part, the result found that number of stations, opinion of passenger, time shifting, time change and discount also significant to the model.

Time changing: The results showed that, when changing travel time from the original time increased, it can affect the declining trend of changing time of ARL passengers to off-peak hours.

Discount: The result indicates that when fare reductions increase, the number of passengers likely to change their times of travel will also increase.

Delay: In evening peak period, passengers tend to change travel times to after peak hours.

Passenger's opinion: Passengers are of the opinion that the off-peak fare reduction can reduce congestion during rush hours. There is a tendency to change the departure time from rush hour when compared with the opposite group.

Number of stations: For passengers traveling more than 3 stations, there is a tendency to not postpone travel time compared to passengers traveling 1–3 stations.

Distance for traveling: when distance of using Airport Rail Link for traveling in evening peak increases, passenger tend not to change travel time.

Passenger card for traveling: ARL passengers who use cards for traveling tend to not change travel times when compared with passengers who bought tickets from automatic machines.

Value of Time (VOT)

The estimated coefficients from the developed models have been used to estimate VOT measures and the overall performance of the ordered logit and the generalized linear mixed model has been found to be superior to the binary logit model.

Time value of passenger traveling from 17.30–19.00 AM: using the data from the model of 548 samples, the model can calculate the value of time based on the formula:

$$\text{Value of Time(VOT)} = \text{Time Coefficient /Cost Coefficient}$$

And reference data will be equal to (−0.01677)/(0.141068) = −0.11 baht per minute.

The time value from the above model shows that the passengers in Evening peak (17.00–19.00) are willing to shift times of travel by 1 min if ARL fare discount is 0.11 baht, or if discount is 6.6 baht they are willing to shift time of travel 1 h.

4.5 Results and Analysis of Scenario 4 (18.00–20.00)

For scenario four, times that passengers arrive at station between 18.00 and 20.00. When checking the correlation coefficient of independent variables, the result found correlation between age twenty-four to forty year and age lower than twenty four, office worker and age lower than twenty-four, income lower than 15,000 baht and age lower than twenty-four, office worker and income lower than 15,000 baht, Purpose of travel to study and office worker, income lower than 15,000 baht and Purpose of travel to study, Purpose of travel to work and income lower than 15,000 baht. The last one found correlation between Purpose of travel to study and Purpose of travel to work. All of this independent variable are check with high Pearson correlation. Age_L24, Occu_3, Inc_L15k and Purp_s were cut from the model for analysis. The definitions of dependent and independent variables for binary logistic regression model of Scenario 4 as shown in Table 9.

Using the binary logistic regression model to analyze the characteristics of the travelers willing to shift travel times from 18.00 to 20.00 to early or later than this period when fare is discounted. Result of some independent variables are the same with the Table (Analysis of scenario 3 in period of 17.00–19.00), such as discount, time change, Opinion of passenger, delay, Number of stations for traveling and ARL passenger card usage (Table 10).

Moreover, the results show that passengers who use ARL for travelling five or more than 5 days per week tend to shift times of travel when receiving fare reductions in off peak hours. And passengers who have time fixed for travel are unlikely to change travel times when compared with passengers with no time fixed for traveling.

In the characteristics of the traveler part, the results found that gender, age, education, status, income and occupation are significant to the model.

Gender of passenger: males tend not to change travel time compared to females.

Age of passenger: Passengers older than 40 years are unlikely to change travel times when compared to passengers aged between 24 and 40.

Education of passenger: passengers with education level at bachelor's degree and higher than bachelor's degree were not likely to shift times of travel compared with passengers with lower than bachelor's degree.

Status of passenger: Single variable was found that has a negative effect on the dependent variable. It means single passengers tend to not shift times of travel compared with passengers who are not single.

Occupation of passengers: the results show that students are not likely to shift travel times when compared with freelance.

Table 9 Definitions of dependent and independent variables for binary logistic regression model of scenario 4 (18.00–20.00)

Variable Type	Variable	Definition	Average
Dependent	Timeshift	1 = Shift travel time, 0 Not shift travel time	0.26
Independent	Male	1 = Male, 0 Otherwise	0.38
	Age_L24	1 = Age lower than 24 years, 0 Otherwise	0.37
	Age_2440	1 = Age between 24 and 40 years, 0 Otherwise	0.54
	Age_M40	1 = Age more than 40 years, 0 Otherwise	0.09
	Edu_LB	1 = Education lower than bachelor's degree, 0 Otherwise	0.27
	Edu_B	1 = Education with bachelor's degree, 0 Otherwise	0.57
	Edu_MB	1 = Education higher than bachelor's degree, 0 Otherwise	0.16
	Single	1 = Single, 0 Otherwise	0.78
	Occu_1	1 = Student, 0 Otherwise	0.19
	Occu_2	1 = Scholar, 0 Otherwise	0.16
	Occu_3	1 = Office worker, 0 Otherwise	0.62
	Occu_4	1 = Freelance, 0 Otherwise	0.03
	Inc_L15K	1 = Income lower than 15,000 baht, 0 Otherwise	0.35
	Inc_15to25	1 = Income between 15,000 and 25,000 baht, 0 Otherwise	0.22
	Inc_M25K	1 = Income higher than 25,000 baht, 0 Otherwise	0.43
	Accom	1 = Travel alone, 0 Otherwise	0.90
	Transfer	1 = Transfer mode with public transportation, 0 Otherwise	0.59
	Purp_S	1 = Traveling for study's purpose, 0 Otherwise	0.29
	Purp_W	1 = Traveling for work's purpose, 0 Otherwise	0.66
	Purp_P	1 = Traveling for personal' s purpose, 0 Otherwise	0.05
	TimeFix	1 = Fixed time of travel, 0 Otherwise	0.77
	Freq5days	1 = Travel 5 days or higher per week, 0 Travel lower than 5 days per week	0.82
	Card	1 = Travel by using passenger card, 0 Buy ticket from the machine	0.63
	Sta_M3	1 = Travel more than 3 stations, 0 Otherwise	0.48
	Dist	Distance for traveling	14.02
	Opinion	1 = Agreed with the price reduction, 0 Otherwise	0.61
	Delay	1 = Travel Delay, 0 Otherwise	0.50
	Tchange	Changing Time (Minute)	60.16
	Discount_baht	Discount (Baht)	6.93

Table 10 Regression analysis results of factors affecting decisions by fare discount of scenario 4 (18.00–20.00)

Independent variable	Model 1 Coef	Model 2 Coef	Model 3 Coef	Model 4 Coef
Male	−0.23727**	−0.22973**	−0.11176	
Age_L24				
Age_2440				
Age_M40	−0.4396**	−0.43743**	−0.28428*	
Edu_LB				
Edu_B	−0.66178***	−0.66965***	−0.88260***	
Edu_MB	−1.52249***	−1.56497***	−1.85442***	
Single	−0.51274***	−0.51976***		
Occu_1	−0.64191***	−0.65083***	−1.09710***	−0.35340***
Occu_2	−0.24107	−0.24720	−0.27496*	−0.21451
Occu_3				
Occu_4				
Inc_L15K				
Inc_15to25				
Inc_M25K	0.02425	0.02344	−0.04205	−0.44817***
Accom	−0.01159	−0.01520		
Transfer	0.12311	0.12247	0.23917***	0.25448***
Purp_S				
Purp_W				
Purp_P	0.02424	0.02490	−0.05753	−0.21001
TimeFix	−0.57047***	−0.56761***	−0.41046***	−0.41881***
Freq5days	0.30968**	0.31689**		
Card	−0.78478***	−0.79275***		
Sta_M3	−0.18924		−0.18093*	−0.30009***
Dist	0.00514	−0.00432		
Opinion	0.46458***	0.45363***		
Delay	0.94930***	0.95026***		
Tchange	−0.02058***	−0.02059***	−0.01429***	−0.01393***
Discount_b ~ t	0.10001***	0.09777***	0.09370***	0.08956***
Summary of statistics				
Observations	410	410	410	410
Log likelihood	−1580.8872	−1581.7226	−1683.9808	−1728.7405

***Indicates significance at the 1% level
**Indicates significance at the 5% level
*Indicates significance at the 10% level

Income of passengers which is more than 25,000 baht show negative effect to the dependent variable when compared with passenger's rate of income between 15,000 and 25,000.

Value of Time (VOT)

The estimated coefficients from the developed models have been used to estimate VOT measures and the overall performance of the ordered logit and the generalized linear mixed model has been found to be superior to the binary logit model.

Time value of passenger who traveling from 18.00–20.00 AM: using the data from the model of 410 samples, the model can calculate the value of time based on the formula:

$$\text{Value of Time(VOT)} = \text{Time Coefficient /Cost Coefficient}$$

And reference data will be equal to (−0.02058)/(0.100014) = 0.21 baht per minute.

The time value from the above model shows that the passengers in Evening peak (18.00–20.00 PM.) are willing to shift time of travel by 1 min if ARL fare discount is 0.21 baht, or if discount is 12.60 baht are willing to shift time of travel 1 h.

5 Conclusions and Recommendations

This study aims to identify factors that influence passengers to shift time of travel to off-peak hours by using fare discounts and value of time of Airport Rail Link passengers. The methodology used in this study was designed to observe the ARL passengers by using the questionnaire survey from the site and online methods. The socioeconomic characteristics of ARL passengers were also obtained by using self-reported questionnaires. This study used State Preference and analyzed by using Binary Logit Model.

The first objective to analyze the characteristics of people willing to shift time of travel. The results of this part show that older passengers tend not to shift times of travel when compared to young passengers. Most passengers who have high income are more likely to not shift times of travel. Passengers with higher degree of education tend not to shift times of travel. Passengers who travel five or more days per week are unlikely to shift the time compared with passengers who travel less. Passengers travelling more than 3 stations are unlikely to change the time. Passengers who travel using ARL passenger cards tend not to shift times of travel. Passengers who have inflexible schedules tend not to shift times of travel. Freelance workers are likely to shift the time compared with students. When considering passengers' attitude towards willingness to shift times of travel, passengers who think that a fare discount in off-peak hour can solve congestion problem in peak hour, tend to shift travel times. In the morning peak hour, the results found that the major reasons are that people are not willing to wake up earlier, people had no flexibility with their working or study hours, and they had family commitments. The reasons why people do not change their times to later is that their working and studying hours are not flexible.

The Second objective is to study the factors that motivate people to shift times of travel. For ARL passengers, higher discount rate will motivate passengers to shift the time to off-peak hours. From statistics data, it was found that most passengers are willing to shift times of travel to avoid congestion and get a fare discount at a rate of 10 percent. Longer travel distances will result in a decrease in passengers' willingness to shift times of travel.

The third objective is to investigate the amount of time that people are able and willing to change for a given discount. The result found that the value of time in Morning peak can be expressed as people are willing to shift time of travel by an hour for ARL fare discount of around eight to nineteen baht. The value of time in Evening peak is that they are willing to shift time by an hour if there are ARL fares discounts of around seven to thirteen baht.

However, this study uses the data of number of passengers counted in Airport Rail Link in 2014 which should be updated if there is a further study, such as number of passengers in rush hour. If there is more time to study, the data collection should be carried out in a way that the number of samples in each station are equal in order to increase the accuracy of data analysis. Online surveys are a good way to distribute surveys and get a quick sample of people and save time for data collection when compared to onsite surveys. Nevertheless, there are some disadvantages such as respondents cannot ask the questions when they become confused with the questionnaire. Therefore, if an online survey is to be used, it should be planned well. Moreover, the fare reduction policy can be applied to solve the congestion problems in the peak hour in the future. There should be several levels of fare reduction at different times for passenger distribution improvement. Most of all, passengers have fixed schedules for travel, therefore, one policy that can support fare reduction policy to reduce overcrowding in rush hour is that people should not have the same time to start and finish work in the same area.

6 Appendices A: Questionnaire

Passenger characteristics and the opinion of Airport Rail Link commuters about the fare discount in off peak hour for spreading peak demand questionnaire

Part 1 General information

1.1 Gender (Choose only one)

- Male
- Female

1.2 Age (Choose only one)

- Lower than 18 years
- 18-23 years
- 24-30 years
- 31-40 years
- 41-50 years
- 51-60 years
- Higher than 60 years

1.3 Educational status (Choose only one)

- Under Senior high school/ Vocational certificate
- Diploma/High vocational certificate
- Bachelor degree
- Higher than Bachelor degree

1.4 Marital status (Choose only one)

- Single
- Marriage
- Divorce/Separated

1.5 Occupation (Choose only one)

- School student
- University student
- Government officer
- Private employee
- Owner/ Freelancer
- Laborer
- Other ………………………………………………………………

1.6 Personal income (Choose only one)

- Less than 15,000 Baht
- 15,000 – 25,000 Baht
- 25,001 – 35,000 Baht
- 35,001 – 50,000 Baht
- 50,001 – 100,000 Baht
- Higher than 100,000 Baht

Part 2 The ARL travel information

2.1 Number of fellow travelers in this trip

- Only you
- 2 persons including you
- Up to 3 persons including you

2.2 To travel to the destination, which mode do you have to use in addition?

- Walking
- Private car
- Bus
- Van
- BTS/MRT
- Hired motorcycle
- Other ....

2.3 The main purpose of trip

- Go to / Come back from school
- Go to / Come back from university
- Go to / Come back from workplace and have a fixed schedule
- Go to / Come back from workplace and have an unfixed schedule (flexible schedule)
- Go to / Come back from doing personal activity for example go to bank, post office and hospital
- Shopping, Recreation, go to restaurant, watching movie, go to gym and party
- Go to airport
- Other ....

2.4 Frequency of using ARL

- 6-7 days per week
- On weekdays(5days)
- 2-3 times per week
- Once a week
- Less than once a week

2.5 Peak hour duration that you usually travel by ARL

- Commute to travel in morning 07.00 – 09.00
- Commute to travel in evening 17.00 – 20.00
- Commute to travel in both morning 07.00 – 09.00 and evening 17.00-20.00
- Avoid travelling in peak hour

2.6 Type of your card (usually used for traveling)

- Buy a ticket from the machine
- General Card/Adult card
- Student Card
- Senior Card

Travel information in Morning Peak hour

ARL origin station that you always go to

……………………………………………………………………………………………

ARL destination station that you always go to

……………………………………………………………………………………………

Number of stations

…………………… stations

Distance

…………………… km

Fare

…………………… Baht

Approximate time to arrive at ARL station in Morning

…………hour………… minutes

Part 3 Morning commuter's opinion (Example of questions in this part)

3.1 If the fare discounts 20% of normal fare from 20 Baht to 16 Baht per trip, would you shift the arrival time at the origin station from 07.00 – 08.30 for getting the discount?

- Interest and can shift the time to earlier 07.00
- Interest and can shift the time to after 08.30
- Interest and can shift the time to both earlier 07.00 and after 08.30
- Not interest

3.1.1 The main reason that you cannot shift the time to earlier 07.00

- Have to wake up early
- Family commitment
- Fixed schedule
- Fixed appointment
- Have to shift the other mode of transportation which has the permanent plan
- Think that the congestion is no different from the current time
- The discount is not significant
- Other ....

3.1.2 The main reason that you cannot shift the time to after 08.30

- Family commitment
- Fixed schedule
- Fixed appointment
- Have to shift the other mode of transportation which has the permanent plan
- Think that the congestion is no different from the current time
- The discount is not significant
- Other ....

Travel information in Evening peak

ARL origin station that you always go to

....

ARL destination station that you always go to

....

Number of stations

…………………… stations

Distance

…………………… km

Fare

…………………… Baht

Approximate time to arrive at ARL station in evening

…………hour………… minutes

Part 4 Evening commuter's opinion (Example of questions in this part)

4.1 If the fare discounts 20% of normal fare from 20 Baht to 16 Baht per trip, would you shift the arrival time at the origin station from 19.00 – 20.00 for getting the discount?

- Interest and can shift the time to earlier 19.00
- Interest and can shift the time to after 20.00
- Interest and can shift the time to both earlier 19.00 and after 20.00
- Not interested

4.1.1 The main reason that you cannot shift the time to earlier 19.00

- Family commitment
- Fixed schedule
- Fixed appointment
- Have to shift the other mode of transportation which has the permanent plan
- Think that the congestion is no different from the current time
- The discount is not significant
- Other ……………………………………………………………………………

4.1.2 The main reason that you cannot shift the time to after 20.00

- Family commitment
- Fixed schedule
- Fixed appointment
- Have to shift the other mode of transportation which has the permanent plan
- Think that the congestion is no different from the current time
- The discount is not significant

- o Cannot leave later 20.00 PM
- o Other ....

Part 5 attitude of passenger

5.1If the fare in off peak period is reduced, do you think the congestion in rush hour (07.00-09.00 and 17.00-20.00) would be decreased?

- o May solve the congestion problem
- o Can solve the congestion problem
- o Cannot solve the congestion problem

5.2 Rate these factors which could reduce the congestion problem by arranging in order of effectiveness from the most effective (4 stars) to the lowest (1 star).

- o Unfixed schedule or flexible schedule
- o Peak time fare surcharge
- o Off peak time fare discount
- o Increase the train frequency

7 Appendices B: Random Questions for Test Fare Discount

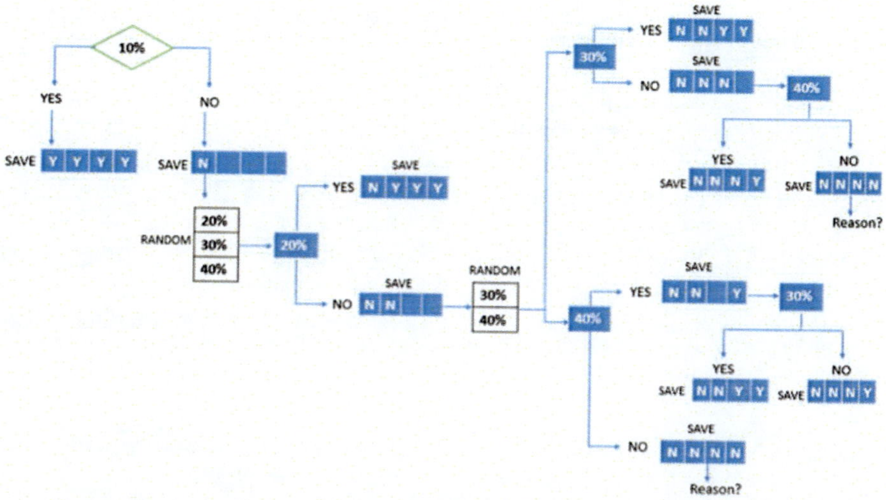

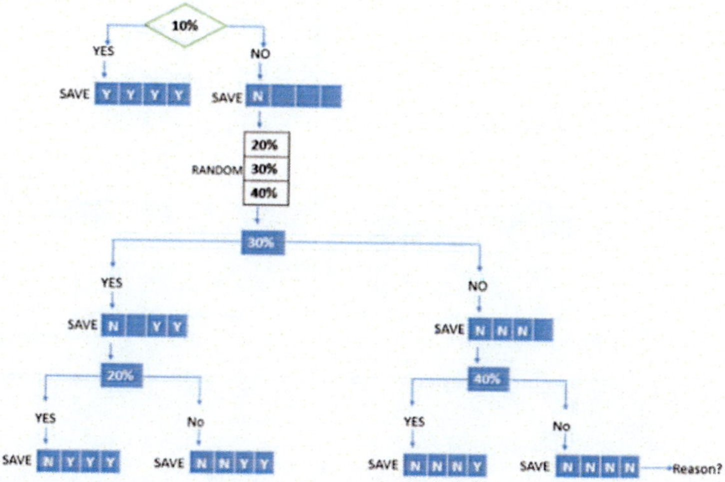

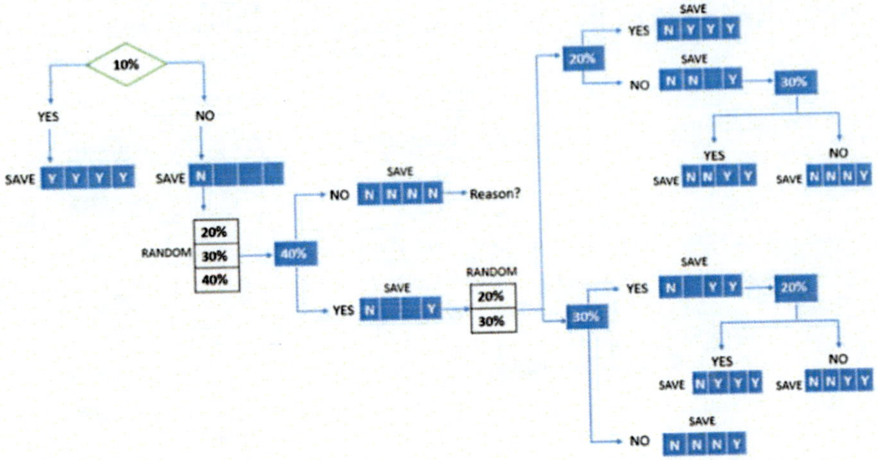
10%
YES
NO
SAVE Y Y Y Y
SAVE N
RANDOM
20%
30%
40%
NO
SAVE N N N N
Reason?
YES
SAVE N Y
RANDOM
20%
30%
20%
YES SAVE N Y Y Y
NO SAVE N N Y
30%
YES SAVE N N Y Y
NO SAVE N N N Y
30%
YES SAVE N Y Y
20%
YES SAVE N Y Y Y
NO SAVE N N Y Y
NO SAVE N N N Y

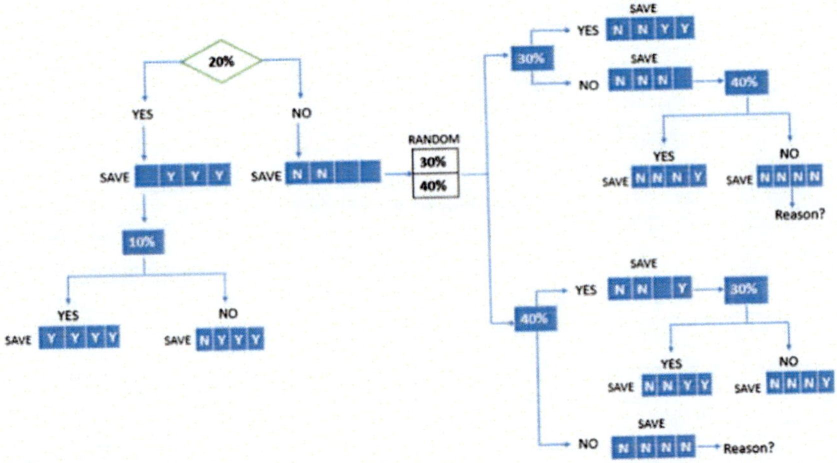
20%
YES
NO
SAVE Y Y Y
SAVE N N
10%
YES SAVE Y Y Y Y
NO SAVE N Y Y Y
RANDOM
30%
40%
30%
YES SAVE N N Y Y
NO SAVE N N N
40%
YES SAVE N N N Y
NO SAVE N N N N
Reason?
40%
YES SAVE N N Y
30%
YES SAVE N N Y Y
NO SAVE N N N Y
NO SAVE N N N N
Reason?

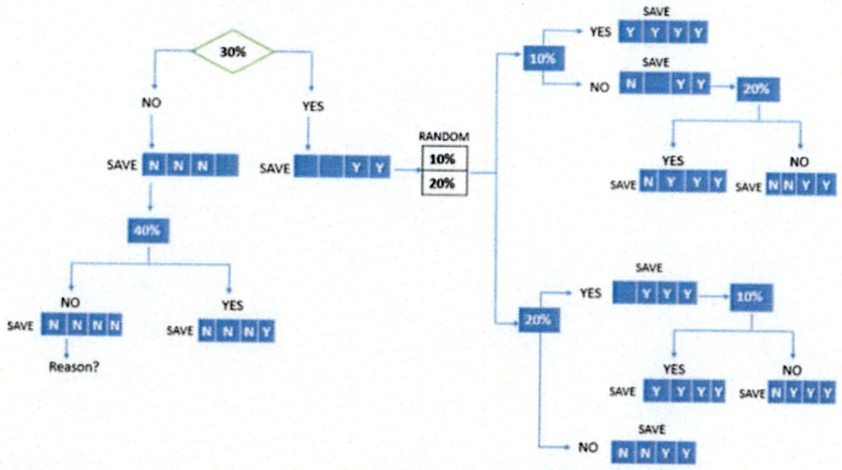
30%
NO
YES
SAVE N N N
SAVE Y Y
40%
NO
SAVE N N N N
Reason?
YES
SAVE N N N Y
RANDOM
10%
20%
10%
YES
SAVE Y Y Y Y
NO
SAVE N Y Y
20%
YES
SAVE N Y Y Y
NO
SAVE N N Y Y
20%
YES
SAVE Y Y Y
10%
YES
SAVE Y Y Y Y
NO
SAVE N Y Y Y
NO
SAVE N N Y Y

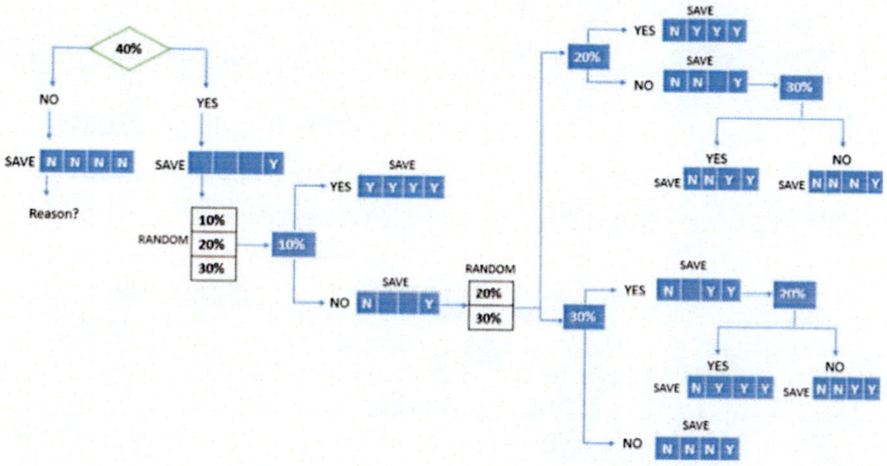
40%
NO
YES
SAVE N N N N
Reason?
SAVE Y
RANDOM
10%
20%
30%
10%
YES
SAVE Y Y Y Y
NO
SAVE N Y
RANDOM
20%
30%
20%
YES
SAVE N Y Y Y
NO
SAVE N N Y
30%
YES
SAVE N N Y Y
NO
SAVE N N N Y
30%
YES
SAVE N Y Y
20%
YES
SAVE N Y Y Y
NO
SAVE N N Y Y
NO
SAVE N N N Y

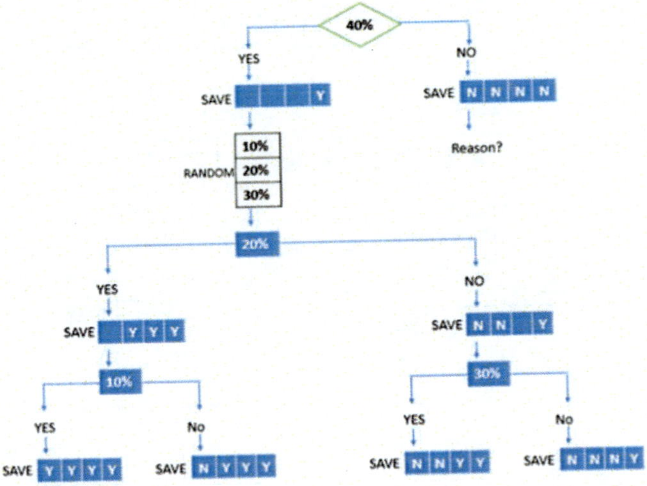

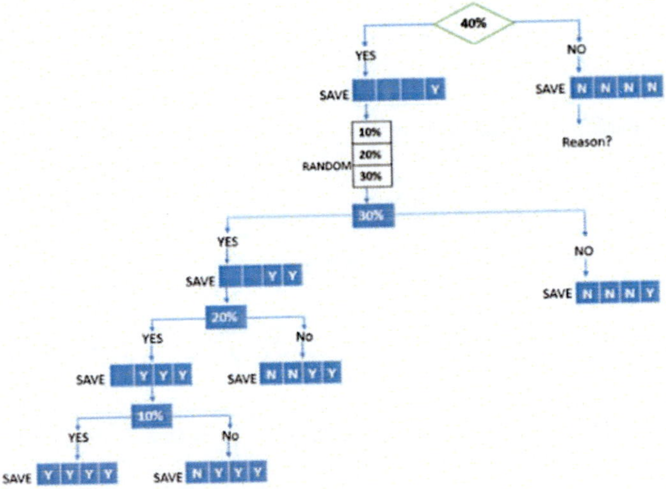

References

Airport Rail Link (n.d.) Retrieved 2017, from https://www.srtet.co.th/index.php/th/about-history/aboutproject.

Cervero R (1986) Time-of-day transit pricing: Comparative US and international experiences. Transp Rev 6(4):347–364

Currie G (2010) Quick and effective solution to rail overcrowding: free early bird ticket experience in Melbourne, Australia. Transp Res Rec J Transp Res Board 2146(1):35–42. https://doi.org/10.3141/2146-05

Edge of morning peak travel—Research findings prepared for Passenger Focus by Consolidated Communications (2006) Retrieved from https://www.transportfocus.org.uk/researchpublications/publications/edge-of-morning-peak-travel-research-findings-prepared-for-passenger-focus-by-consolidated-communications/

Faber M (2007) Demand management techniques: peak spreading. Final Report to Department for Transport, Transport for London and Network Rail. (Tech.)

Nature (2009) An investigation into a variable ticket pricing strategy for train travel Melbourne, Metlink

RailCorp (2008) SmartSaver trial online survey results December 2008 Sydney

Whelan G, Johnson D (2004) Modelling the impact of alternative fare structures on train overcrowding. Int J Transp Manag 2(1):51–58. https://doi.org/10.1016/j.ijtm.2004.04.004

Metro Performance Indicators for Service Operations in Thailand

Jirapan Liangrokapart

Abstract With a large number of people all over the world, the transportation modes of choice for public to travel at the same time within urban areas include buses, trains, trams, boats, and mass rapid transit. This research uses the word "Metro" to represent the mass rapid transit systems which are electric railway systems provided on designated lines between stations in Bangkok Metropolitan region in Thailand. Currently, there are 5 lines in operation including the BTS green line, BTS light green line, MRT blue line, MRT purple line and ARL. Interestingly, the metro lines are operated by different operators under different contracts and requirements. Therefore, it is extremely difficult for the Ministry of Transport to monitor the operations of each operator and compare the performance of each metro line. The objectives of this research are to identify the appropriate key performance indicators to monitor each operator's performance and to suggest avenues for performance improvements. Extensive literature review analysis and focus discussion groups are among the research method used in this study. The main finding from this study is a consensus of six dimensions for measuring metro performance including "Transport volume", "Punctuality", "Reliability", "Availability", "Resource Utilization", and "Customer satisfaction".

Keywords Performance measurement · Metro performance · Service operations · Mass rapid transit

1 Introduction

With growing numbers of people in the world, transportation modes of choice for the public to travel at the same time within urban areas include buses, trains, trams, boats, and mass rapid transit. Mass rapid transit may be known as Metro, Subway, Underground, Tube, U-Bahn, or others. This research uses the word "Metro" to represent the mass rapid transit system which is an electric railway system provided

J. Liangrokapart (✉)
Faculty of Engineering, Mahidol University, Nakornprathom 10730, Thailand
e-mail: jirapan.lia@mahidol.ac.th

W. Weerawat et al. (eds.), *Urban Rail Transit*, Lecture Notes in Mobility,
https://doi.org/10.1007/978-981-15-5979-2_5

on designated lines between stations. The service operations studied in this research are in the Bangkok Metropolitan region in Thailand. The major metro operations stakeholders consist of project owners, metro operators, policy makers and regulators, and research and development agencies.

The common problems found during preliminary survey and field observations on metro service operations include insufficient information for trip planning, various ticketing systems among different metro operators, long queues for ticketing services, inconvenience of entrances to and exits from the stations, large volumes of passengers during peak hours, confusion of station names, etc. The problems may vary from one service operator or one metro line to others. It is because each has different practice and adheres to different standards. Currently, there are 5 lines in operations including the BTS green line, BTS light green line, MRT blue line, MRT purple line and ARL. Each line is operated with different terms of contract. It is very difficult to compare the performance of each metro operator in order to know how well each of them is doing and which functions or operations are in need of improvement.

Therefore, two research questions have emerged: (1) What are the key performance indicators (KPI) for measuring the metro service operations in Thailand? (2) How can the operators make improvement for public benefit? The objectives of this research are to identify the appropriate key performance indicators to monitor each operator's performance and to suggest avenues for performance improvement.

2 Framework for Performance Measurement

Performance measurement describes the feedback or information on activities with respect to meeting customer expectations and strategic objectives. It reflects the need for improvements in areas with unsatisfactory performance. In this research, Metro Performance Indicators from various literatures were reviewed.

2.1 CoMET and NOVA

The most common metro performance indicators applied by many metro operators all over the world are the ones specified by CoMET (Community of Metros) and NOVA Group of Metros which are the world's metro benchmarking groups. One of their objectives is to share knowledge and identify best practices in a confidential environment. Currently the two consortia are made up of 38 large and medium sized metro systems from 36 cities around the world. Examples of CoMET members are Beijing Subway, Hong Kong MTR, London Underground, Singapore MRT, Taipei Metro, while examples of Nova include Bangkok MRT, Newcastle Tyne & Wear Metro, Metro Rio and Sydney Trains (CoMET and Nova Metro Benchmarking Groups 2018).

According to Baron (2016), the Key Performance Indicator (KPI) system is the basis of the benchmarking process, and is designed to compare performance of different organizations. CoMET and Nova use the KPI system with approximately 30 top-level indicators, which are designed to measure the overall performance of the organization in six distinct areas: Growth, Learning and Innovation; Financial; Customer; Internal Processes; Safety and Security; and Environment. The KPI system is comprehensive and represent all of the different parts of the metro business. The data of each indicator is collected from all CoMET and Nova members on an annual basis. The KPIs used in CoMET and Nova are shown in Fig. 1.

2.2 EN 13816 and EN 15140

Operators in public passenger transport provide their services directly to the customer. Therefore, the service quality perceived by the customer will have a significant effect on the selection of the provider. The quality of public passenger transport is measured based on European standards EN 13816 and EN 15140.

The EN 13816 is the standard for evidencing quality capability of transport providers in public passenger traffic. The standard is classified into eight categories (European standard (EN 13816: 2002)).

1. Availability
2. Accessibility
3. Information
4. Time
5. Customer care
6. Comfort
7. Security
8. Environmental impact

The EN 15140 standard has basic requirements and recommendations for public transport operators to measure delivered service quality. Public transport providers must understand what is important to the customer. The standard provides recommendations on service quality measurement in the framework of EN13816 mentioned above. The focus is on advising operators and authorities how to set measurement processes, formulate specific indicators and set clearly defined targets.

Growth, Learning & Innovation	• % change network size & passenger journeys • % change operated capacity km. & car km. • No. of training hrs/1000 staff hrs • Non-fare commercial revenue/Fare revenue & /Passenger journey
Financial	• Total commercial revenue/operating cost • Operating cost/car km • Investment cost/ car km • Operating cost/passenger journey & km • Fare revenue/passenger journey & km
Customer	• Capacity km/route km • Passenger km/capacity km • Passenger Hrs' delay/passenger journey • Passenger journeys on time/passenger journey • Trains on time/total trains scheduled & /total trains actual • Hrs of Train delay/train hrs operated • %stations with step free access
Internal Process	• % cars available / % cars used in peak hrs • Car km/hrs between incidents • Passenger journeys/staff+contractor hrs • Capacity/staff + contractor hrs • Car km/staff+contractor hrs • Train hrs/driver hrs • % employee absenteeism • Traction energy consumed/car km • total energy consumed / passenger journey • total energy consumed /passenger km
Safety & Security	• Total fatalities/passenger journeys • Incidence of crimes/passenger journeys • Staff lost time through accident/staff hrs
Environment	• CO2 per passenger km.

Fig. 1 KPIs used by CoMET and Nova Benchmarking Groups. Adapted from Baron (2016) Baron

Table 1 The NRPS satisfaction measures (The National Rail Passenger Survey, 2016)

Station facilities	Train facilities
1. Overall satisfaction with the station 2. Ticket buying facilities 3. Provision of information about train times/platforms 4. The upkeep/repair of the station buildings/platforms 5. Cleanliness 6. The facilities and services 7. The attitudes and helpfulness of the staff 8. Connections with other forms of public transport 9. Facilities for car parking 10. Overall environment 11. Personal security whilst using the station 12. The availability of staff 13. The provision of shelter facilities 14. Availability of seating 15. How request to station staff was handled 16. The choice of shops/eating/drinking facilities available	1. Overall satisfaction with the train 2. The frequency of the trains on that route 3. Punctuality/reliability (i.e. the train arriving/departing on time) 4. The length of time the journey was scheduled to take (speed) 5. Connections with other train services 6. The value for money of the price of ticket 7. Upkeep and repair of the train 8. The provision of information during the journey 9. The helpfulness and attitude of staff on train 10. The space for luggage 11. The toilet facilities 12. Sufficient room for all passengers to sit/stand 13. The comfort of the seating area 14. The ease of being able to get on and off 15. Your personal security on board 16. The cleanliness of the inside 17. The cleanliness of the outside

2.3 *Metro Service in Other Countries*

2.3.1 UK

In the UK, the National Rail Passenger Survey (NRPS) had conducted satisfaction surveys of the passengers who travelled by London Metro, Regional trains and Long-distance trains twice a year. The satisfaction measures in the survey focus on station and train facilities (see Table 1).

2.3.2 Italy

The Railway Service of Northern Italy (RSI) had developed a set of KPIs for measuring their performance. Eboli et al. (2017) had developed a performance measurement system from Eboli and Mazzullai (2008). The KPIs are classified into Safety, Cleanliness, Comfort, Service, Additional Services, Information, and Personnel (see Fig. 2).

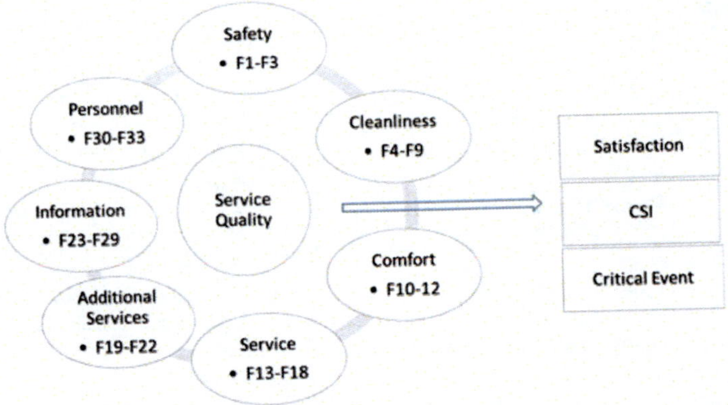

Fig. 2 RSI criteria. Adapted from Eboli et al. (2017) Eboli et al.

2.3.3 Australia

In Australia, Metro Trains Melbourne (MTM) has adopted four KPIs for measuring their service operations which include Punctuality, Service Delivery, Total Kilometer Scheduled, and Customer Satisfaction Index. In addition, Transport for New South Wales reported its customer satisfaction index based on the following 8 dimensions (Transport for New South Wales 2018):

1. Timeliness
   - This train turning up on time
   - Frequency of this train service
   - Journey time given the distance travelled
   - Time to connect to other transport services
2. Safety and security
   - Feeling safe at the train station
   - Feeling safe while on the train
3. Ticketing
   - Ease of purchasing my ticket
   - Ease of using my card
4. Accessibility
   - Ease of accessing the train station
   - Ease of getting on/off the train
   - Usefulness of signs to help you find your way
5. Comfort

- Comfort at the trains trop
- Seat availability on this train
- Seat comfort on this train
- Temperature on this train
- Personal space on this train

6. Cleanliness

- Cleanliness of the train stop
- Cleanliness of the train

7. Information

- Availability of arrival information for the train
- Availability of next stop information for the train
- Availability of information about service delays
- Ease of finding information about routes, stops, and timetables

8. Customer service

- Willingness of train staff to help
- Knowledge of train staff
- Presentation of train staff

In New Zealand, two main concerns for the metro service operations are Punctuality (3 min delay allowed in Wellington and 5 min delay allowed in Auckland) and Service Delivery (Auckland Transport Board Meeting Report 2014).

2.3.4 Singapore

There are 3 Metro operators including SMRT Trains, SBS Transit, and Light Rail Transit. In general, the performance is measured based on CoMET and Nova criteria with additional measure on the customer satisfaction once a year. According to the SMRT Corporation Ltd Annual Report 2016, the measures include:

1. Train Arrival Punctuality (within 2 min of schedule weekly)
2. Train Departure Punctuality (within 2 min of schedule weekly)
3. Train Service Delivery (weekly)
4. General Ticketing Machine Reliability (weekly)
5. Passenger Service Machine Reliability (weekly)
6. Automatic Fare Gates Reliability (weekly)
7. Escalator and Conveyors Reliability (monthly)
8. Lift Reliability (monthly)
9. Passenger Injury Rate (monthly)
10. Train-kilometers between delays of more than five minutes
11. Number of delays lasting more than 30 min

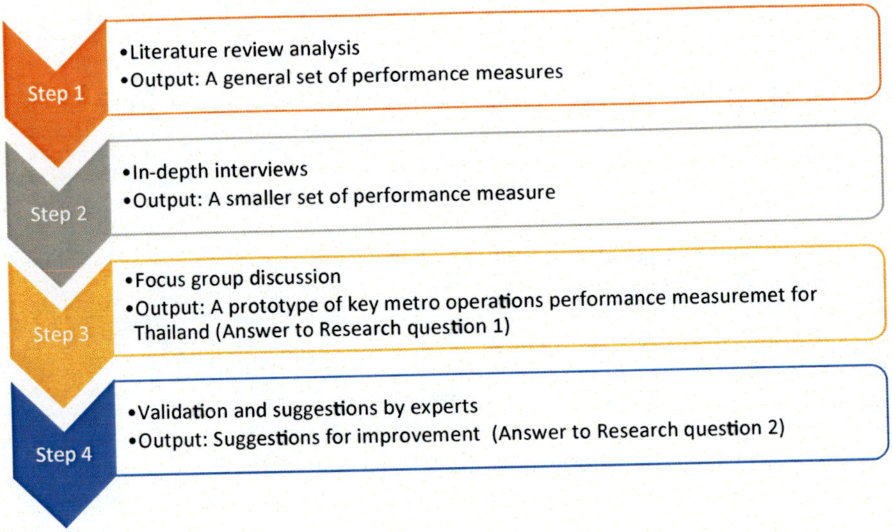

Fig. 3 Steps of research methodology

From the review of literature, it was found that different metro services have different performance measurement criteria. Other measures used in major metro operators in London Underground in UK and MTR HongKong are also reviewed. The literature review analysis presented a list of performance measures and criteria for metro operations measurement. Then the output was presented to each of the target participants from metro operators, metro project owners, and policy makers. They were asked to select the measures based on their own judgement.

3 Our Methodology

In order to meet the research objectives, the method of the study includes literature review analysis, several rounds of in-depth interviews with relevant stakeholders, and focus group discussion. Finally, validation and suggestions were done by experts. The steps of methodology is shown in Fig. 3.

4 Findings

Step 1. Literature review analysis.

This step started from the review of previous literature, theories, annual reports and practices in other countries with relevance to the metro service operations. The output of the extensive review was a list of key performance indicators used in the literature. The list would be used as a seeding for next step.

Step 2. In-depth interviews.

Stakeholders involve with metro operations services in Thailand were identified. The stakeholders included project owners, metro operators, policy makers, and research and development agencies who have a direct mission in the support of rail operations. Executives, directors, and managers from Mass Rapid Transit Authority of Thailand (MRTA), Bangkok Expressway and Metro Public Company Limited (BEM), BTS Group Holdings Public Company Limited, National Science and Technology Development Agency (NSTDA), Office of Transport and Traffic Policy and Planning (OTP), Krungthep Thanakom PCL (KT) Airport Rail link PCL (ARL), SRT Electrified Train Limited (SRTET), Bangkok Metropolitan Administration (BMA), and relevant researchers from universities were interviewed and gave their views on the current and future metro operations performance measurement. The questions are "what are the KPIs you used for measuring metro operations? and which KPIs should be used for monitoring metro operations in Thailand?" The time for interview varied from 1 h to a few hours. The output from this step is a smaller set of key measures as shown in Table 2.

Step 3. Focus group discussion.

This step was to endorse the selection of important key performance indicators for metro operations in Thailand using the focus group method. Twenty-four panelists from metro operators, metro project owners, policy makers, and research and development agencies participated in the focus group. Most of the panelists were senior decision-making level executives who were nominated by the organizations. The discussions were conducted in detail considering each selected KPI one by one and open for free discussion until the consensus was reached. The total time spent for the focus group discussion was a bit longer than 3 h. The final list of metro operations performance indicators for Thailand is shown in Table 3.

Step 4: Validation of selected KPIs and suggest avenues for improvement.

The list of selected KPIs from Step 3 were confirmed by another group of experts who have at least 10 year experience working in the metro service. The list was confirmed and each KPI was defined clearly, and the frequency of data collection were also agreed. Then the recommendations for improvement were prepared for overall public benefits.

A set of key performance indicators (KPIs) for metro operations has been listed for performance measurement as well as efficiency and quality enhancement. The KPIs will be used as a mechanism to raise operators and stakeholders' awareness of their performance. It is expected that the results from performance comparisons will encourage improvements, developments and acceptance among operators and policy makers. The definition of 9 KPIs and how to improve each KPI can be summarized as follows:

Table 2 Output summarized from the interview

Dimension	No.	KPI	Unit	Definition	Frequency
1. Growth, learning ands innovation	1	% change passenger journeys	%	(No. of passenger journey of fare paying customers this year-last year) * 100/last yea	Yearly
	2	No. of training hrs/1000 staff hrs	h	Hours of staff training/1000 staff hours	Yearly
2. Financial	1	Operating cost/passenger journey (or passenger km)	USD	Total operating cost in USD/No. of passenger journey (or passenger km)	Yearly
	2	Operating cost/car km	USD	Total operating cost in USD/car km	Yearly
	3	Maintenance cost/car km	USD	maintenance cost/car km	Yearly
	4	Fare revenue/passenger journey (or passenger km)	USD	Fare revenue in USD/passenger travel(or Km total passenger travel)	Yearly
3. Customer	1	Train service punctuality	%	No. of trains operated on time within 5 min delay*100/total trains	Monthly
	2	Hours of train delay/train hours operated	%	Hours of train delay within 5 min.*100/train hours operated	Monthly
	3	Train service availability	%	No. of Train-Trips actually operated in a Month/No. of scheduled Train-Trips agreed in contract	Monthly
	4	No. of >30 min delay	times	No. of > 30 min Delays excl. delays caused by external factors	Monthly
	5	% Station with step-free access	%	No. of stations with step-free access*100/total no. of stations	Monthly

(continued)

Table 2 (continued)

Dimension	No.	KPI	Unit	Definition	Frequency
4. Internal process reliability, availability, efficiency	1	% cars available and used in peak hours	%	Toal cars/cars in use during peak period	Monthly
	2	Mean distance between delays > 5 mins	car km	Car km between incidents causing a delay of 5 minutes or more	Monthly
	3	Traction Energy consumed /car km	KWh/km	Traction energy used in Kwh/car km	Monthly
	4	Total energy consumed/passenger journey	KWh/km	Total energy purchased in Kwh/passenger journey	Monthly
	5	Passenger journeys/staff hours	trips/h	Total number of passenger trips/staff hour	Monthly
5. Safety and security	1	Deaths from accidents/passenger journeys	Cases	No. of cases died from accidents/passenger journeys	Monthly
	2	Incidence of crime/passenger journey	Cases	No. of crime incidents/passenger journey	Monthly
6. Environment	1	CO_2 per passenger km	grams	CO_2 emission in grams/million passenger km.	Monthly

Table 3 Key performance indicators for measuring metro operators in Thailand

Dimension	KPI
1. Transport Volume	1. Number of passenger journeys
2. Punctuality	2. Train service punctuality
3. Reliability	3. Hours of train delay/train hours operated 4. Number of delays of more than 30 min delay 5. Mean distance between delays of more than 5 min
4. Availability	6. Train service availability
5. Resource utilization	7. Percent of cars used in peak hours 8. Passenger journeys/staff hours
6. Customer satisfaction	9. Customer satisfaction

- KPI 1 Number of passenger journeys (frequency: monthly)

It is a global policy to encourage people to use mass rapid transit services and increase the number of passenger journeys. It is a great challenge to make urban rail systems more attractive and comfortable. Below are some ideas.

1. Increase channels of ticket sales for passenger convenience.
2. Initiate discounted fare structures for different target groups.
3. Increase accessible physical connections and adequate parking space.
4. Provide information available for passengers both in Thai and English.
5. Add other facilities to serve passengers' needs.

- KPI 2 Train service punctuality (frequency: monthly)

On-time service is one of the most important performance indicators which reflects the availability of system and system management and operations. It is suggested that to improve train service punctuality, the operators should:

1. Ensure that equipment is ready to use and have proper maintenance.
2. Monitor station equipment including lifts, escalators, automatic fare gates, platform screen doors, ventilation systems to be in good condition.
3. Train the staff to follow the correct operating procedures
4. Respond to unusual events professionally.
5. Conduct risk analysis and management regularly for unexpected event preparation.

- KPI 3 hours of train delay/train hours operated (frequency: monthly)

This indicator measures the reliability of the service, based on the percentage of hours served on time, allowing a delay of up to 5 minutes. Root cause analysis is recommended for service improvement.

1. Monitor and upgrade equipment regularly.
2. Modify traffic management, scheduling management, human resource management, and system administration.
3. Apply standards compliance and regulations.

- KPI 4 Number of delays longer than 30 min (frequency: monthly)

This indicator measures the reliability of the service by counting the unusual delays longer than 30 min. Cause of delays should be identified and corrections should be made to prevent future delays. There is a need to measure both the actual delay to train service and the impacts of train delays on customers. It is important to report delay as quickly as possible and a proper refund policy should be announced.

- KPI 5 Mean distance between delays of more than 5 min (frequency: monthly)

This indicator measures the reliability of the service by measuring the distance in kilometers between the two delays. The greater the number means that the train is more reliable. Some of the analysis tools that should be used include cost-benefit analysis and life cycle costing management.

- KPI 6 Train service availability (frequency: monthly)

It is a measure of the ability to comply with the terms of the contract. The indicator is subject to change depending on the number of passengers in each period of the contract. Incentives and penalties should be applied.

- KPI 7 Percent car available and used in peak hour (frequency: monthly)

It is a measure of cars available in peak periods. In a good practice. it is necessary to have some spare trains or having more than 100 percent car available in case of failure.

- KPI 8 Passenger journeys/staff hours (frequency: monthly)

The productivity of the staff is another indicator for metro performance. The appropriate number of staff helps to improve the quality of service and enhance the efficiency.

- KPI 9 Customer Satisfaction Index (frequency: monthly)

Customer satisfaction is one of the major concerns of metro operators. The passenger satisfaction survey should be conducted at least twice a year. Surveys should include facilities and accessibility, information provided, customer service, responses to complaints, handling abnormal situations, and passenger comfort. Increasing passenger satisfaction scores implies the sustainability of metro services in the long run.

5 Conclusion

This research is a practical example of research focusing on the identification of appropriate key performance indicators to monitor metro operator performance in Thailand and suggestions for performance improvements. A set of key performance indicators (KPIs) for metro operations has been identified for metro operators to monitor their operations performance. A consensus of six dimensions for measuring metro operators' performance included "Transport volume", "Punctuality", "Reliability", "Availability", "Resource Utilization", and "Customer satisfaction." Nine KPIs summarized from this research are: (1) Number of passenger journeys (2) Train service punctuality (3) Hours of train delay/train hours operated (4) Number of delays more than 30 min (5) Mean distance between delays more than 5 min (6) Train service availability (7) Percent car used in peak hour (8) Passenger journeys/staff hours and (9) Customer satisfaction.

It is noted that the suggested metro performance indicators from this research are limited to nine indicators which is considered a good number for the initial stage.

Too much emphasis on the measurement of train service performance with too many indicators can be at the expense of operators and may lead to unfocused elements of service quality and the bias of actual customer experience. The nine indicators were a consensus from relevant stakeholders and validated by industry experts in Thailand. Suggestions for each KPI improvement were also made with an aim to improve the service operations and satisfy customers.

In the future, data collection for measuring metro performance will be much easier as technology-driven machines create new data sources. The digital data will help the metro operators to analyze their performance in the focused areas. With Business Intelligent and Business Analytic tools, it will be much easier and faster for operators to access their performance, be notified of the problem areas, identify the causes of the problems, and find proper solutions. More KPIs may be identified in addition to the nine KPIs specified from this research. Other KPIs related to environmental issues may be added in the future as the issue of energy consumption and carbon credit environmental savings are more important these days.

Acknowledgements This project is funded by the Office of Transport and Traffic Policy and planning (OTP), Thailand Ministry of Transport with partially supported by the European Union's Horizon 2020 Research and Innovation Programme RISE under grant agreement no. 823759 (REMESH) and the Newton Fund: Industry-Academia Partnership Program with grant reference IAPP18-19/100.

References

Auckland Transport Board Meeting Report (2014) Bus reliability and punctuality performance, 26 Aug 2014. Available at: https://at.govt.nz/media/679083/Item-11ii-Bus-reliability-and-punctuality-performance.pdf. Accessed 1 Mar 2019.

Bangkok Expressway and Metro Public Company Limited (2015) A strong foundation for the limitless potential of the future integrated transport system, 22–62. Available at: https://metro.bemplc.co.th/?lang=en. Accessed 12 Aug 2019

Barron A (2016) World class metro systems: What can be learned from the global metro community? Washington Council of Governments Symposium, Washington DC

CoMET (the Community of Metros) and Nova Metro Benchmarking Groups. Available at: https://cometandnova.org/. Accessed 11 July 2018

Eboli L, Fu Y, Mazzulla G (2017) Railway service quality in northern Italy: A multilevel synthetic assessment. Adv. Mech. Eng. 9(3).

Eboli L, Mazzullai G (2008) A stated preference experiment for measuring service quality in public transport. In.: Transportation planning and technology, vol 31, No. 5. Routledge Publisher, member of the Taylor and Francis Group, 2008. ISSN 0308-1060, pp. 509–523

European Standard (EN 13816 : 2002) (2002) Transportation—logistics and service—public passenger transport—service quality definition, targeting and measurement. Available at: https://ec.europa.eu/eip/ageing/standards/city/transportation/en-138162002_en. Accessed 12 Aug 2019

European Committee for Standardization (2006) EN 15140:2006 allows public transports operators to measure the quality of its service. Available at: https://www.cen.eu/news/brief-news/Pages/NEWS-2019-030.aspx. Accessed 2 Feb 2019

Konečný K, Kostolná M (2014) The possibilities of standardizing the quality level in public passenger transport from customer point of view. Perner's Contacts 2(9). Available at: https://pernerscontacts.upce.cz/35_2014/Konecny.pdf. Accessed 1 Aug 2019

National Rail Passenger Survey (2016) Spring 2016 main report. Available at: https://www.transportfocus.org.uk/research-publications/publications/national-rail-passenger-survey-nrps-spring-2019-main-report/. Accessed 21 July 2019

SMRT Corporation Ltd. Annual Report (2016). Twin tracks to performance and growth. Available at: https://www.smrt.com.sg/Portals/0/InvestorRelations/Annual%20Report/2016/SMRT%20Annual%20Report%202016_LR.pdf. Accessed 1 Mar 2019

Transport for New South Wales (2018) Customer satisfaction index, Nov 2018. Available at: https://www.transport.nsw.gov.au/news-and-events/reports-and-publications/customer-satisfaction-index. Accessed 2 Feb 2019

The Metro Network Extension in the West Midlands: A Socio-economic Impact Assessments

Andrea Gualtieri and Marin Marinov

Abstract This paper provides an evaluation of the socio-economic impacts of the metro network expansion on the West Midlands. The principal municipalities of this English region are already involved in the metro service, but the Midland Metro Alliance is aiming to extend the service to more towns and suburbs in the region. The metro network expansion project has been segmented, analysed and explored using a variant of the classic decision-making matrix, which enabled the assessment of the values of 8 factors responsible for the social and economic development of the West Midlands within the metro network extension itself. The analysis contains both quantitative and qualitative data. A forecasting activity intending to extend the data gathered during the socio-economic impact assessment, considering a 22-year medium-term horizon is organised. The activity has been undertaken using both proper software and different data analysis functions in order to project the socio-economic impact of the metro extension in the West Midlands.

Keywords Metro · West Midlands · Value management · Impact assessment · Decision-making matrix · Data analysis forecasting

1 Introduction

The West Midlands is currently experiencing significant demographic and economic growth that is driving businesses and transport infrastructure networks to further expand (Stubbs and Riley 2018). In particular, the West Midlands Metro extension plan is ambitious and relevant for the development of the entire region (MidlandsMetro 2019d), and it has been met with enthusiasm by citizens as well as by Midland Metro Limited (Midland Metro Alliance 2018).

Exploring a set of socio-economic factors this paper aims to evaluate and forecast the impacts of the metro system expansion on the West Midlands region.

A. Gualtieri · M. Marinov (✉)
Aston University, Birmingham, UK
e-mail: m.marinov@aston.ac.uk

W. Weerawat et al. (eds.), *Urban Rail Transit*, Lecture Notes in Mobility,
https://doi.org/10.1007/978-981-15-5979-2_6

1.1 Motivation

Several metro network extension plans have been announced and approved, and these plans will impact areas across the West Midlands from 2019 to 2026. A large amount of information is publicly available about the technical characteristics of the future metro network, the location interested in the new network, the interconnections available with other public transport systems and strategic infrastructure (Midlands Metro 2019d).

Metro systems are expensive to build and costly to maintain, therefore running them under utilised is not acceptable. Studies on metro system utilisation and new service design include (Kelly and Marinov 2017; Brice et al. 2015; Dampier and Marinov 2015; Wales and Marinov 2015; Darlton and Marinov 2015; Motraghi and Marinov 2012; Toal and Marinov 2019; Marujo et al. 2019), due to the nature of these publications we shall not discuss them further. Over the last two years, passenger engagement in the West Midlands Metro service has not met managerial expectations (Shoaf and Hewitt 2019). Consequently, research on the impact of a further network expansion may contribute to understanding the extent to which the metro service operations are positively affected. Although documents on the quantitative aspect of the expansion are publicly available, there is no research that aims to forecast the social and economic results within the areas included in the West Midlands Metro service. The paper therefore has the objective of measuring and forecasting the impact that a further-extended network will have within the region, assessing the potential benefit of the metro network expansion.

1.2 Objective and Methodology

The socio-economic impact assessment is useful for informing the value that the West Midlands Metro provides in the entire region. In addition, this paper aims to provide an overview of the socio-economic impact of the current metro network by considering some factors that have also been used in the medium-term forecast. For this purpose, secondary data regarding the economic variables have been properly gathered and refined through the usage of visual tools.

The methodology to assess the socio-economic impact of the metro network extension was chosen from a set of available options that were analysed for their efficacy in and suitability for the specific activity needed to meet the paper's aims. In the light of this, the Decision-Making Matrix (Pugh 1981) is a tool that can furnish better consistent information for the case, and as a result it has been selected for the analysis.

All of the information contributing to the development of the research has been extracted from public domain resources. In particular, data have been gathered from metro-specific annual reports communicated by the Department of Transportation (Transport for West Midlands 2017) and from other sources (Stubbs and Riley

2018). Qualitative information has been elaborated from sources such as multimedia, specialist websites and applications (MKT Transport Media 2019; MidlandsMetro 2019c), online archives and published articles (Expressandstar.com 2017; Khamkar 2018).

The factors that have been used to develop the socio-economic assessment tables have been precisely elaborated by exploiting a rich database (Bell 2019) and specific Microsoft Office package functions and applications. The forecast itself has been made possible thanks to the use of a working paper that, despite targeting a different issue within the metro sector, uses a methodology that enables trend projections to be made in the mid to long term span (Wood and Mysore 2019).

1.3 Outcome

The research showcases 22-year projections of the social and economic impact the metro network extension has on the West Midlands. In particular, the major contributors to the impact are visible in graphs and tables favouring point-to-point observations as well as a cumulative overview of the expected trends. Observation of the value of the existing metro service and of the further extension plan have been exhaustively provided and summarised considering both the value addition over time and the transformation of the lines.

Finally, the results have been summed up and appropriately commented, and some insight for further studies that might emerge from this research has been given.

1.4 Paper Organisation

The paper has been organised into 6 sections.

Section 2 describes the layout of current properties and provides a brief overview of the past metro activity in interested locations in the West Midlands. The metro network expansion plan, along with its description, is presented.

Section 3 focuses on the motivations and drivers that have pushed managers to pursue the massive extension of the existing metro network.

Section 4 assesses the socio-economic impact of the network transformation on the West Midlands according to specific weighted factors that differently add value in each extension according to the selected methodology as showcased in tables and graphs.

Section 5 extends the socio-economic assessment in a 22-year window. The impact trend as a whole, along with the focus on the factors change, has been shown. In this section, comments on potential challenges, opportunities and relevant observations have been made.

Section 6 displays the conclusion of the paper, along with avenues for developing further studies on the topic being studied are given.

2 The Current Layout and Further Expansion

The metro as a public transport system in the West Midlands has been in use since 1872 in the city of Birmingham. The service was designed for passenger transport only, and the initial metro line linked Hockey Brook with Dudley Port. Over the years, several lines have been opened, exploiting the advancements in technology and engineering systems as well as the industrial expansion of the area. The consequences of World War 2, along with the local council resources shortage and business closures, have started a decline in the metro network expansion, since lines have been progressively closed. Following the trend, the metro transport mode was terminated in 1953, while bus transport was heavily promoted (Harris 2019).

The metro system's revival began in 1999 with the activation of Line 1. It includes 26 stations from Birmingham to Wolverhampton in a highly populated area in the West Midlands that has approximately 1.5 million inhabitants, which is almost half of the region's total population (Ukpopulation.org 2019). The Line 1 layout recalls the pattern of the Great Western Railway main line pattern, a train line that was closed in 1972. The metro network, called West Midlands Metro, is now the property of Transform for West Midlands, which is directed by the West Midlands Combined Authority, while Midlands Metro Ltd. is in charge of running the service (Metroalliance.co.uk 2018; Transport for West Midlands 2017).

The consortium of the Midland Metro Alliance is in charge of designing, developing and maintaining future metro network expansions.

The current Metro line connected Birmingham New Street Station with Wolverhampton St. George after a 2015 implementation's project that has since linked Snow Hill Station to the Birmingham city centre shopping area and the central railway station, Grand Central (Expressandstar.com 2017). Along the current line, passengers can easily access additional services such as bus stops, railway stations, parking for motorised vehicles and bicycles (MKT Transport Media 2019). Figure 1 shows an edited current West Midlands Metro network Line 1 map.

The service has been supplied with more than twenty Urbos 3 brand vehicles in order to secure comfort and high performance during the journey (Technology 2019). However, the metro utilisation level has not always met the managers' expectations, as Table 1 shows.

Passengers have skyrocketed since the Birmingham City Centre expansion in 2016. The network, which previously had Snow Hill Station as the Line 1 terminus point, has been stretched to Grand Central. The increase in passengers is linked not only to access to the Birmingham central railway New Street Station but also to High Street and businesses and important shopping malls present in the area (Expressandstar.com 2017).

West Midlands metro, which is classified as a light rail public transport system, permits passengers to travel across the region faster and cheaper than other transport modes. The ticket can be acquired by both traditional contact-less modes starting from the price of £1 per a single short-hop-on ticket. Additionally, the Swift Card was introduced, following the London Oyster Card's example for London transportation

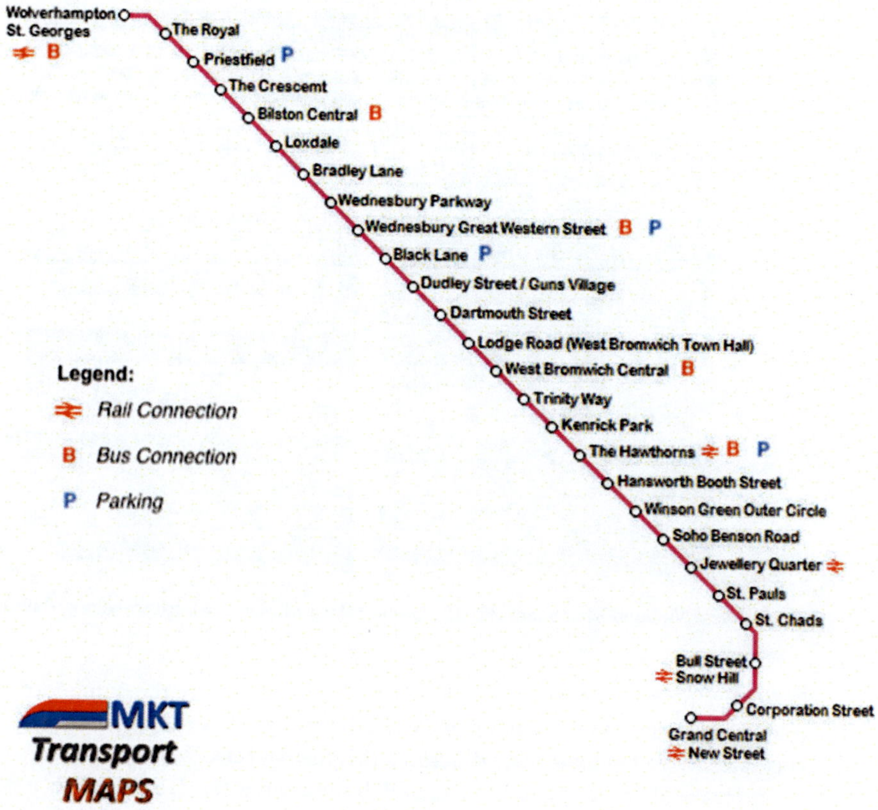

Fig. 1 Current metro network map. *Source* MKT (2019)

Table 1 Expected passengers (2016–2020).

Year	Current passengers (in millions)	Expected passengers (in millions)	Gap (in millions)
2015–2016	5	7.9	−2.9
2016–2017	6.6	7.9	−1.3
2017–2018	6.2	7.9	−1.7
2018–2019	7	7	0
2019–2020	7.3	7	+ 0.3

Source Re-adapted from Expressandstar.com (2017)

(Transport for London 2019). In Fig. 2, the current West Midlands fares table is shown.

The West Midlands Metro objective, along with partners in the Midland Metro Alliance, is to create an integrated regional transport system that can promote advancements for the entire West Midlands area. The ambitious objectives consist

Cash/Contactless

Adult

Single	Return	Off-peak Return
£1.00	£2.00	£2.00
£2.50	£3.20	£3.00
£3.20	£5.00	£3.80
£4.00	£5.50*	£4.00*

*Day Ticket

Child

Single	Return	Off-peak Return
£0.50	£1.00	£1.00
£1.30	£1.60	£1.50
£1.60	£2.50	£1.90
£2.00	£3.20	£2.50

swift

Adult

Single	Return	Off-peak Return
£0.90	£1.80	£1.80
£2.40	£3.10	£2.90
£3.10	£4.90	£3.70
£3.90	£5.40	£3.90

Child

Single	Return	Off-peak Return
£0.45	£0.90	£0.90
£1.25	£1.50	£1.40
£1.55	£2.40	£1.80
£1.95	£3.10	£2.40

Fig. 2 West Midland Metro fares (2019). *Source* West Midlands Metro website (2019c)

of improving the current metro network of 34 km to enlarge the number of stops and interchanges, involving new towns and suburbs in the network, and facilitating the accessibility of strategic infrastructures (MidlandsMetro 2019c).

In terms of the current Line 1, which comprises the Birmingham City Centre extension, the approved expansion programme includes the following 5 macro areas.

1. *Birmingham Westside*—The service will involve, at the end of 2019, the two very central locations of Victoria Square and Centenary Square. Moreover, in 2021, the Line 1 design will include the Edgbaston area through Brindley Place and Five Ways.
2. *Wolverhampton City Centre*—The project, due to open at the end of 2020, aims to link the existing St. Georges stop to Piper's Row and Wolverhampton Central Station.
3. *Birmingham Eastside*—From 2022, West Midlands Metro will have Line 2. The passengers will be able to utilise a brand new catenary-free metro line that connects Snow Hill Station to the area of Digbeth (Technology 2019). It consists of 4 new stops that will allow the passengers to exchange to the High Speed 2 train station (HS2) in Canal Street and to the coach station in Deritend High Street.
4. *East Birmingham to Solihull*—From Deritend High Street, Line 2 will extend towards Solihull, which will enable important infrastructure, such as Birmingham International Airport and the HS2 Birmingham Interchange, to be reached. This project, due in end of 2026, aims to implement the Birmingham Eastside network in adding strategic stations that will open access to users in the vital area of

Fig. 3 Metro extension map. *Source* Readapted from Westmidlandsmetro.com (2018)

Greater Birmingham, which includes Heartland Hospital, Birmingham Business Park and the NEC.

5. *Wednesbury to Brierley Hill*—Planned to open in 2023, this new line aims to link Wednesbury, Dudley and Brierley Hill (Shoaf and Hewitt 2019)

Figure 3 presents a map of the current metro network system and the metro extension programme.

3 Reason for the Expansion

Section 2 provided an idea of the complexity of the extension programmes. Doubtlessly, several resources have already been spent on the development of the metro network, the expansion programme has been undertaken for specific reasons. As anticipated, the West Midlands Metro management plan is to build an integrated transport network that can exploit several existing transport systems and strategic infrastructure across the region (MidlandsMetro 2019c).

Indeed, management has underlined the objective of creating the first integrated transport system outside London. This process has been supported by side decisions, such as designing a modern and attractive new logo that is standard between the trains, metro and buses within the region (MidlandsMetro 2019b).

Metro Line 1 has been built on the existing Great Western Railway (GWR) main route. Accordingly, the presence of railway tracks that are currently not in use, such in the Dudley area, has increased the interest in developing a modern network utilising part of the existing infrastructure (Technology 2019). Including the metro network in areas in which the light rail has already been underlines the need of the population to be connected with a principal public transportation system a need that has been suspended in the last century (Harris 2019). The network aims to re-connect those suburbs and promote connectivity through a renewed metro service. In addition, the Midland Metro Alliance programme enables users of areas distant from the main centres to move across the region utilising an alternative to the motorised vehicles, which causes several disadvantages in term of cost and traffic congestion.

The demographic growth in the region has driven the development of the metro network expansion plan. All seven principal municipalities of the West Midlands have registered an incremental demographic growth in the last two decades, and as a consequence, a higher value of metro users is expected in the next years (Brinkhoff 2019). Figure 4 shows the increasing population in the West Midlands.

If a large demographic growth might bring important benefits to an area, in large centres, it might also cause challenges such as traffic congestion and pollution that a state-of-the-art metro system can help avoid. For instance, the Birmingham population grows approximately 3% every year, and improving and making the metro system adequate and an efficient integrated public system in the long term are vital (Ukpopulation.org 2019) In addition, the metro light rail improvement may fulfil the tourist demand, along with occasional presence of visitors that is expected in the West Midlands due to prestigious international events such as the 2021 Commonwealth Games that Birmingham will host (MidlandsMetro 2019a).

Moreover, some important business areas on the west side of the Birmingham city centre have previously been reachable by bus and by train, only by utilising the service in Five Ways Station (Metroalliance.co.uk 2018). The presence of a single

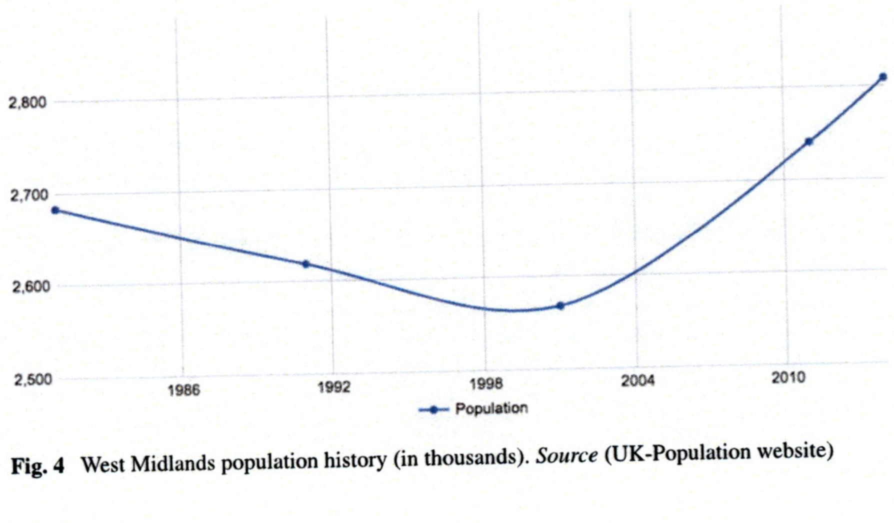

Fig. 4 West Midlands population history (in thousands). *Source* (UK-Population website)

railway station in the area creates discomfort for workers in terms of accessibility. In addition, utilising private transportation methods can be a disservice due to the heavy traffic congestion in peak hours. As a consequence, implementing the metro extension facilitates the passengers to reach their workplace compared with other transportation methods. The area included in the metro extension plan programme will benefit by having a higher level of potential users and will improve the general value and attractiveness of those areas when considering the factors that are the core of the developments in this paper.

4 Socio-economic Impact Assessment

The expansion of metro light rail system across the entire West Midlands region is ambitious and demanding. However, it will accelerate the progress and prosperity of the whole West Midlands; it will promote services that may stimulate a higher level of commerce and benefits for the entire region. Given the nature of the metro extension plan, the technical and financial feasibility have also been evaluated with the consideration of variables such as expected capacity, infrastructure cost, maintenance and return on investment (Shoaf and Hewitt 2019).

However, the analysis elaborated so far has only been reported from the West Midlands Metro viewpoint. Indeed, data about the metro light rail extension implication in economic and social terms have not yet been collected. For this reason, a specific method for assessing the socio-economic impacts on the West Midlands has been researched and implemented. For this purpose, a rather traditional method to evaluate the benefits of the expansion has been taken into consideration. In particular, a survey, project management approach, risk assessment and decision-making matrix are the methods that have been analysed in terms of their compatibility with the socio-economic impact assessment.

The survey was the first method examined. Surveys can be appropriate for assessing factors such as the perception of cleanliness, security or punctuality (Giles 2018). In the specific case of the Midlands Metro Alliance project, this method is not suitable for measuring any socio-economic impact. Indeed, interviewing stakeholders and passengers could have given information about their satisfaction level or about the advantages they personally perceive to gain from using the metro, but it would not have provided quantitative values (Giles 2018). The assessment requires a number of quantitative data in order to proceed in developing its medium-term forecast (Wood and Mysore 2019).

The project management approach is widely used in the transportation industry and may incorporate several applications depending on the project's complexity. However, such an approach is not optimal in this specific case (Online Project Management 2019) because its outcomes may be excessive in numerical terms and it may not be suitable for measuring the socio-economic impact in the areas considered.

An attractive alternative is the risk assessment method, which offers a wide range of layouts and high adaptability according to the object. The issue with this specific method is the high dependability on probabilities, which are challenging to obtain, considering the data sources that are available and the specific project's complexity (Praxiom.com 2015).

4.1 Decision-Making Matrix

The methodology review was conducted with a decision-making matrix, which was considered the most suitable tool for measuring the West Midlands Metro network extension's socio-economic impact. This method was originally credited to Benjamin Franklin in 1772, and although it was originally used in studies concerning the employment field, it is effectively adaptable to a broad range of issues. Named by its creator as "Moral and Prudential Algebra", the method consists of assigning a score to a limited set of options or decisions while considering the sum of the selected variables according to their specific weight (ScarboroughCivitelli 2019). The model has been used in diverse fields in which assessing uncertain, not easily quantitatively assessable objects is required (Edwards 2018).

The method suits the central objective of the paper, which aims to furnish qualitative and quantitative data on both social and economic aspects on the metro extension plan while maintaining the dynamism of the factors considered.

The model has been adapted to the West Midlands Metro extension case, and two sets of factors, *dynamic* and *standard,* have been considered. Each *dynamic factor* weight is specific according to the value it has in terms of the overall value, while *standard factors* are not affected by any modification. The final outcome of the decision-making matrix, after multiplying the weights of the dynamic factors and the value assigned to each of them, gives a measure of the socio-economic impact of the extension being considered.

4.2 Dynamic Factors

To proceed with the socio-economic metro network assessment, the dynamic factors of the household, business, accessibility, interchange, tourism, mobility, entertainment and prices have been selected and described. Each feature has been assigned a certain weight, from a minimum of 0% to a maximum of 100%, depending on the extent to which the factors contribute to the overall impact (Fig. 5).

According to Fig. 5, the most valuable factors are *interchange* and *mobility.* The first factor refers to the number of bus hubs, railway stations, HS2s and transportation services in the line. This indicates that the users may have easily switch between one type of public vehicle to another. In addition, the importance of providing interchange stations has also been highlighted by the management, who are aiming to create an integrated transportation system for the West Midlands (Metroalliance.co.uk 2018).

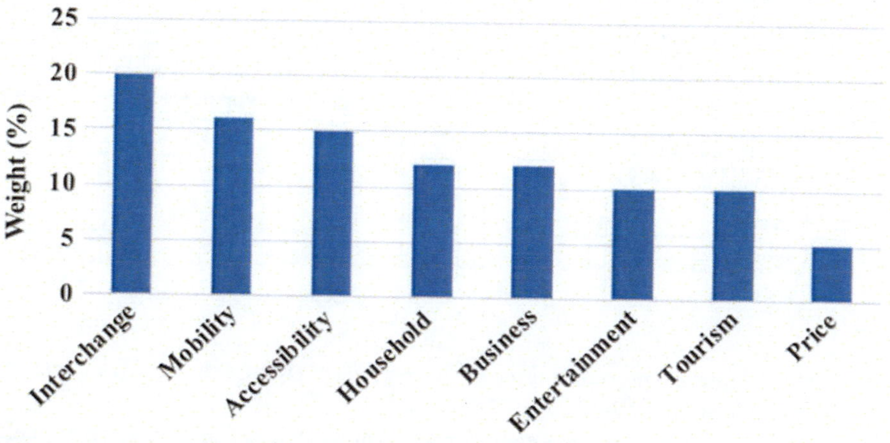

Fig. 5 Weight of each dynamic factor. *Source* (Own illustration)

Mobility is a valuable factor since the region has been abandoning its light train development for decades, thus indirectly encouraging roads (Harris 2019). The network expansion in high-traffic areas might drive owners of motorised vehicles to choose the metro alternative, resulting in saved time and resources from a user point of view as well as a better flow of the traffic stream.

Household, business and *accessibility* are also particularly relevant factors. Household and business measure the potential number of users, considering the areas in which they live and their workplace locations. The first factor is specifically considered to be household concentration, while the second is intended to be business density within an area.

Accessibility refers to the movement of people or goods that, through the metro extension, are able to travel to areas, sites of interest and institutions.

The measurement of this factor considers users reaching public institutions and buildings related to education, such as universities and libraries. The realization of an effective metro network might increase the average number of users more frequently visiting installations such as hospitals and governance structures as well as buildings that the current metro system does not permit visitors to reach.

Tourism and *entertainment* refer to attractiveness to tourists and the accessibility of entertainment-focused locations, respectively, which have a moderate impact because of the elastic number of users they might attract. However, the metro can promote the exploration of museums, historic buildings, parks and suggested sites for tourists and can boost traffic in areas dedicated to nightlife and events.

Finally, the less influential factor is *price,* which changes according to the trip length. In particular, the shorter the journey is, the higher the price factor value is because the fare for the users is more likely to increase with travel distance. Given the significance of the plan, price is not considered a powerful factor, as the service demand is static because a metro network usable throughout the entire region would completely transform the mobility in the area. The users may not be concerned

Table 2 Standard factors per postcode

2021—line 1, Birmingham Westside			
Station	Postcode	Population	Household
Victoria Square	B1 1	4207	2245
Centenary Square, Brindley Place	B1 2	3026	1503
Five ways	B15 1	2264	1003
Edgbaston	B16 8	6937	3624
Total		16,434	8375

about paying a premium price if the services overcome the disadvantages of private transportation methods.

In total, 6 out of 8 *dynamic variables* presented in Fig. 5 have been qualitatively evaluated using publicly available sources (2018).

However, the scores for *household and business* has been mathematically determined to exploit an online database, which contains information on the population and household numbers for each postcode (Bell 2019).

For each of the dynamic factors, a score was assigned, and their sum results in the socio-economic impact of the given metro extension.

4.3 Standard Factors

The value of the variables of population (UK Census Data.com 2012) and business density (Wright 2018) are directly reported with no modification because they contribute to the assessment of dynamic household and business factors. In particular, the household score is considered to be the sum of the values of household numbers on population in each location, while business density is a ratio of the business density in the West Midlands to the population in each location. As a result, Table 2 shows the methodology undertaken, using the Birmingham Westside extension as an example.

The same data collection and elaboration exercises were undertaken for the other extensions using similar tables, and they have been useful for building the table and graphs below.

4.4 Assessment of the Lines

To facilitate the understanding of the socio-economic evaluation, the analysis is presented in the following structure:

1. Line 1 assessment from 2019 to 2021.

2. Line 2 assessment from 2022 to 2026.
3. Dudley line assessment from 2023 on.

Each line segment has been assessed using the decision-making matrix model, as Table 3 shows.

Each extension has its own table, which is also represented using charts.

Figures 6, 7, 8, 9, 10 and 11 show the value each factor delivers on an isolated metro segment and give an overview of the comparison of these segments within each line. Specifically, Fig. 6 shows the current situation.

Line 1 Assessment from 2019 to 2021

Line 1 includes 3 additional extensions. In particular, it is formed by the current Line 1 layout, the additional improvement in Wolverhampton City Centre and the Birmingham Westside extension. The West Midlands are divided into two phases:

Table 3 Line 1 decision-making matrix

2019—current line 1: Wolverhampton St. Georges—Birmingham New Street		
Factors	Weight (%)	Impact score
Interchange	20	0.40
Mobility	16	0.5
Accessibility	15	0.7
Household	12	0.4
Business	12	0.0
Entertainment	10	0.4
Tourism	10	0.7
Price	5	0.4
Total	100	44.3

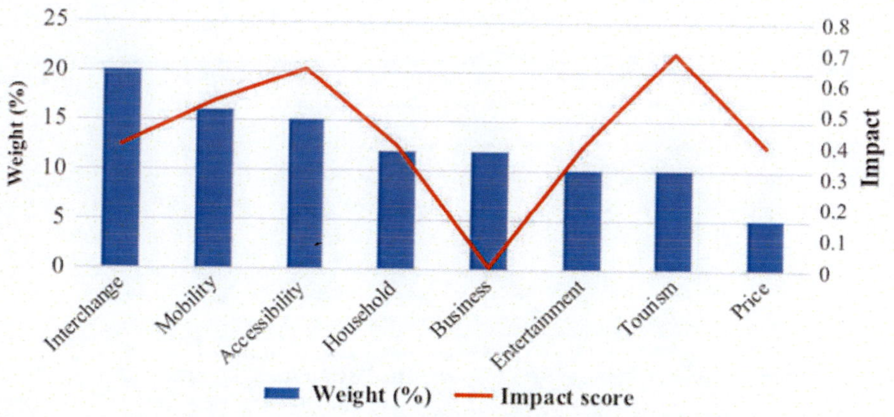

Fig. 6 Current metro network impact. *Source* (Own illustration)

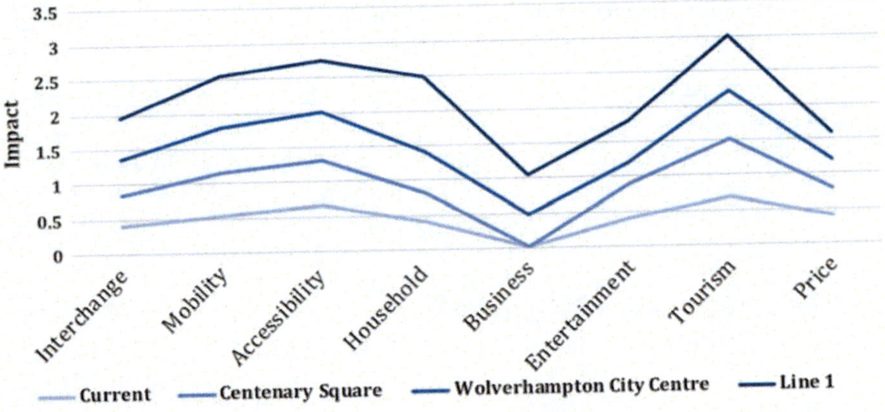

Fig. 7 Impact from the factors of line 1. *Source* (Own illustration)

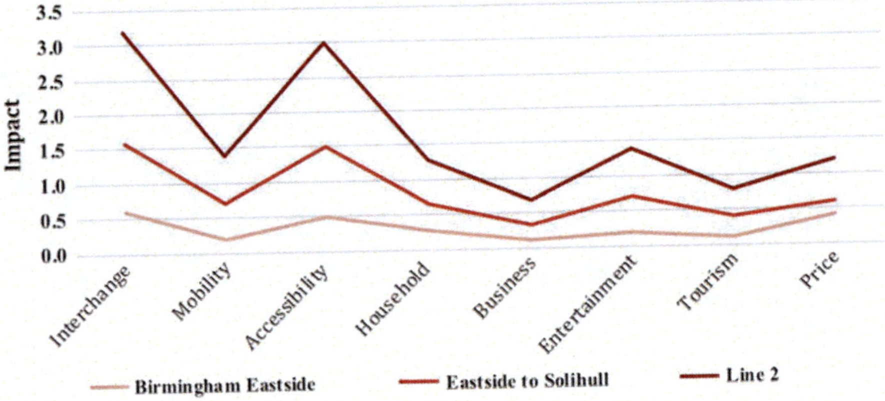

Fig. 8 Impact from the factors of line 2. *Source* (Own illustration)

Fig. 9 Impact from the factors of Dudley line. *Source* (Own illustration)

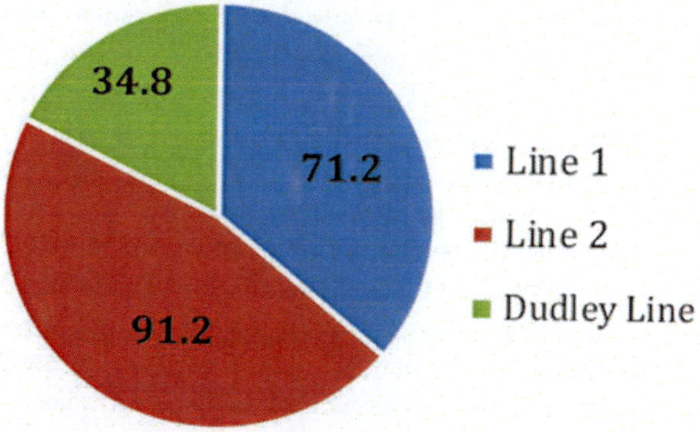

Fig. 10 Impact from the factors of all the lines. *Source* (Own illustration)

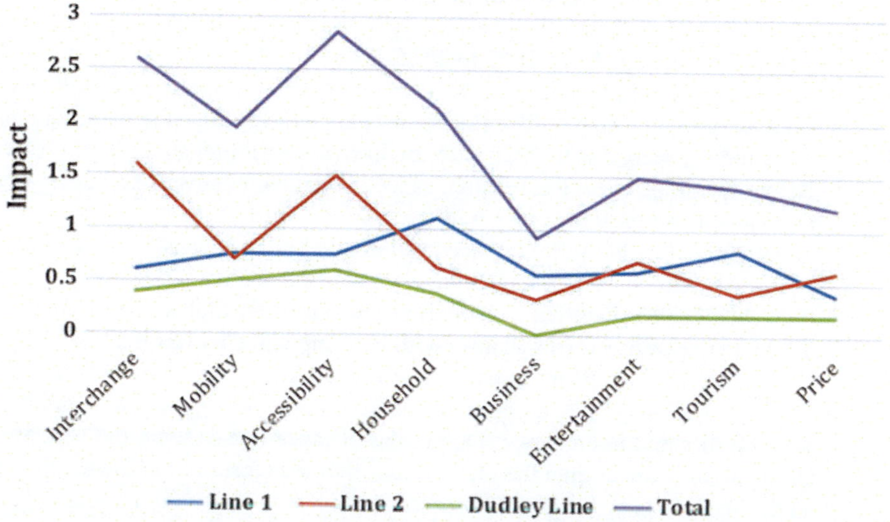

Fig. 11 Value of dynamic factors in the overall impact of the lines. *Source* (Own illustration)

the extension towards Centenary Square, due to open in 2019, and the extension towards Edgbaston, planned to open at the end of 2021. The first chart shows the result of the socio-economic impact evaluation on the current Line 1 metro network from Wolverhampton St. Georges and Grand Central.

Line 1 is very relevant for the community since it brings a high number of public transportation interchanges with the 7 bus stations and 5 railway stations across the segment. Accessibility is fostered because the network connects the two largest West Midlands cities, Birmingham and Wolverhampton, in which a high number of households and level of infrastructure are spread across. The low business density level is compensated for by the value given by tourist attractions and entertainment, which are further improved by the extension to Centenary Square because of the massive presence of shopping centres, museums and monuments.

The socio-economic impact assessment was also assessed for the Line 1 current extension plan, which, with Westside Birmingham, aims to connect the town centre to Edgbaston. The cumulative value, including Wolverhampton city centre and the Birmingham Westside extension, is presented in Fig. 7.

The Line 1 impacts will skyrocket after the 2020 Wolverhampton City Centre Extension, which will enable the involvement of a higher number of households and businesses condensed in the Wolverhampton Central Station area.

When the 2021 Westside extension will be terminated, the cumulative score of Line 1 will be very positively affected by the number of households and businesses served by the network, and there will be the added improvement of solving traffic congestion in sites such as the Five Ways area. The impact on the entertainment level in the area is very high since users can visit the suggested areas of the canals and dedicated nightlife locations, such as Broad Street, using the metro service.

Line 2 Assessment from 2022 to 2026

The second brand new line is, at the moment, under construction and will be available to passengers in 2 stages. The first is in 2022 with the Eastside Birmingham extension, and the second is in 2026 with the Eastside Birmingham to Solihull extension. Below, the graph of each extension, as well as the cumulative Line 2 socio-economic assessment, are presented in Fig. 8.

Despite a low household concentration and business density levels, the Eastside extension reaches a similar score in terms of accessibility and interchange as Line 1 did. This result is justified, considering the extremely high value of connecting the HS2 and the Deritend Bus Station with Birmingham City Centre.

The presence of Birmingham International Airport and the second High-Speed Railway Station (HS2), along with the secondary rail connections, greatly fosters accessibility and interchange. Due to the distance users travel, the benefits in term of the price is lower than the one on lines that are closer to the towns, as is the case in the Wednesbury–Brierley Hill extension.

Dudley Line

Due to open in 2023, the extension aims to link several of Dudley's areas with the existing Line 1 in Wednesbury. Below, the socio-economic impact of the line is shown (Fig. 9).

The 2023 metro extension enables the population of Dudley to use the West Midlands Metro service to travel across the region along the path of the first line ever built in 1872, thus improving the benefits in terms of accessibility. Additionally, this factor is improved upon by taking into account the infrastructure of the town, such as the railway station in Dudley Port and the bus hub in the town centre.

The cumulative factor of value and the proportion of the metro lines are represented in the pie diagram below (Fig. 10) along with their total value score. Line 2 is responsible for almost half of the overall economic impact on the region. This line performance derives from the high value of its factors, particularly, as Fig. 8 highlights, in terms of accessibility and interchange.

Figure 11 sums up the values of the factors according to each line. It is important to assess the present findings to proceed in the forecasting process.

5 Socio-economic Impact Forecasting

The data gathered in Sect. 4 are sufficient for undertaking a 22-year forecast in line with the aim of the paper. As a consequence, a forecasting model suitable for the decision-making matrix assessment was selected after research on the available methodology.

Several studies on demand forecasting have been the object of research, as have the researchers who developed the Passenger Demand Forecasting Handbook, which, in its last version, explains the mode of assessing demand forecasting through a mathematical formula [(Balcombe et al. 2004)]. The forecast formula could have been adapted to the West Midlands case and its factor value projections. However, although the source might have been adapted to the network expansion-specific case, the complexity of the model, which includes different variables that are not available for minor-size areas such as the West Midlands, has caused the model to be discarded.

The forecasting methodology has followed the method of an industry-relevant study utilising the Excel forecasting function (Wood and Mysore 2019). In particular, the linear and exponential function has been considered in designing the socio-economic value contribution for each factor considered.

The line charts show the trend of the dynamic factors as a result of the forecasting activity along the axis of time and the socio-economic impact of the metro extension. The forecast results are shown in Figs. 12 and 13.

In both linear and exponential contexts, the factors follow a common trend. Until 2026, data have been manually elaborated, while from 2026 to 2041, the scenario has been simulated by the forecasting software. More precisely, since no metro network has been planned to be opened in the years 2024 and 2025, the value for those years has been artificially elaborated with an average function. The metro operation does not alter the forecast because the time interval between the missing data and the available data is very short compared to the 22-year analysis window.

According to the graphics shown in Figs. 12 and 13, a series of observations have been made:

- Accessibility has a peak in 2022 that matches the opening of the first stage of the Birmingham Eastside extension, which links Line 1 with the HS2 in New Canal Street. The socio-economic contribution grows over time, enabling the accessibility factor to be the greatest value-added factor across the network for at least 7 years.
- The interchange factor is important for the overall value even though its growth is relatively stable without significant changes. This trend is justified by the wide distribution of train and bus stations within the metro lines. The strategic infrastructure will be constantly enriched to meet the aim of building an integrated

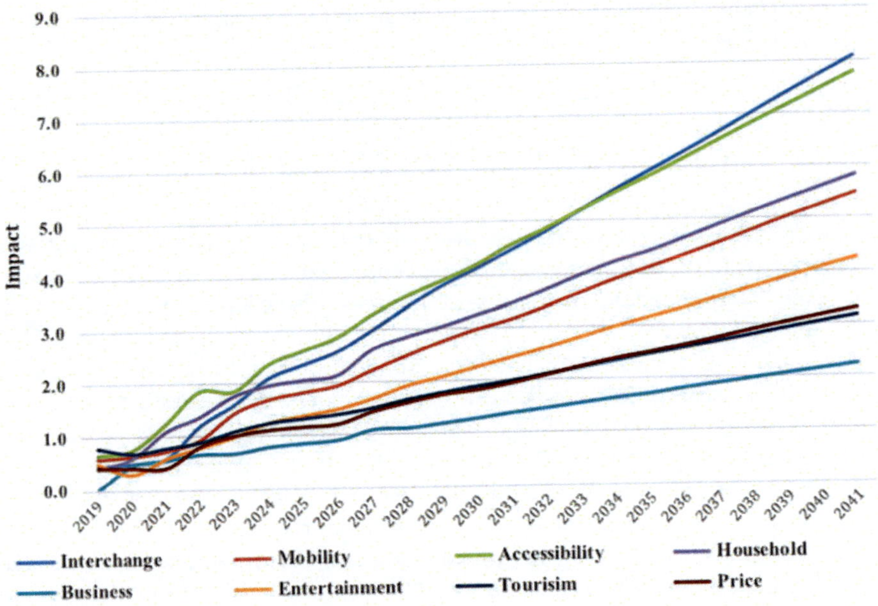

Fig. 12 Linear forecasts of the trend of the factors. *Source* (Own illustration)

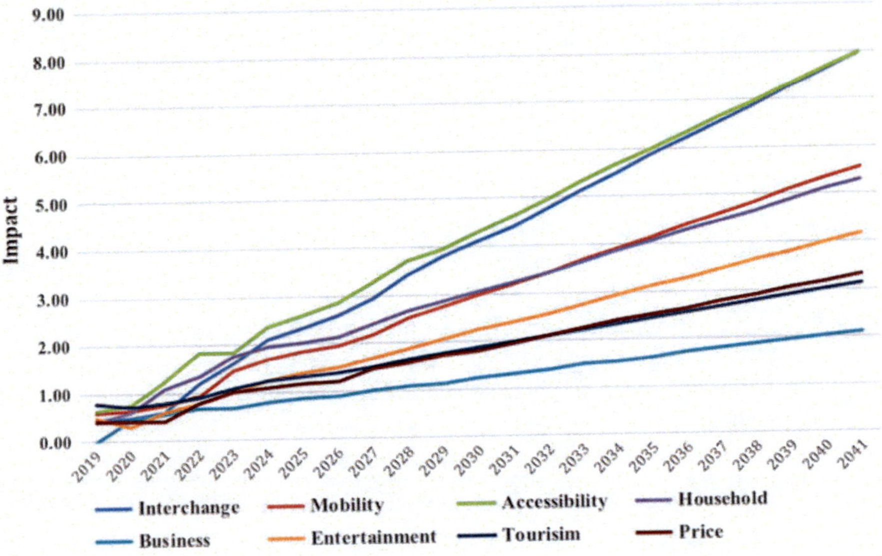

Fig. 13 Exponential forecasts of the trend of the factors. *Source* (Own illustration)

transport system. This action will enable the interchange value to overlap with the accessibility factor in diverse points over the years and finally become the major value contributor.
- The data for the substantial parity between household and congestion reduction shows that in the long term, the West Midlands Metro expansion can contribute to reducing traffic and, at the same time, improve the number of users. These data are particularly powerful since the household factor derives from the mathematical elaboration that focuses on the population in postcode districts. This means that the metro development may drive better mobility in several areas of the West Midlands, rather than in the city centres only. However, the household value increases with the Westside extension finalisation, given the massive household concentration in areas such as Five Waysor Edgbaston.
- Price and tourism factor advancements have the same impact over time. It is an encouraging figure since the value linked to the price is positive in terms of economic impact. As a reminder, price negatively depends on the trip length. The only small variation in price value is given in 2026 with the finalisation of Line 2, which results in longer journeys and potentially higher prices because of the presence of the airport and 2 HS2 stations. However, this variation is only temporary, since the trend comes back on its existing growth rate.
- Despite their important contributions to socio-economic value, overall in Line 1, tourism and business do not register any particular change during these years.
- The overall trend, considering the data available, shows the higher relevance and volatility of the accessibility and interchange value levels compared to the other variables. The growth gap between the factors was weighted at 20%, and the lighter variables, configuring a reliable but static forecast, were characterised by dynamism only until the year 2031.

The linearity may depend on the "fixed factors" that contributed to the creation of the household and business levels that have been static, not considerable changes over time. As an example, an adjustment of the population per postcode district considering the demographic growth rate every year would have reduced the errors in the future projections. Additionally, the same static factors have not been included in to support dynamic factor developments, and they may have given a different shape to the forecast charts.

Additionally, the information about further expansion is not available until 2026, and management is cautious to announce another extension proposal at the moment.

The tourism factor is particularly static because in this research, the estimation of incoming visitors due to international sports events has not been included, and other factors, such as Brexit, may influence tourist behaviour. The macro-variables need to be analysed carefully and with more advanced tools since they can have a critical impact on the social and economic aspects in the long term.

Adding up the value of the factors for each year of the forecast enables the calculation of the value of the overall impact when considering all the metro network segments. Figure 14 shows the progression of the socio-economic development of the network expansion over the years.

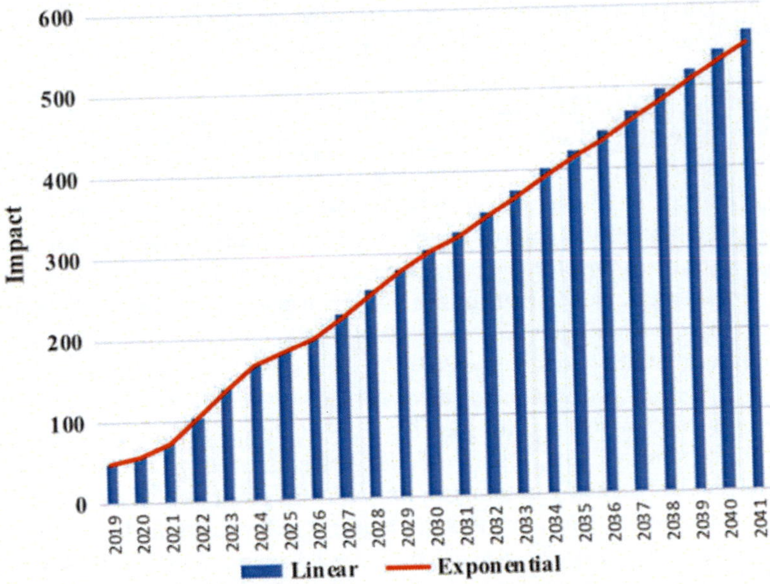

Fig. 14 Forecast of the socio-economic impact trends. *Source* (Own illustration)

The socio-economic impact is progressive over the years, and the introduction of the new metro lines starting in 2021 significantly improves the value. Considering the socio-economic impact trends derived by both the linear function and the exponential function, a justification of the linearity may be given because the analysis does not consider some additional factors, such as the potential substitution of machines and systems as well as the presence of further international events that the West Midlands may host.

6 Conclusion

The method of a decision-making matrix has been extremely useful for assessing the recent challenge of evaluating the socio-economic impact of the metro network expansion in a highly populated and high-growth region such as the West Midlands. The forecast shows the trend of 8 variables in a 22-year timeline, and the key contributors in the long-term socio-economic value have been identified as the accessibility and interchange factors.

The majority of the variables have been evaluated qualitatively to guarantee dynamism in the research, in a field in which the literature does not have specific non-mathematic tools yet faces the challenge of a project such as that of the West Midlands Metro. Nevertheless, increasing the number of analytical variables would have produced more precise data and, as a consequence, a more solid forecast.

Further research could be undertaken to assess various aspects of the metro network expansion that will certainly transform the West Midlands public transport system. The project involves several stakeholders of different industries and fields that may be interested in the nature and outcomes of the metro network expansion. Additionally, an analysis of the appropriateness of developing a 100% surface metro system rather than integrating underground systems could be added after seeing the challenges that a wholly over-ground system may have on accessibility and traffic. The usage of more quantitative data and a different methodology that enable more dynamic trends and more precise outcomes may be suggested because such usage can deepen the data research on the socio-economic impact as well as open new avenues to further explore and assess the West Midlands extension plan.

References

Balcombe R, Mackett N, Paulley N, Preston J, Shires J, Titheridge H, Wardman M, White P (2004) The demand for public transport: a practical guide. [online] Trl.co.uk. Available at: https://trl.co.uk/sites/default/files/TRL593%20-%20The%20Demand%20for%20Public%20Transport.pdf. Accessed 21 Jun 2019

Bell C (2019) UKPostcode districts. [online] Doogal.co.uk. Available at: https://www.doogal.co.uk/PostcodeDistricts.php. Accessed 21 Jun 2019

Brice D, Marinov M, Rüger B (2015) A newly designed baggage transfer system implemented using event-based simulations. Urban Rail Transit 1(4):194–214

Brinkhoff T (2019) West Midlands (United Kingdom): boroughs—population statistics, charts and map. [online] Citypopulation.de. Available at: https://www.citypopulation.de/php/uk-westmidlands.php. Accessed 21 Jun 2019

Dampier A, Marinov M (2015) A study of the feasibility and potential implementation of metro-based freight transportation in Newcastle upon Tyne. Urban Rail Transit 1(3):164–182

Darlton A, Marinov M (2015) Suitability of tilting technology to the Tyne and wear metro system. Urban Rail Transit 1(1):47–68

Edwards C (2018) What is a decision matrix? Definition and examples. [online] Business News Daily. Available at: https://www.businessnewsdaily.com/6146-decision-matrix.html. Accessed 21 Jun 2019

Expressandstar.com (2017) Midland metro numbers jump by a third after Birminghamextension. [online] Available at: https://www.expressandstar.com/news/transport/2017/03/18/midland-metro-numbers-jump-by-a-third-after-birmingham-extension/. Accessed 21 Jun 2019

Giles R (2018) Tram passenger survey (TPS)—Midland Metro Autumn 2017 results. [online] Transport for West Midlands. Available at: https://www.tfwm.org.uk/media/3318/wbhe-f16-tram-passenger-survey-midland-metro-autumn-2017-results-march-2018-transport-focus.pdf. Accessed 21 Jun 2019

Harris J (2019) A history of BirminghamTrams—Midland Metro Alliance. [online] Metroalliance.co.uk. Available at: https://metroalliance.co.uk/a-history-of-birmingham-trams/. Accessed 21 Jun 2019

Kelly J, Marinov M (2017) Innovative interior designs for urban freight distribution using light rail systems. Urban Rail Transit 3(4):238–254

Khamkar F (2018) Light rail and tram statistics, England: 2017/18. [online] Assets.publishing.service.gov.uk. Available at: https://assets.publishing.service.gov.uk/government/uploads/system/uploads/attachment_data/file/720315/light-rail-and-tram-statistics-england-2018.pdf. Accessed 21 Jun 2019

Marujo LG, Blanco EE, Mota DO, Leite JMLG (2019) The use of public railway transportation network for urban intermodal logistics in congested city centres. Sustain Rail Transp 3:187–207

Metroalliance.co.uk (2018) About us—Midland metro alliance. [online] Available at: https://metroalliance.co.uk/about-us/. Accessed 21 Jun 2019

Midland Metro Alliance (2018) Midland metro alliance. [online] Available at: https://metroalliance.co.uk/. Accessed 21 Jun 2019

MKT Transport Media (2019) Maps—MKT transport media. [online] Available at: https://mkttransport.co.uk/maps/. Accessed 21 Jun 2019

Motraghi A, Marinov M (2012) Analysis of urban freight by rail using event-based simulation. Simul Model Pract Theory 25:73–89

Online Project Management (2019) Project management approach to successful projects | Sinnaps. [online] Available at: https://www.sinnaps.com/en/project-management-en/project-management-approach. Accessed 21 Jun 2019

Praxiom.com (2015) Risk assessment methods and procedures. [online] Available at: https://www.praxiom.com/risk-assessment.htm. Accessed 21 Jun 2019

Pugh (1981) Concept selection: a method that works. In: Hubka V (ed) Proceedings international conference on engineering design. Review of design methodology. March 1981, Rome. Zürich: Heurista, blz, pp 497–506

ScarboroughCivitelli J (2019) How to use a decision-making matrix-VocationVillage.com. [online] VocationVillage.com. Available at: https://www.vocationvillage.com/how-to-use-a-decision-making-matrix/. Accessed 21 Jun 2019

Shoaf L, Hewitt P (2019) Metro investment programme briefing. [online] Governance.wmca.org.uk. Available at: https://governance.wmca.org.uk/documents/s2553/Metro%20Investment%20Programme%20Briefing%20V8%20-%20Final%20TDC%20version.pdf. Accessed 21 Jun 2019

Stubbs E, Riley R (2018) Birmingham economic review 2018. [online] Greaterbirminghamchambers.com. Available at: https://www.greaterbirminghamchambers.com/media/621749/ber-2018-ch-3-infrastructure.pdf. Accessed 21 Jun 2019

Railway Technology (2019) Midlandmetro—railway technology. [online] Available at: https://www.railway-technology.com/projects/midland/. Accessed 21 Jun 2019

Toal J, Marinov MA (2019) Check-in and bag drop service on-board light rail vehicles for passengers travelling to the airport, sustainable rail transport, vol 3, pp 1–82

Transport for London (2019) Visitor Oyster card. [online] Available at: https://tfl.gov.uk/travel-information/visiting-london/visitor-oyster-card. Accessed 21 Jun2019

Transport for West Midlands (2017) TfWM to take direct control of MidlandMetro services. [online] Available at: https://www.tfwm.org.uk/news/tfwm-to-take-direct-control-of-midland-metro-services/. Accessed 21 Jun 2019

UKCensus Data.com (2012) West Midlands—UKCensus data. [online] Available at: https://www.ukcensusdata.com/west-midlands-e12000005#sthash.Firh4P8V.nyJwTuAu.dpbs. Accessed 21 Jun 2019

Ukpopulation.org (2019) BirminghamPopulation 2019 | Population UK. [online] Available at: https://www.ukpopulation.org/birmingham-population/. Accessed 21 Jun 2019

Wales J, Marinov M (2015) Analysis of delays and delay mitigation on a metropolitan railway network using event based simulation. Simul Model Pract Theory 52:55–77

Westmidlandsmetro.com (2018) West Midlands metro expansion plan. [online] Available at: https://westmidlandsmetro.com/wp-content/uploads/2018/08/MetroExpansion.pdf. Accessed 21 Jun 2019

West MidlandsMetro (2019a) About Midland Metro Limited (MML) [online] Available at: https://westmidlandsmetro.com/about/mml-about/. Accessed 21 Jun 2019

West MidlandsMetro (2019b) Transport For West Midlands reveals a new public transport identity—West MidlandsMetro. [online] Available at: https://westmidlandsmetro.com/2018/07/20/new-public-identity/. Accessed 21 Jun 2019

West MidlandsMetro (2019c) Where we are going—West MidlandsMetro. [online] Available at: https://westmidlandsmetro.com/about/expansion-programme/where-we-are-going/. Accessed 21 Jun 2019

West MidlandsMetro (2019d) Expansion programme—West MidlandsMetro. [online] Available at: https://westmidlandsmetro.com/about/expansion-programme/ Accessed 21 Jun 2019

Wood B, Mysore K, Marinov M A study on greenhouse gas emissions, analysing the potential of five rail passenger systems for meeting EU targets, IF (Benjamin Wood, Kaushik Mysore & Marin Marinov, 2019 *accepted for publication*)

Wright O (2018) Business population estimates for the UKAND regions 2018. [online] Assets.publishing.service.gov.uk. Available at: https://assets.publishing.service.gov.uk/government/uploads/system/uploads/attachment_data/file/746599/OFFICIAL_SENSITIVE_-_BPE_2018_-_statistical_release_FINAL_FINAL.pdf. Accessed 21 Jun 2019

Rail Careers in ASEAN: Employers Search for Talents, Skills and Knowledge

Janene Piip

Abstract Investment in rail projects in ASEAN (Association of South East Nations) will increase mobility for local populations and improve access to education, work and social development. In Thailand, multinational companies are part of this expansion. Their quest to hire local employees identifies education and training must consider employability or 'soft' skills, and work skill development that leads to sustainable career pathways. While the Thai government has identified a qualifications framework and 13 professional qualifications for the rail industry, employability skills or 'soft' provide graduates with a balance of talents to ensure they stay in jobs. Employability skills, 'soft' or work skills, are less well defined in rail qualifications. Our paper explores qualifications frameworks, qualifications in TVET (Technical and Vocational Education and Training) and Higher Education and professional recognition processes in Thailand with commentary from a global consulting firm based in Bangkok.

Keywords Qualifications in ASEAN · Work · Skill development · Rail industry

1 Introduction

New rail projects are proposed or under development in ASEAN countries as local populations grow and become more affluent (Allurentis Ltd. 2019). Global agencies and governments realise access to mobility is an essential enabler of individual and local economic productivity and wealth, facilitating access to jobs, markets, social activities and education (United Nations Development Programme 2018).

Currently, Thailand is expanding its metro network from 100 to 500 km, with two new monorail lines and various other new, and extension, projects (Pwc 2019). It is expected more staff will be needed to operate these networks. While there may be existing skills and talent reports on workforce development needs for rail projects in Thailand, these documents were unable to be located during this research study.

J. Piip (✉)
JP Research and Consulting, PO Box 2614, Port Lincoln, SA 5606, Australia
e-mail: janenepiip@gmail.com

W. Weerawat et al. (eds.), *Urban Rail Transit*, Lecture Notes in Mobility,
https://doi.org/10.1007/978-981-15-5979-2_7

However, local media articles state worker shortages are placing planned rail projects in doubt (Apisitniran 2019). Other cities in Thailand, for example, Phuket and Khon Khaen, are also considering rail expansion through light rail (tram) services. In addition, two High Speed Lines are being discussed, one with China and another with Japan.

As a workforce development researcher, my interests relate to the rail industry, qualifications frameworks, the composition of vocational and higher education qualifications, particularly the 'soft' skills components and their importance in the future of work. More than 630 million people, 94% of whom are literate, and 50% of whom are under 30 years of age, live in the ASEAN region (Brink News 2018). In the next ten years, disruption of traditional industries, adapting to the future of work, artificial intelligence and robots all point to low skilled jobs being replaced by highly skilled roles. The importance of training and education is an enabler for workers to 'transition into jobs requiring softer, customer-oriented or interactive skill sets' (Brink News 2018).

For new rail projects in developing countries, frameworks provide regulation and guidance to ensure education and training meets industry needs (European Centre for the Development of Vocational Education and Training 2011). In the last thirty five years, employability, generic, 'soft' or core skills have been added to qualifications frameworks (Wibrow 2011; Bateman and Coles 2013; Bateman and Liang 2016) because graduates must be able to apply their talents, skills and theoretical knowledge to a work situation. The World Economic Forum (Leopold et al. 2016) highlights how existing jobs are changing at a rapid rate, requiring new and different skill sets. Employees must be able to continually adapt and change to environmental conditions with new skill sets that facilitate employment success in an environment of continual change.

Leopold et al. (2016, pp.21–22) describe the growing demand for complex problem solving skills in more than 36% of all jobs compared to physical abilities or strength in only 4% of jobs traditionally found in manual roles. Other skills common across most jobs into the future will be social skills (19%), process skills such as active listening and critical thinking (18%), systems skills (17%), cognitive abilities such as creativity and mathematical reasoning (15%), resource management skills (13%), technical skills (12%), content skills (10%). Figure 1 highlights soft skill requirements in the future.

This paper, therefore, provides an opportunity to investigate the research problem relating to the talent, skills and knowledge needed for new rail projects in the ASEAN context, particularly Thailand, considering local qualifications frameworks and industry skill needs, particularly 'soft' or employability skills.

1.1 Structure

This paper is structured as follows. After the Introduction in Sect. 1, the paper investigates the literature in Sect. 2. Section 3 explains the methodology applied to the data

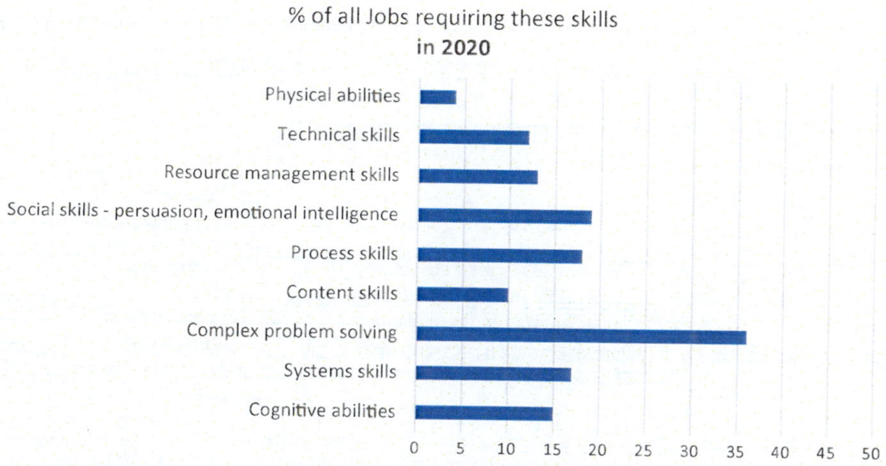

Fig. 1 Jobs requiring employability skills by 2020. *Source* Based on Leopold et al. (2016, p. 22)

collection and provides an industry commentary on talent, skills and knowledge for Thai rail projects. Section 4 highlights the challenges for organisations in sourcing talent, skills and knowledge for rail projects in Thailand. In Sect. 5, conclusions and recommendations are outlined, while finally, Sect. 6 lists the limitations of the study and suggests the next research steps.

2 Background

2.1 Workforce Development

Harris and Short (2014) describe workforce development as both a human resource development process and collection of workplace practices that educate, train, develop and build the competences of employees. Industries such as the rail industry are confronted by never-ending challenges such as market competition, labour mobility, skills shortages, a mismatch of talent, emerging economies, global currencies and digital disruption. The importance of developing the workforce with the talent, skills and knowledge to predict, stay ahead of the competition and solve these challenges, is what makes the difference between companies that succeed or fail in engaging local workforces with relevant skills.

Furthermore, Roche (2001, p. 6) proposes that workforce development requires a 'broad, comprehensive and multifaceted focus'. In other words, workforce development is more than education and training—it is a system wide approach to addressing skill needs. Areas requiring consideration in workforce development are outlined in Table 1.

Table 1 A systematic approach to workforce development

Level	Workforce development descriptor	Examples
Level 1 Systems	Involves creating environments and systems that support the full range of workforce development strategies	Legislation Policy Funding Recruitment and retention Resources Support mechanisms
Level 2 Current workers	Encompasses methods of improving individual professional functioning, ensuring that individuals have opportunities to develop skills, knowledge and attitudes	Formal education Training Workplace training Mentoring On-the-job learning e-Learning Best practice guidelines
Level 3 Future workforce	Involves ensuring a sufficient pool of skilled workers for the future with a range of planning and strategies	Recruitment strategies Education and training Affordable and accessible education and training Adequate funding to employ staff Support and facilitate policies

Source Based on Roche (2001)

2.1.1 Workforce Development in Developing Countries

The issues with workforce development in emerging economies such as those in the ASEAN region are evident at many levels (Tan et al. 2016). While workforce development is an ongoing undertaking, reports by global agencies (World Bank 2019) cite a mismatch of skills demand and supply of talent, skills and knowledge. Individual ASEAN countries are at differences in their stage of development, economic structure, and growth strategy with different skill mixes need for different industries (Tan and Tang 2016). Increasing investment in education by ASEAN countries has lifted the outcomes for graduates, but many are unable to find jobs equal with their expectations and theoretical knowledge while employers complain of skills shortages and talent mismatches (Tan et al. 2010). Often, employers describe 'deficiencies in soft skills (e.g., problem-solving, communication, effectiveness in team work, etc.) in the workforce, skills that are considered essential for business productivity in today's inter-connected and technologically-driven global economy' (Tan et al. 2010, p. 1). On a broader scale, skills mismatches are making it difficult for countries to innovate and move into more lucrative areas of economic activity. With new rail projects on the horizon; skills, talents and knowledge of potential employees, matched with employer demands, is essential for ASEAN countries. Tan et al. (2010, 2016) write that workforce development efforts in most low- and middle-income countries are not as effective as it could be hoped, further describing similar issues related to systems, current workers and the future workforce, as outlined in Table 1 by Roche (2001).

While ASEAN countries have lifted their graduate outcomes in the vocational and higher education sectors, the importance of 'soft' skills should not be underestimated. Global economies are characterised by businesses working across time zones, and around the clock, so employees need to be agile in their ways of working to interpret and participate in different workspaces. 'Soft' skills are the glue that helps graduates find and maintain work.

While the literature discusses the need for graduates having 'soft' skills, leaders and managers also require problem solving, leadership and communication skills for getting work done (Gropello et al. 2011). The demand for skills goes beyond merely learning theory in books. It means employees need to exercise a degree of judgement in their work by communicating with others, thinking through problems and being confident to trust their skills. The report by Gropello et al. (2011) describes the need for young people to develop these skills early as work is becoming more informal and individuals need to think of themselves as a flexible entrepreneur—finding their own work as one stream dries up, and another offers a new opportunity.

2.2 *Qualifications Frameworks*

In 2007, to address these strategic workforce issues, ten ASEAN leaders signed documentation to 'develop human resources through closer cooperation in education and life-long learning and in science and technology, for the empowerment of the peoples of ASEAN and for the strengthening of the ASEAN Community' (Association of Southeast Asian Nations 2007, pp. 4–5). The purpose of the unified approach was to recognise the importance of labour mobility between jobs in various nations as local economies grew, and the fact that different countries have different workforce development approaches to education, training and learning (Bateman and Liang 2016). Varying approaches by country to recognition of individual competence and skills were not conducive to a united ASEAN, as countries' needs for skilled labour, knowledge and talent increases (Association of Southeast Asian Nations 2016a). Therefore, a combined approach was conceived.

Qualifications frameworks, therefore, are viewed as 'an instrument for the development and classification of qualifications (e.g. at national or sectoral level) according to a set of criteria (e.g. using descriptors) applicable to specified levels of learning outcomes' (European Centre for the Development of Vocational Education and Training 2011, p. 82). In general, qualifications frameworks provide direction on competence at different levels in the vocational and higher education qualifications and are endorsed by industry and national training boards of each country (Bateman and Liang 2016). The composition of qualifications frameworks varies by country with skill and qualification recognition processes agreed upon by the ASEAN association (Association of Southeast Asian Nations 2016a) to facilitate integrated approaches to workforce development. A comprehensive account of frameworks for each country is provided in the report by Bateman and Liang (2016). In Thailand, each qualification is comprised of three domains, including knowledge, skills and attributes (Bateman and Liang 2016, p. 10).

2.2.1 Employability Skills

Along with the skills and competences required at each level, many qualifications frameworks address employability skills, or 'soft' or generic skills required for work. A definition from the Internal Labour Office about employability skills is as follows (Brewer 2013, p. iii):

> the skills, knowledge and competencies that enhance a worker's ability to secure and retain a job, progress at work and cope with change, secure another job if he/she so wishes or has been laid off and enter more easily into the labour market at different periods of the life cycle. Individuals are most employable when they have broadbased education and training, basic and portable high-level skills, including teamwork, problem solving, information and communications technology (ICT) and communication and language skills. This combination of skills enables them to adapt to changes in the world of work.

Furthermore, employability, 'soft' or core work skills 'help individuals to understand the labour market, make more informed choices about their options in education, training, wage employment, self-employment, cooperatives. They also help them become better citizens and contribute to their communities and societies' (Brewer 2013, p. 5).

Employability, 'soft' or core work skills are, therefore, an integral set of skills to help individuals remain engaged in the workforce, navigating seamlessly from study to employment and being prepared to embrace new challenges throughout life as they arise. Theoretical skills are important, but employability, 'soft' or core work skills are the balance of the whole person, often embedded in practical training and assessment in qualifications.

In early societies, talents (currency or coins) were used as instruments of trade and exchange (Piip 2014). The term 'talent' has evolved from this early time to mean an item of value or person of value to an organisation (Michaels et al. 2001). The talents of people, including hard and soft skills, make up valuable skills or the whole person—and this what employers are looking for in employees in the twenty-first century (Fig. 2).

2.2.2 Soft Skills

While the terminologies soft and hard skills are readily separated, the components or attributes of soft skills are more difficult to define. The government in Australia recognises this debate, attempting to define the differences. In reality, 'employability, generic, transferable, or more recently, twenty-first century skills, soft skills encompass a range of non-technical skills (as opposed to technical, job-specific skills) that,

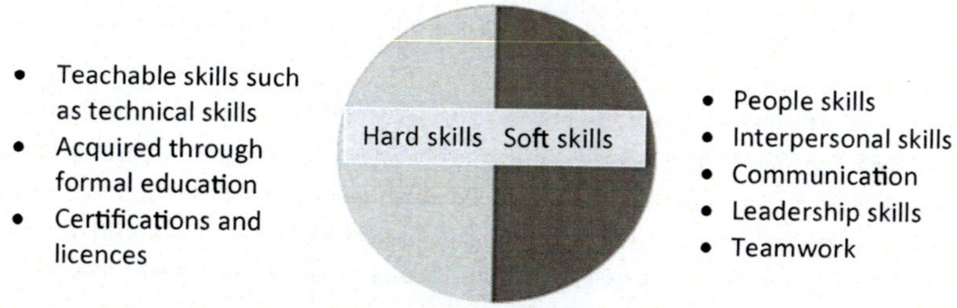

Fig. 2 Hard and soft skills. *Source* Based on Piip (2014)

increasingly, employers require in potential employees. These skills include communication skills, teamwork and collaboration skills, creativity, problem solving and analytical skills, entrepreneurial and digital skills' (Ncver 2018).

Lucas and Hanson (2016, p. 5) report soft skills are 'sometimes used as a synonym for non-cognitive skills' being a 'set of useful personal and social attributes such as problem-solving, communicating, collaborating and so on. These are 'soft' in contrast to 'hard' or technical skills, such as 'operating specialist equipment'. Furthermore, they report that soft skills can be associated with an individual's personality and can be defined broadly under three categories: attitudinal skills, life skills, and transferable skills.

Personal character that drives conduct in the workplace, resilience and perseverance, a positive attitude and 'can do' approach comprises aspects of soft skills highly valued by twenty-first century employers (Hanson 2010). These attributes are difficult to measure as each person is diverse. As a result, soft skills terminologies have been developed further into employability skills frameworks in many countries (Ncver 2018). In Australia, an employability skills framework guides the delivery of vocational education and training and is incorporated into competency-based training to specify what is required in the workplace in terms of attitudes and behaviours. These developments have occurred because employers want work ready graduates who can be employed directly from the training environment to the workplace, providing immediate value to their business.

In the vocational training sector in Australia, the development of workplace skills has been commonplace, especially regarding apprenticeships, where students gain employability and workplace skills while learning and being employed. As well, NCVER (2018) highlight that employability or soft skills frameworks in Australia provide the opportunity to create an environment where the formal assessment of soft skills has been incorporated into the certification process of competency based training.

In the Higher Education learning environment, simulating work environments during degrees to ensure graduates are both employable and work ready has become more common recently. Graduates have to be employable (possessing a 'set of valued and valuable skills which are necessary but not sufficient for gaining employment') as well as work ready ('a set of skills including employability' as well initial skills to commence employment') (Sachs et al. 2016, p. 8; Bca 2016).

2.3 Qualifications

The qualifications framework in Thailand includes a professional qualifications system focused on assessing competencies of individuals following occupational standards (Thailand Professional Qualification Institute 2014; Unesco-Unevoc International Centre for Technical and Vocational Education and Training 2015).

As an example, the Qualifications Framework in Thailand includes three components:

- Basic education (primary and secondary education)
- Technical and Vocational education
- Higher education, and
- A system for assessing professional qualifications gained in the workplace or other countries.

In this paper, we are concerned with TVET and Higher education qualifications, as these are the qualifications that prepare people for work. The content of these qualifications is the responsibility of the provider in line with the Qualifications framework of the country.

TVET or technical vocational education and training is post-secondary education, defined by UNESCO as skills for work, stating: 'The current focus is increasingly upon preparing knowledge workers to meet the challenges posed during the transition from the Industrial Age to the Information Age, with its concomitant post-industrial human resource requirements and the changing world of work' (United Nations Education Scientific and Cultural Organization 2017).

TVET in Thailand includes nine specialisations of (1) Trade and industry; (2) Arts and crafts; (3) Home economics; (4) Commerce and business administration; (5) Tourism industry; (6) Agriculture; (7) Fishery; (8) Textile industry; and (9) Information and Communications Technology (ICT) (United Nations Education Scientific and Cultural Organization 2017).

2.4 Academic Qualifications in Higher Education

Academic qualifications have six levels in the Thai qualifications framework (Ministry of Higher Education 2006), including:

Level 1. Advanced Diploma.
Level 2. Bachelor.
Level 3. Graduate Diploma.
Level 4. Master.
Level 5. Higher Graduate Diploma.
Level 6. Doctor.

Accordingly, academic qualifications comprise five domains relating to knowledge, skills and attributes—or personal qualities, working habits or behaviours. The domains are as follows:

1. '*Ethics and moral*—related to habits of acting ethically and responsibly in personal and public life.
2. *Knowledge*—the ability to understand, recall and present information including knowledge of specific facts, knowledge of concepts, principles and theories and knowledge of procedures.
3. *Cognitive skills*—the ability to apply knowledge and understanding of concepts, principles, theories and procedures.
4. *Interpersonal skills and responsibilities*—the ability to work effectively in groups, exercise leadership; to accept personal and social responsibility, and to plan and take responsibility for their own learning
5. *Numerical analysis and communication and information technology skills*—the ability: to use basic mathematical and statistical techniques, communicate effectively in oral and written form, and to use communications technology' (Bateman and Coles 2013 p. 18).

Having a workforce with the right skills, stronger literacy, analytical reasoning, and problem solving skills can thus help Thailand move up the value-added ladder to a more knowledge-based economy (Sondergaard et al. 2016). These skills are essential for new rail projects in the region.

Dacre Pool and Sewell (2007) suggest graduates would integrate more seamlessly to work if courses at TVET level and University considered the following components:

- Active learning where students are involved with real life problem solving
- Integration of practice and theory in the classroom
- Workplace awareness including knowledge about types of industries and employers
- Career development learning about how to develop one's career
- Workplace experience where the student undergoes 'real' work in various workplaces
- Awareness and importance of employability or 'soft' skills and emotional intelligence with the opportunity to develop these skills.

2.5 Skills Needs of Industry

There is a lack of information on specific workforce needs for new rail projects in ASEAN and Thailand. However, rail workforce development has been categorised by the Thai government broadly in three sectors to address company and workforce needs. The sectors are: 'transportation and railway policy, railway education and research, and railway operation and maintenance and manufacturer' (Rail Professional 2019).

Considering Thai strategic direction and the opportunity to recognise workers existing skills, the TPQI (Thailand Professional Qualification Institute 2014) has assigned Kasetsart University to define 13 professional qualifications for the rail industry (Kasetsart University Rail Engineering Center 2019). The qualifications are as follows:

1. RST (Rolling Sstock) Engineer—Level 5
2. RST (Rolling Stock) Technician (Mechanical)—Level 4
3. RST (Rolling Stock) Technician (Electrical)—Level 4
4. AFC (Automatic Fare Collection) Engineer—Level 4
5. AFC (Automatic Fare Collection) Technician—Level 4
6. Signalling Engineer—Level 5
7. Signalling Technician—Level 4
8. Mainline Signalling Technician—Level 4
9. Operation Controller—Level 4
10. Station Master—Level 4
11. Train Controller—Level 4
12. Commuter Train Controller—Level 4
13. Operation Investigator—Level 5.

The standard has seven levels for each profession, from Level 1, or Basic Skilled; to Level 7, Consummate Specialist. The seven levels includingsho Vocational and Higher Education Qualifications are aligned with Professional Qualifications.

2.6 Skill Competence

The level descriptors of Thai rail qualifications comprise levels of complexity of learning outcomes. The level descriptors include the notion of 'competence' or the ability that extends beyond the possession of knowledge and skills (Association of Southeast Asian Nations 2016b, p. 15) including:

1. 'cognitive competence involving the use of theory and concepts, as well as informal tacit knowledge gained experientially
2. functional competence (skills or know-how), those things that a person should be able to do when they work in a particular area
3. personal competence involving knowing how to conduct oneself in a specific situation, and
4. ethical competence involving the possession of certain personal and professional values.'

While industry qualifications, including rail qualifications, have been prescribed, Tan and Tang (2016, p. 9) report a shortage of workers with sufficient science and engineering, and technical skills as critical skills challenges for Thailand. Weaknesses in soft skills and English proficiency also present challenges as the workforce faces increasing global competition in sectors such as the rail industry where industries are developing rapidly. More effort is needed to develop 'soft' or employability skills is recommended, notably:

- Communication skills for graduates
- Creative/critical thinking
- Analytical and problem-solving competencies
- Ability to work independently
- Time management skills
- Teamwork
- Foreign language skills primarily English proficiency (Tan and Tang 2016, p. 47)
- Entrepreneurial skills
- Flexibility.

More than 80% of survey respondents in Tan and Tang's (2016, p. 47) study, believed the 'university curricula should be revised to reflect the realities in the labour market and 90% of them felt more practical training should be provided for students'.

Tan and Tang (2016, p. 146) highlight the difficulty in collecting data on the type and demand for soft skills. They assert soft skills such as 'creativity', 'empathy' and 'flexibility' are difficult to define and measure. However, they believe there is a lower value placed on teaching soft skills, and these skills are often neglected in national education systems in ASEAN countries despite their inclusion in national qualifications frameworks. Furthermore, soft or employability skills have more importance in highly competitive labour markets (Dacre Pool and Sewell 2007). Graduates of rail engineering and those with professional rail qualifications are highly sought after in Thailand as new rail projects emerge. Mahidol University, Thailand also confirms the importance and development of knowledge, skills and attributes as 'hard skills will make you get knowledge, but soft skills will make you get a job' (Mahidol University 2019).

2.7 *Measuring Employability in Higher Education Courses*

Following on from the discussion about employability and work readiness in Sect. 2.2.1, employability means 'being ready to participate, make suggestions, accept new ideas and constructive criticism, and to take responsibility for outcomes' (Blades et al. 2012 p. 10). A critique of the literature identified many graduates of higher education courses have limited workplace experience and inability to apply theoretical knowledge to a new workplace role (Tan and Tang 2016, p. 108). Higher education courses typically assess theoretical knowledge but not employability skills in a valid and ethical way.

Work integrated learning (Sachs et al. 2016) is one method of developing employability skills and work readiness in students who are undertaking theoretical degrees. Many students come directly from secondary school and have limited knowledge of the workplace, what is expected of them in the workplace and limited confidence in their own abilities. Work integrated learning should also include formal learning about the workplace, practical experience in the workplace and formal study modules about the workplace that are incorporated into the assessment process of the degree to ensure students participate in the learning and are formally assessed.

In many countries, a common language around employability skills and the establishment of clear benchmarks for addressing them, with distinctions between levels of skill, have been developed into frameworks that connect the gap between TVET, university and work.

Bandaranaike and Willison (2014) developed a Work Skills Development Framework comprising five levels of autonomy, starting with the initial work placement while at university to autonomy or readiness for work. The levels from 1 to 5 are described as follows:

1. *Prescribed Direction*—Highly structured directions and guidance from the supervisor
2. *Bounded Direction*—Boundaries set by and limited direction from the supervisor
3. *Scaffolded Direction*—Works independently and within provided guidelines
4. *Self-Actuated Direction*—Develops own abilities and works with limited guidance.
5. *Open Direction*—Works within self-determined guidelines appropriate to context.

In this example framework, developed for an environmental science degree, work skills are embedded in the assessment structure of a subject on professional placement (Bandaranaike and Willison 2010). The importance of work skills development is:

> in the recognition and accreditation of life-engaged learning outside of the classroom; the search for knowledge outcomes from university teaching, together with the skills and knowledge required to operate in public and private enterprise. (Bandaranaike and Willison 2010, p. 3)

3 Industry Commentary on Talent, Skills and Knowledge for Thai Rail Projects

In summarising this literature review, we identified disconnects between the policy of the qualifications frameworks, the need for high level skills in new rail projects, and employability skills for industry in ASEAN countries. In Thailand, there is a potential shortage of workers with the best skills match for rail projects, a need for English proficiency for the workplace, and the teaching, development and measurement of soft skills in TVET, Professional and Higher Education qualifications.

We were interested in discussing how education providers' courses are meeting industry needs, graduate employability and 'soft' skills, whether 'soft' skill gaps are affecting organisations and graduate organisational programs in the workplace, with a view to future research. The literature had provided articles and reports from 2008 to 2016 so by interviewing an industry expert; we sought to identify whether workforce development problems related to skills, talent and knowledge had changed in recent years. We acknowledge that this paper is an initial investigation, that it is exploratory and qualitative, and requires deeper focus on particular issues.

We posed five questions to a senior manager at rail consulting firm, ABC (not the real company name) Bangkok, (2019), a global professional services firm with offices in Thailand involved in the development of rail projects in Thailand and other global locations. The manager was recruited through a call on LinkedIn for rail professionals with an interest in workreadiness of graduates.

The participant who was interviewed has lived and worked in Thailand for:

> For two decades now. I speak three languages fluently, English, German and Thai. So, I can skip between the different worlds. Sometimes it's very difficult for me to wrap my German head around Thai culture. And I also have the same thing vice versa, when we have people joining our projects from abroad.

Across the world, ABC employs 46,000 people.

The questions were provided before the discussion and responses recorded and then transcribed. The questions are provided below.

1. Do you have a comment about the inability of educational institutions such as universities and vocational providers to meet industry demands, especially in relation to your organisation?
2. What do you believe are the gaps of soft skills in professionals and new graduates?
3. Can you describe the soft skill gap that has the most impact on your organisation?
4. What do think are the reasons for these gaps?
5. Is your organisation actively addressing these gaps with programs or other measures?

4 Results

Our interview with ABC highlighted challenges aligned with the literature.

Educational Organisations Role in Preparing Graduates for Workplace Practices

ABC indicated a gap in graduates' ability to apply theory to practice in the workplace. This phenomenon is also widespread globally as graduates often have limited experience of the workplace and limited understanding of what is required in a new job. ABC has a graduate program in other locations but in Thailand has a strategy to hire professionals with experience. ABC described the study to the workplace gap in Thailand as follows:

> The major issue is I work in the consulting industry, and we usually don't deal that much with graduates… we're usually hiring people that have already some experience.
>
> Our organisation hires contractors and from what I can tell is that most of the graduates in Thailand that have had some sort of university degree, lack the application knowledge. So, they come out fresh from university, they know how to do the math and how to do this, how to do that, but they lack how to apply that knowledge to the work that they would be doing.

Soft Skills Gaps of Professionals and Graduates

Educational organisations are preparing graduates with theoretical knowledge according to the Thailand Qualifications Frameworks. However, ABC identified cultural expectations of parents, the ability of graduates to be independent and trust their own judgement as barriers for new employees, along with other skills of personal leadership and showing initiative at work to name just a few gaps. Soft gaps comprise the subsequent challenges:

> So basically, they come with their theoretical knowledge, but then how to apply that theoretical knowledge to an actual construction site or actual work site, that is very difficult for them.
>
> I think that as a first, it's that the Asian culture generally. Many students are sponsored by their parents. So, they haven't learned yet to be independent and stand on their own two feet, so when they come into the work environment, they're still looking for other people to lead, other senior people to lead. So, that's I think the major problem here.

Impact of Soft Skill Gaps

ABC identified graduates and professionals with a university qualification are trialled in companies for an initial period. If their skills, talent and knowledge are not suitable for company requirements, employees may be retained in laboring rather than professional roles. This gap means graduates and professionals are missing out on employment opportunities, and university qualifications are not fully utilised. Outcomes after comprehensive study programs that do not result in long term employment are an issue for ASEAN countries, described also as a key issue by ABC:

> To be really honest, not much, because as I said, we don't deal directly with graduates. However, I know that the contractors organisation, they keep hiring people, and then the ones that they are confident with that they can work with, they will try to retain, and the ones that they are not that confident with, they try to get rid of in the end.
>
> So, they look, they're just fishing, and basically, around 20 to 30% are the people that they retain, and the others they either get rid of them or they don't continue their contracts, or they just resign by themselves, or they keep them on as just labor, basically.

Reasons for the Mismatch of Individual Skills and Company Requirements

ABC indicated that graduates and professionals have a sound understanding of theoretical concepts but lack experience applying theory to real workplace problems. Further opportunities to participate in workplace experience is recommended as well as support structures to develop employability and 'soft' skills. The mismatch of skills relates partly back to educational institutions as described by ABC and would benefit from more focus on this area of skill development as well as career counselling and workplace experience for graduates as demonstrated in this comment:

> I think also there's a bit of a problem in the ways that the universities teach. They don't emphasize a lot of the application cases, so they emphasize a lot of theory, so it's very difficult for them to connect the dots from theory to the practical application.

Ways ABC is Addressing Skill Gaps

Mentoring is utilised at ABC to enable new employees to 'learn the trade' or the culture of the organisation. New employees are paired with a senior person in an informal learning approach that can take between one and four years. After this stage, formal learning may be offered to employees which comprises access to the company university through an online learning platform with courses from around the world. Additionally, ABC supports employees to work on real life problems and scenarios. These situations build confidence and allow professionals to work on problems supported by more experienced staff. ABC has three workforce skills gaps strategies in other global locations that have graduates, but not in Thailand. ABC described these ongoing learning strategies as follows:

> So, We Have Three Ways of Dealing with Graduates.
>
> First of all, we put them together with senior members of our team or other mentor. So, we put them together with senior members of our team who will teach them, basically, the trade. They just tag along for a year or two or three or four, just during the process, and see how it all evolves, and that's their first hands-on experience, and then they change to a new project. So, in the new project, they will accept a more senior role, so that would be one thing.
>
> Now, for people who are already quite skilled and interested, we have what is called our company university. It's an online learning platform where you can go and you can get courses from basically all over the globe, online courses, and you can download them and you can do them, and you get actually credit for them within the organization.
>
> And then, the third thing that we have going is what we do with our people, is we put them into situations where there are problems and try to encourage them to figure out solutions for those problems, and then somebody helps them along. So, that would be basically the second step after they have one full project exposure already.

5 Conclusions and Recommendations

Talent, skills and knowledge are needed to develop new rail projects in Thailand. Our research has outlined measures governments are taking in the area of qualifications frameworks, and TVET, Higher Education and Professional qualifications to ensure the hard skills of the rail industry are met. On face value, the pipeline of rail talent from TVET, Higher Education and professional recognition processes is being directed towards employment in organisations in ASEAN and Thailand. However, our exploratory investigation has identified further work is required to specify employability, 'soft' or generic skills that provide graduates and professionals with work skill development to pursue a career pathway in their chosen area of work, particularly for rail industry projects. The literature highlights how ASEAN countries are investing heavily in education and training, but these efforts will be affected if graduates do not pursue career pathways in their chosen field of study. In the rail industry, the demand for work ready graduates is increasing.

6 Limitations and Next Steps

As mentioned previously, this research explored the skills, talents and knowledge required for new Thai rail projects. Despite all efforts to make this research as feasible as possible, there are some deficiencies which influence this paper. The critical limitation of the approach includes the dialogue with only one rail industry organisation. If other companies had been included, the results might have been different. However, to start the discussion on talents, skills and knowledge required for rail projects in Thailand, a broadbrush approach instigated a starting point for further discussion. Our next steps are to develop further research on barriers to soft skill development, programs addressing work skill development, and ways educational institutions are bridging the gap between education and training and work.

References

Allurentis Ltd. (2019) Investing in ASEAN 2019 2020 [Online]. Available: https://asean.org/storage/2019/06/Investing-in-ASEAN-20192020-App.pdf. Accessed 6 Oct 2019

Apisitniran L (2019) Worker shortage casts doubt on rail projects [Online]. Bangkok Post, Thailand. Available: https://www.bangkokpost.com/business/1748729/worker-shortage-casts-doubt-on-rail-projects. Accessed 7 Oct 2019

Association of Southeast Asian Nations (2007) Charter of the association of Southeast Asian Nations [Online]. Available: https://asean.org/asean/asean-charter/charter-of-the-association-of-southeast-asian-nations/

Association of Southeast Asian Nations (2016a) ASEAN guiding principles for quality assurance and recognition of competency certification systems [Online]. ASEAN Secretariat, Jakarta. Available: https://asean.org/wp-content/uploads/2018/03/Guiding-Principles-for-Quality-Assurance-and-Recognition-of-Competency-C....pdf

Association of Southeast Asian Nations (2016b) ASEAN qualifications reference framework [Online]. ASEAN Secretariat, Jakarta. Available: https://asean.org/wp-content/uploads/2017/03/ED-02-ASEAN-Qualifications-Reference-Framework-January-2016.pdf

Bandaranaike S, Willison J (2010) Work skill development framework: an innovative assessment for work integrated learning. In: Campbell M (ed) Work integrated learning—responding to challenges: proceedings of the 2010 ACEN national conference, September 29–October 1, 2010. Curtin University, Perth

Bandaranaike S, Willison J (2014) Work skills development framework [Online]. Adelaide, Australia. Available: https://www.adelaide.edu.au/rsd/framework/frameworks/WSD-2009_2014.pdf. Accessed 15 Aug 2019

Bateman AJ, Coles M (2013) Qualifications frameworks and quality assurance of education and training [Online]. The World Bank. Available: https://olc.worldbank.org/sites/default/files/Qualifications%20frameworks%20and%20quality%20assurance%20of%20education%20and%20training_final.pdf. Accessed 7 Aug 2019

Bateman AJ, Liang X (2016) National qualification framework and competency standards: skills promotion and job creation in East Asia and Pacific (English). World Bank Group, Washington

Bca. 2016. *Being work ready: a guide to what employers want* [Online]. Australia: Business Council of Australia. Available: https://d3n8a8pro7vhmx.cloudfront.net/bca/pages/4101/attachments/original/1527731958/Being_Work_Ready_Guide.pdf?1527731958. Accessed 12 Oct 2019

Blades R, Fauth R, Gibb J (2012) Measuring employability skills: a rapid review to inform development of tools for project evaluation [Online]. National Children's Bureau, London. Available: https://hdl.voced.edu.au/10707/319995. Accessed 10 Aug 2019

Brewer L (2013) Enhancing youth employability: what? Why? and How? Guide to core work skills [Online]. International Labour Office, Geneva. Available: https://www.ilo.org/skills/areas/skills-for-youth-employment/WCMS_213452/lang--en/index.htm. Accessed 10 Aug 2019

Brink News (2018) How can ASEAN prepare for the future of work? [Online]. Marsh & McLennan Insights, Inc. Available: https://www.brinknews.com/how-can-asean-prepare-for-the-future-of-work/. Accessed 10 Aug 2019

Dacre Pool L, Sewell P (2007) The key to employability: developing a practical model of graduate employability Educ + Train 49:277–289

Centre E, for the Development of Vocational Education and Training, (2011) Development of national qualifications frameworks in Europe. Publications Office of the European Union, Luxembourg

Gropello ED, Kruse A, Tandon P (2011) Skills for the labor market in Indonesia: trends in demand, gaps, and supply. The World Bank Group, Washington, DC

Hanson S (2010) Gender and mobility: new approaches for informing sustainability. Gender Place Cult 17:5–23

Harris R, Short T (2014) Workforce development : perspectives and issues. Springer, New York

Kasetsart University Rail Engineering Center (2019) Professional qualifications for the Thai rail industry [Online]. Available: https://kurail.org/. Accessed 17 Aug 2019

Leopold TA, Ratcheva V, Zahidi S (2016) The future of jobs: employment, skills and workforce strategy for the fourth industrial revolution. The World Economic Forum, Geneva, Switzerland

Lucas B, Hanson J (2016) Learning to be employable—practical lessons from research into developing character. City and Guilds, London, England

Mahidol University 2019. *Professional and personal skills development* [Online]. Available: https://www.grad.mahidol.ac.th/softskills/. Accessed 13 Aug 2019

Michaels E, Handfield-Jones H, Axelrod B (2001) The war for talent. , Harvard Business School Press, Boston

Ministry of Higher Education S, Research and Innovation (2006) National QUALIFICATIONS FRAMEWORK FOR HIGHER EDUCATION in Thailand [Online]. Thailand. Available: https://www.mua.go.th/users/tqf-hed/news/FilesNews/FilesNews8/NQF-HEd.pdf. Accessed 10 Aug 2019

Ncver (2018) Focus on soft skills, employability and education [Online]. National Centre for Vocational Education Research (NCVER), Australia. Available: https://www.voced.edu.au/focus-soft-skills-employability-education. Accessed 7 Oct 2019

Piip J (2014) Exploring leadership talent practices in the Australian rail industry. Doctor of Philosophy, Australia

Pwc (2019) Thailand infrastructure news [Online]. Thailand. Available: https://www.pwc.com/th/en/deals/assets/infrastructure-news/Issue8.pdf. Accessed 6 Oct 2019

Rail Professional (2019) Human resource development in the rail industry [Online]. Available: https://www.railprofessional.com/magazine/human-resource-development-rail-industry.

Roche A (2001) What is this thing called workforce development? Australia, National Centre for Education and Training (NCETA), Adelaide

Sachs J, Rowe A, Wilson M (2016) 2016 good practice report: work integrated learning (WIL). Department of Education and Training , Canberra

Sondergaard LM, Luo X, Jithitikulchai T, Poggi C, Lathapipat D, Kuriakose S, Sanchez Martin ME, Reungsri T, Mohib SA, Chavapricha R, Peamsilpakulchorn P, Arin T (2016) Thailand—systematic country diagnostic : Getting back on track - reviving growth and securing prosperity for all [Online]. World Bank Group, Washington DC. Available: https://documents.worldbank.org/curated/en/855161479736248522/Thailand-Systematic-country-diagnostic-getting-back-on-track-reviving-growth-and-securing-prosperity-for-all. Accessed 7 Aug 2019

Tan J-P, Lee KH, Flynn R, Roseth VV (2016) Education and Skills for Growth in Emerging Economies. Workforce development in emerging economies: comparative perspectives on institutions, praxis, and policies

Tan J-P, Mcgough R, Valerio A (2010) Workforce development in developing countries: a framework for benchmarking. Human Development Network, World Bank, Washington, DC

Tan KS, Tang JT (2016) New skills at work: managing skills challenges in ASEAN-5 [Online]. Research Collection School Of Economics, Singapore. Available: https://ink.library.smu.edu.sg/soe_research/1891. Accessed 8 Aug 2019

Thailand Professional Qualification Institute (2014). Thailand. Available: https://www.tpqi.go.th/home.php. Accessed 18 Aug 2019.

Unesco-Unevoc International Centre for Technical and Vocational Education and Training (2015) World TVET Database Thailand. UNESCO-UNEVOC International Centre for Technical and Vocational Education and Training, Bonn, Germany

United Nations Development Programme (2018) Human development indices and indicators: 2018 statistical ipdate. In: Steiner A (ed). United Nations Development Programme, New York

United Nations Education Scientific and Cultural Organization (2017) Technical vocational education and training [Online]. Available: https://unevoc.unesco.org/go.php?q=What+is+TVET. Accessed 14 Augs 2019

Wibrow B (2011) Employability skills at a glance [Online]. NCVER, Australia. Available: https://www.ncver.edu.au/__data/assets/file/0013/2407/2404.pdf. Accessed 8 Aug 2019

World Bank (2019). Available: https://www.worldbank.org/. Accessed 14 July 2019

The Development of a Railway Track Inspection System for Track Maintenance

Korntham Sathirakul, Suradate Suratisak, Jatesada Borsub, and Sompat Praeknokkeaw

Abstract The railway-track inspection system for track maintenance being developed in this research project consisted of 2 sub-systems, namely (1) Railway-track Locomotion sub-system, consuming electric energy to propel itself along the track in the inspection operation, with light gross weight, operated by one person, and (2) Railway-track Inspection and Data Recording sub-system with high-speed and high-resolution image capturing capability to suit the requirements of railway-track inspectors' field operations. The 2 parts were researched, designed and developed such that the railway-track inspection system had simple construction and operation, and could be built with locally available supplied parts and equipments, leading to low manufacturing costs. A prototype of the system was built and tested on a mock-up test track simulating a real operation. The test results were used for later improvements of the system to yield an appropriate performance for real railway-track operation. The prototype had been planned to be tested on a real track in a real field operation in a near future. This developed system had been expected to put into real inspection and maintenance service of the railway operators to help ensuring safe tracks for railway passengers safety and comfort.

Keywords Railway-track inspection · Track maintenance · Railway-track vehicle

1 Introduction

Nowadays, railway transportation in Thailand has been improved and developed extensively, to provide safe, reliable, efficient and convenient services, especially the on-going dual-track construction for the regional (long-distance) services and the new network of metro transit systems in urban cities, e.g., Bangkok. One important thing that will come with such extensive development of the railway systems is their efficient maintenance. Railway inspection, one of the key activities in railway infrastructure maintenance, must be efficiently achieved. Our research project had

K. Sathirakul (✉) · S. Suratisak · J. Borsub · S. Praeknokkeaw
Department of Science Service, Innovative Robotics and Automation Group, Engineering Materials Division, 75/7 Rama 6 Road, Ratchatewi, Bangkok 10400, Thailand
e-mail: korntham@dss.go.th

W. Weerawat et al. (eds.), *Urban Rail Transit*, Lecture Notes in Mobility,
https://doi.org/10.1007/978-981-15-5979-2_8

been conducted to develop a system that assists railway-track inspectors to carry their duties. Traditionally, inspectors must walk along railway tracks to visually inspect the tracks for irregularities, e.g., missing fasteners, broken fishplates, cracked rails. This process is laborious and tiresome for inspectors and prone to imperfect results. Moreover, track inspection with high-technology inspection car, although yielding detailed and accurate inspection data, is very expensive. The developed railway-track inspection system in this work aims to fill the gap between the two.

1.1 *Objective of the Developed System*

The research project is aimed at developing a system that efficiently and economically assists railway inspectors to successfully achieve their work.

1.2 *Scope of the System*

The system was meant to fulfill the needs of the track inspectors' routine operations, as well as to meet all necessary safety regulations and constraints. The system was not intended to substitute the expensive imported high-technology inspection cars, like EM80 or EM120 currently in service.

2 Background

2.1 *Railway Track Inspection*

Both the on-going dual-track construction for the regional (long-distance) services and the new network of metro transits in urban cities are currently contributing to a rapid expansion of the Thai railway network. The investment in such infrastructure is enormous. The quality of the railway transportation services can be obtained, partly but significantly, from good maintenance of the infrastructures. Railway track inspection and maintenance are crucial activities. These need to be done very effectively and economically.

There are two main methods of railway track inspection. First is done by the inspectors, daily walking the track to visually inspect the condition of the track sections they are assigned to be responsible for, usually 25–30 km long for each individual inspector. The inspectors look for missing fasteners, broken fishplates, cracked/broken rails, etc., and they sometimes fix the problems right at the scene with hand tools. Locations of other irregularities are marked and recorded, waiting for further maintenance crews to return to the scene with heavy tools and equipment to service and repair. This method of inspection requires many competent and well trained inspectors on duty to cover the entire railway network.

The other way is to inspect the track with railway-track inspection cars. These inspection cars are well equipped with expensive high-tech sensors and instruments to measure track parameters, such as, track geometry, rail profiles, rail wear, as well as, multiple-axis vibrations, real-time ultrasonic testing for rail flaw detection, and video recording of the rails and track components. Although these inspection cars are capable, they are a substantial investment, and required specialists to operate and service. The Thai railway operators have only a few such expensive cars which may not be sufficient for the rapid expansion of the future railway network.

Our research team developed a system that assists the inspectors to carry out their tasks more effectively and conveniently. The system can be used for both ballasted track, e.g., the State Railway of Thailand (SRT), and non-ballasted track, e.g., BTS and MRT. The developed system can fill the gap between the two means of inspection.

2.2 *Requirements and Specifications of the System*

The requirements of the railway-track inspectors had been explored, as well as, the rules and regulations for the use of the equipment on the railway track of the SRT, especially the light railcar for track service and maintenance purposes had been studied. The key requirements and specification of the developed system are listed as follows.

2.2.1 The inspection system shall be a light rail vehicle that can carry two persons together with the necessary inspection instruments. Furthermore, it can be operated by one person.

2.2.2 The operation of the system shall not interfere with the train operations, i.e., the railway signaling and control shall not be interrupted.

2.2.3 The operation range shall at least cover 25–30 km along the railway track, and the necessary operation duration. It shall be able to travel at an appropriate speed for efficient inspection operations, but shall not be over 50 km per hour limit (according to the SRT regulations).

2.2.4 The weight of the system shall be light enough that it can be hauled off the track by two persons, when necessary.

2.2.5 The system shall be equipped with a sub-system with high-speed camera that can capture the images/videos of the parts of the railway track under inspection. The speed of the camera system shall be high enough to obtain clear high-resolution images while the system travels at operating speed. It shall also have a monitor screen that shows the images of the track sections/parts being inspected for the inspectors to view during the operation.

2.2.6 The system shall be equipped with a sub-system that can identify and record its precise location, matched with the captured images mentioned in 2.2.5, for the purpose of indicating the locations of track irregularities for later repairing/maintenance services.

2.2.7 The system shall be equipped with necessary obvious marks, e.g., red flags, and lamps, to clearly identify its existence and current location, during both day and night, to avoid potential accidents. A sound warning signal, like horn, is also necessary.

2.2.8 The system shall be affordable in terms of investment.

All of the above listed information was thoroughly considered as inputs to the design process of the prototype of our railway inspection system. The design and development of the prototype of the inspection system will be presented in 3.

2.3 *Related Literature Reviews*

Railway track needs to be periodically inspected and well maintained in order to obtain a safe and comfortable ride. The railway track inspection procedure needs to be done by complying with the operator's rules and regulations, and to appropriate standards (SRT 2006; US Department of Defense 2008). The overview of the advancement of railway track inspection technologies was presented by Innotrack (2008), and later by Barragan et al. (2011). Sawadisavi et al. (2008) introduced a computer machine-vision technique for railway track inspection. Berry et al. (2008) utilized a high speed video camera and an advanced image-processing technique for the inspection of rail joint bars. Furthermore, a system that was able to visually recognize elastic rail clips was developed (Hsieh et al. 2007). Such recognition used an artificial intelligence technique. Resendiz et al. (2013) developed a laboratory-scale prototype of an automated visual inspection for railway tracks. There are quite a few research works aimed at developing competent railway inspection and maintenance systems. These published technologies can be carefully studied, adopted, modified and appropriately applied to the case of Thailand's railway system.

3 Design and Development of the System

The railway-track inspection system for track maintenance in our research work had been designed and developed such that it consisted of 2 sub-systems, namely (1) Railway-track Locomotion sub-system, and (2) Railway-track Inspection and Data Recording sub-system. The 2 sub-systems were researched, designed and developed according to the derived requirements and specifications mentioned in Sect. 2.2. Furthermore, the system was developed such that the railway-track inspection system had simple construction and operation, and could be built with locally available Thai supplied parts and equipment, leading to low manufacturing costs.

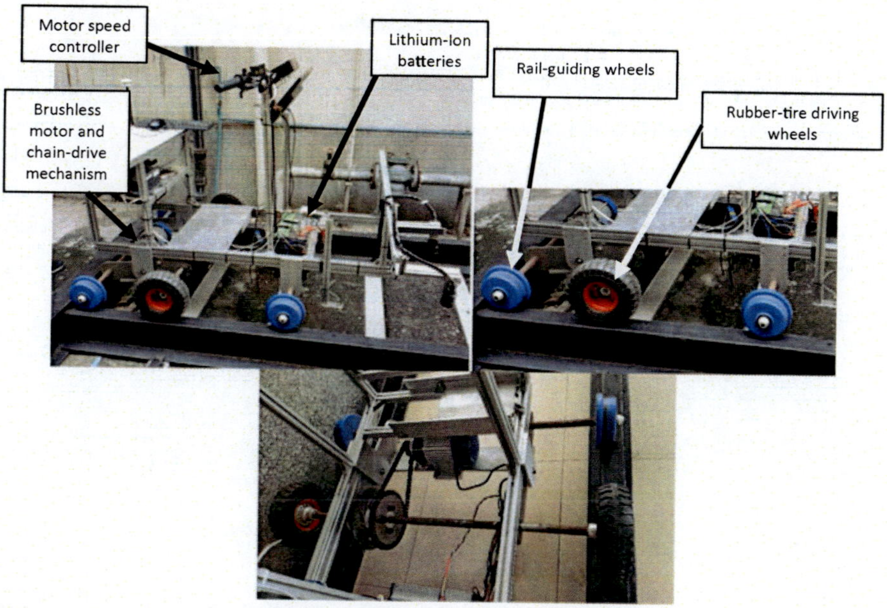

Fig. 1 The Railway-track locomotion sub-system

3.1 Railway-Track Locomotion Sub-System

This sub-system is the one that propels the whole system, traveling along the inspected railway track. The body frame is made from an aluminum extruded profile. There are 2 pairs of rail-guiding wheels, made of strong non-conductive material, e.g., nylon. There is a pair of rubber-tired driving wheels, powered by a 72 V 750 W brushless DC motor, through a chain driven shaft. There is a controller unit with a hand throttle for controlling speed and direction of the motor. The set of motor and controller unit is widely commercially available in Thailand. Six 12 V Lithium-Ion deep cycle rechargeable (40 Ah) batteries are used for the electrical source. A set of motorcycle disc brakes is adopted for the system braking system. Figure 1 shows the arrangement of the sub-system.

3.2 Railway-Track Inspection and Data Recording Sub-System

The developed system has a sub-system that can capture high resolution images of the sections of the track under inspection. In our research project, the visual inspection of the conditions of the rail fasteners was emphasized. Two high speed and high resolution cameras were installed for each rail, with a lighting system, and

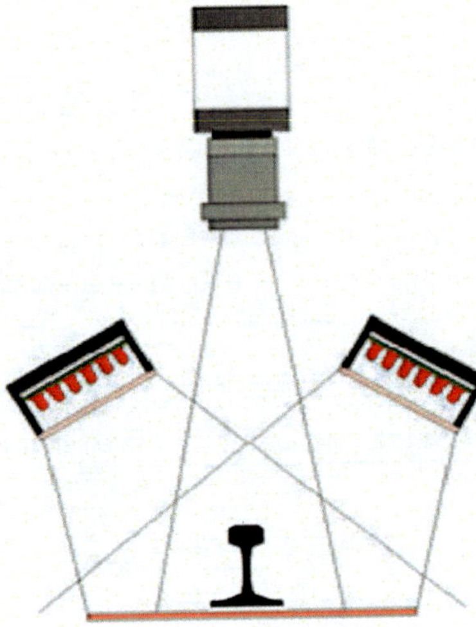

Fig. 2 The arrangement of the cameras and their lighting system

in such a way that high quality images were obtained. There was a proximity (laser type) switch designed to detect the existence of the track sleepers. The operation of the cameras was designed such that they simultaneously captured images once the proximity switch was triggered by each sleeper detection. The captured images were then stored in the system hard disk space together with the concurrent location data (the system localization). Figure 2 shows the arrangement of the cameras and their lighting system.

The sub-system is also equipped with a global positioning system (GPS) as one of the means for its localization. Since GPS has limitations on accuracy within 3 m of radius and is unavailable in some situations, two other means of localization are made available in parallel, namely an odometer from its wheel revolutions, and track sleeper counting. The utilization of these three means of identifying the location of the system together effectively increase accuracy of the localization, which is crucial for identifying the locations of the irregularities for later repairing/maintenance services.

For the time being, the sub-system is only capable of capturing and recording images and localization data of the rail and fasteners on each and every sleeper along the inspected track. The evaluation of the track condition, i.e. the existence of complete fasteners, needs to be post-processed. This can be done manually or automatically using image processing software. A monitor screen that shows real-time images of the track sections/parts being inspected is also provided for the inspectors to view during operation. An on-board industrial-grade PC is used to control this sub-system. Figure 3 shows the schematic diagram of the sub-system operations. Figure 4 shows the arrangement of the Railway-track Inspection and Data Recording sub-system.

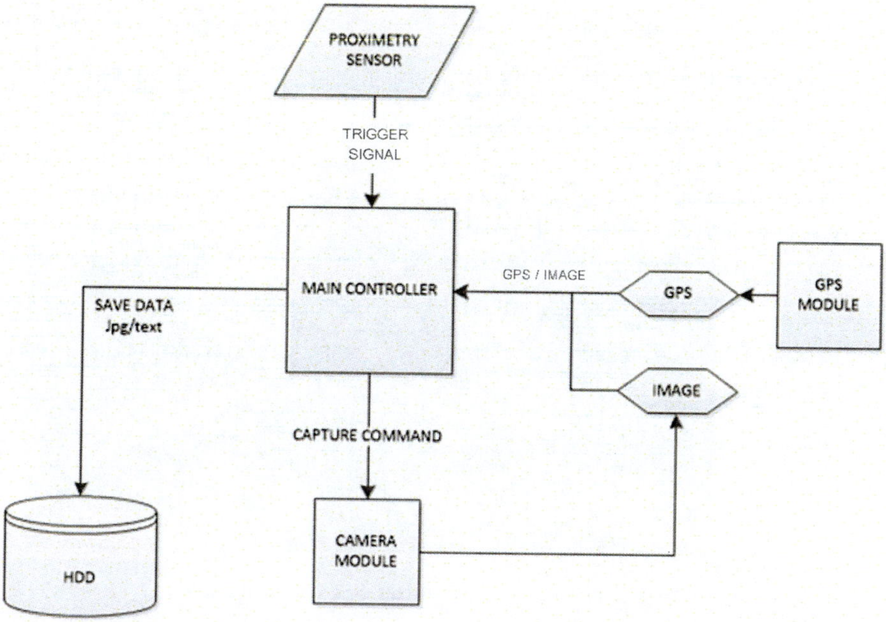

Fig. 3 The Schematic diagram of the railway-track inspection and data recording sub-system

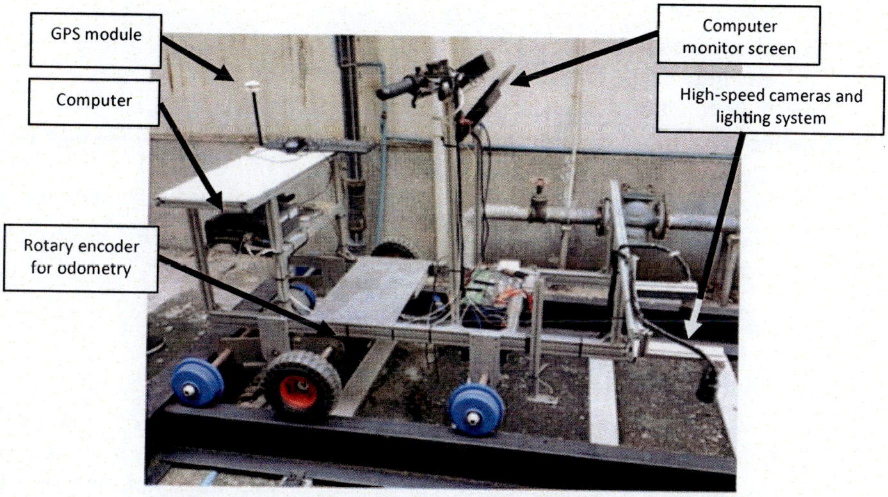

Fig. 4 The arrangement of the railway-track inspection and data recording sub-system

Fig. 5 The Testing of the developed system on the mock-up test track

4 System Testing

The developed railway-track inspection system was extensively tested on a mock-up railway track of 18 m in length, in a straight line. The two sub-systems were verified for their functions. Because of the limited space, the system could only be tested on the 18 m mock-up test track, traveling forwards-and-backwards repeatedly for a duration of approximately 10 h. Testing on a real railway track will be arranged in the near future. Figure 5 shows the testing of the developed system on the mock-up test track. Figure 6 shows some examples of the captured images of the track under inspection.

5 Results and Conclusions

In this research project, the prototype of the railway-track inspection system was designed, developed, and tested. The system met all the requirements. The tested operation speed was 15 km per hour. The maximum speed was around 25 km per hour. With 40 Ah 72 V Lithium-Ion batteries, the system operation could be achieved within a 20 km round-trip distance. The railway-track inspection and data recording sub-system adequately performed its functions. Clear images of all the fasteners of the inspected track (the mock-up test track) were obtained, together with their location data. The evaluation of the rail fasteners' conditions still needed to be manually post-processed. It should be noted here that the sub-system worked better with the

(Left-side of rail) (Right-side of rail)

Fig. 6 Some examples of the captured images of the track under inspection

concrete slab track (non-ballasted), because of its well-organized environment. Using the system with ballasted track, e.g., SRT track, may be prone to faults/missing triggering signal from the proximity switch, due to the irregularities of the ballast. The system weighed approximately 45 kg.

In conclusion, the developed system that consists of the railway-track locomotion sub-system, and the railway-track inspection and data recording sub-system can be operated as designed. Its performance is acceptable to some degree. There is still room for improvements, e.g., extending operating range, implementing image-processing software for post-processing or real-time operation, upgrading the sensors to go beyond visual inspection, increasing the accuracy of its localization, reducing its weight while maintaining functionality, etc. Above all, the system needs to be tested within the real environment and situation. This will open up more research ideas. In terms of cost, the developed system took roughly 300,000 Baht. A more thorough re-consideration of its bill of materials and manufacturing costs will be needed when carrying it to commercialization processes, to ensure the significant cost reduction.

6 Limitations and Future Works

The developed railway-track inspection system, currently, can assist the inspectors by providing an electrically propelled vehicle for traveling along the track, and recording high-resolution photo images of rail sections and their fasteners together with location data for identifications. The inspectors can evaluate the track conditions by manually post processing those recorded photo images. Our future research will be extensively involved with the development of image processing software to automatically post process the images, and then later development of a real-time automatic railway track inspection system that will collaborate and integrate other types of sensors and track measurements.

References

Barragan A, Cembrero P, Caceres N, Schubert F (2011) D2.1 Report on railway inspection and monitoring techniques-Analysis of different approaches. ACEM-Rail. Available at: https://www.acem-rail.eu/documents.html. Accessed 5 June 2014

Berry A et al. (2008) High speed video inspection of joint bars using advanced image collection and processing techniques. In: Proceedings of the 2008 World Congress on Railway Research, Seoul, Korea. Available at: https://www.uic.org/cdrom/2008/11_wcrr2008/pdf/I.3.1.3.2.pdf. Accessed 5 June 2014

Department of Defense, United States of America (2008) Unified facilities criteria (UFC): railroad track maintenance & safety standards, UFC 4-860-03, 13, Feb 2008. Available at: https://www.wbdg.org/ccb/DOD/UFC/ufc_4_860_03.pdf. Accessed 5 June 2014

Hsieh HY, Chen N, Liao CL (2007) Visual recognition system of elastic rail clips for mass rapid transit systems, JRC/ICE 2007-40080. In: Proceedings of the ASME/IEEE Joint Rail Conference & Internal Combustion Engine Spring Technical Conference, Pueblo, Colorado, USA, pp 319–325

Innotrack (2008) D4.4.1-Rail Inspection Technologies. TIP5-CT-2006-031415, University of Birmingham, UK. Available at: https://www.innotrack.net/IMG/pdf/d441.pdf. Accessed 5 June 2014

Resendiz E, Hart JM, Ahuja N (2013) Automated Visual Inspection of Railroad Tracks. IEEE Trans Intell Transp Syst 14(2):751–760

Sawadisavi SV et al. (2008) Machine-vision inspection of railroad track. University of Illinois at Urbana-Champaign, Nov 2008. Available at: https://ict.uiuc.edu/railroad/CEE/pdf/Events/TRB09/Sawadisavi%20TRB%20Paper.pdf. Accessed 5 June 2014

The State Railway of Thailand (SRT) (2006) Rules and regulations for train operations, pp 338 (in Thai language)

Spatio-temporal Retail Competition Factors Accessibility in Hakata Station, Japan

Somsiri Siewwuttanagul and Takuro Inohae

Abstract Retail accessibility refers to the attractiveness of place especially in the transport hub that serves a number of passengers each day. Retail activities are crucial for providing a variety of space utilizations which gives more options to the transport hub's users. A change of Spatio-temporal accessibility needs to be analyzed in order to identify the spatial accessibility potential in transport hubs where the density of people depends on the activity and the time of the day. In this paper, the study presents a new approach to measure competition factors of accessibility by emphasizing centrality analysis tools that help to simulate potential demand as well as implement the reverse proportional of conventional competition factors accessibility measures for determining the level of competition factors accessibility of retail activities. The results illustrated the Spatio-temporal accessibility performance of each type of retailer in the Hakata station. The consequent impacts regarding infrastructure efficiency, circulation network, and the time temporal in the day affects the accessibility and the way people perceive the activity's attractiveness. This study could help decision-makers identifying the dynamic of the attractiveness of spatial retail activity throughout the day for assisting in their spatial planning work.

1 Introduction

Retail activities play a crucial role in transit stations for efficiently utilizing the space in the transport hub as well as providing shopping options for passengers. The accessibility of retailers can interpret not only how to reach the desired destination from the origin location but also the attractiveness of retail location (Hansen 1959; Dalvi and Martin 1976; Handy and Niemeier 1997). Accessibility assessment is usually conducted on a large scale such as district or community scale mainly to explain the

S. Siewwuttanagul (✉)
Graduate School of Science and Engineering, Saga University, 1 Honjo, Saga 840-8502, Japan
e-mail: jamessomsiri@gmail.com

T. Inohae
Department of Civil Engineering and Architecture, Saga University, 1 Honjo, Saga 840-8502, Japan

W. Weerawat et al. (eds.), *Urban Rail Transit*, Lecture Notes in Mobility,
https://doi.org/10.1007/978-981-15-5979-2_9

connection between land use, transportation, and labor markets (Caschili and Montis 2013). But there are a few researches that have been done using the conventional location-based accessibility measures for assessing smaller spatial analysis. Therefore, in this study, the interaction between land use and transportation was focused at the micro-scale as an intrinsical pedestrian network inside a transport hub building where non-motorized trips are conducted in relevance with the public transportation infrastructures.

The attractiveness of the specific activity is usually identified based on the rationale of location-based accessibility measures which emphasized the competition factors accessibility measure. This kind of accessibility measurement investigates the relationships between pedestrian volume (demand) and retail space (supply) that may affect the potential attractiveness of the place. Therefore, the importance of the place competitiveness is also associated with the route-choice decision of public transport users. This can guide the spatial planning and development by demonstrating the dynamics of trip distribution of public transportation users in the study area.

Consider the time period in the day and the change of character of spatial in transport hub site, the performance of retail accessibility may vary due to the time of the day. It is temporal of potential accessibility for which the level of retail competitiveness in each location may not be permanent. The routine of daily activity can help explain the characteristics of passengers in terms of the time that the station is likely to be crowded, the type of retail that people visit the most or can even tell the passenger's shopping behavior. In this case study, the retail activities and public transportation node are investigated for explaining the Spatio-temporal accessibility changing dynamic in a transportation hub which demonstrated the micro-scale of spatial land use and transportation. These aspects are aimed at helping improve the intrinsical accessibility of pedestrian networks in public transport as well as the retail investment that would benefit both passengers and station operators.

2 Literature review

2.1 *Retail Accessibility*

In general, accessibility can be interpreted for reaching a destination from the origin location (Ingram 1971) but from the accessibility measure theoretically point of view, it could be measured for the specific purpose (Bach 1981) in such retail attractiveness which is mainly discussed in this paper. First of all, the space utilized in particularly for the retail activity in the transit station is included for one single unit in the urban economic movement. A transport hub site is considered as an urban public space and passengers or users are considered as pedestrians. In the scale of the transit station, the small unit of retail space is welcomed for the occupancy for attracting the public transit commuters. In this particular environment, commercial activities are the majority of transport hub space utilization. It could differ in a smaller transit station because of the size and complexity of the transport hub that was designed to service more users than a single transit station which has no interchange or is not located in the center of the city's economy.

This phenomenon has been explained as mentioned in the report of ITDP (2017), the retailer attracts pedestrians to come and enjoy the services and in return pedestrians can also utilize the retail location vitality as well. This is mutualism of place that is an association among the urban form's elements (Dempsey et al. 2010). It is also associated with the central place theory expressed through centrality in order to explain the accessibility features of the place. Centrality determines the relevance of distance which refers to cost for reaching the destinations. It is dynamically associated with urban development, dynamics of real estate, and any other social welfares (Porta et al. 2009).

The location of retailers is systematic. There is an interaction between densities, grid structure, and the use of space that shape the retail's location pattern (Hillier 1996). According to the study of Kang (2016), the consideration of accessibility to retail activities enhances the vitality of the urban space. It is also one of the crucial strategies for encouraging pedestrian-friendly environments. Then the study of Dolega et al. (2016) emphasizing the patronage probabilities of retail activity related to the spatial interactions between potential demand as interpreted as the potential customers and the retail supply in the retail center. This work allows us to understand the location effects and agglomeration form of retail activity.

2.2 *Review of Temporal Accessibility Change*

With regards to the belief of space–time dynamic (Tan et al. 2017), the change of Spatio-temporal accessibility during the day refers to the level of potential accessibility in the specific time and also its effect on the individuals in activity (Wee et al. 2013). The perspective of space–time behavior identifies the differences accessibility of each location, the activity's opportunity, and person matter. This means that the level of accessibility varies depending on the time of the day because time determines the concentration of potential demand which is mostly concluded as a pattern of activity.

The study of Spatio-temporal change of accessibility provides spatial information of retail activities which determine the relationships between retail types and the temporal specific or time-sensitivity through the manner of location-based accessibility measure (Lee et al. 2017). According to Jiang et al. (2012), the clusters of individuals are represented in the pattern of daily activity that ran upon the space–time rational and essentially predicted a daily routine of each cluster. Furthermore, in Tribby and Zandbergen (2012) the Spatio-temporal accessibility benefits the transport hub: not only the retail location-allocation guidance but also the transportation services associated with the retail activities around the transport node. The contribution of Hu and Downs (2019) inspired and guided scholars to study a temporal change of spatial accessibility. Supply, demand, and spatial barriers are taken into account for identifying the fluctuation of Spatio-temporal accessibility dynamics in hourly distribution. Therefore, the review of these literatures allow us to consider the fluctuating temporally of retail activities in the complex transport hub site.

2.3 *Review of Competition Factors Accessibility Measure*

The location-based accessibility with competition factors measure is a type of accessibility measure that has been developed in order to consider the interaction of demand and supply to identify the potential between locations. In order to apply the competition factors measure, the gravity model needs to be mentioned for its structure to help clarify the use of accessibility analysis that considers the travel function for defining the potential of opportunities for interaction between locations (Hansen 1959).

The gravity model plays an important role in the competition factors measure with the belief that the destinations are not equal in terms of accessibility, so the weights of the destination have been considered for this measure (see Eq. 1) which are opposed to the cumulative opportunity measure that does consider the weight or size of a node of the destination. Therefore, the gravity model is appropriate for assessing the complexity of the urban network.

$$A_{i,m} = \sum_{j} O_j f\left(C_{ij,m}\right) \tag{1}$$

where

$A_{i,m}$	Is an accessibility at location i travel with mode m to location j;
O_j	Is the activity's opportunity at location j;
$f\left(C_{ij,m}\right)$	Is the travel impedance function of travel mode m from location i to location j.

In the large and complex circulation network of a transport hub which consisted of several activities and numerous routes, the potentials of location have differed regards to the spatial configurations in the station area. The gravity model measures the potential of location on accessibility that is considered on cost or time spent to reach the destination. Therefore, the competition factors accessibility measure is employed in an indoor and semi-indoor network in the station areas. The function of the competition factors measure improves the realism of the gravity measure for accessibility to a particular activity in the site (Cerdá 2009). This model consequently considered the interaction between supply opportunity and the demand of people visiting the opportunity which is able to estimate impacts of changes in the built environment on travel behavior study (Handy et al. 2002) and hence expressed the potential accessibility in a subjected location and able to assist in identifying the location that has a significant impact on transit-oriented development planning.

In regard to the demand for available opportunities that are uniformly distributed across space, these opportunities have no capacity limitations as stated by Shen

(1998), who then redefined accessibility for the competition factors measure which incorporated demand potential based on the gravity model as Eqs. 2 and 3;

$$A_{i,m} = \sum_{j} \frac{O_j f(C_{ij,m})}{D_j} \tag{2}$$

By

$$D_j = \sum_{k} P_k f(C_{kj,m}) \tag{3}$$

where

D_j	Is a demand potential for location j;
P_k	Is the number of people at location j;
$f(C_{kj,m})$	Is the travel impedance function of travel mode m from location k to location j.

3 Methodology

This study presents the competition factors accessibility measure of each time of the day in order to observe the Spatio-temporal change of retail potential accessibility. The information used in this study is mainly obtained from counted pedestrian numbers throughout the station and measurement of the retail space in order to compute the competition factors accessibility of each retail activity. The methodology of this study includes three elements as a supply of opportunity, potential demand, and mathematical foundations of competition factors accessibility.

3.1 Study Area

Hakata station, which is the center of public transportation of Fukuoka prefecture, is the largest public transport hub in the southern part of Japan (Kyushu region). Over 100,000 passengers travel through Hakata station each day. The number of users in Hakata station is the majority in railway services followed by urban bus and non-travel trips that come to the station area in order to pass by just for shortening their walking distance or for shopping purposes.

Hakata station is not only a center of public transportation but also performs as a commercial center with the space utilization of the station services for business and retail purposes as well as the pathway connection between the station area to surrounding office buildings and department stores. Figure 1a illustrates the perspective of Hakata station and Fig. 1b shows the layout of Fukuoka City showing walking distance from the center of Hakata station.

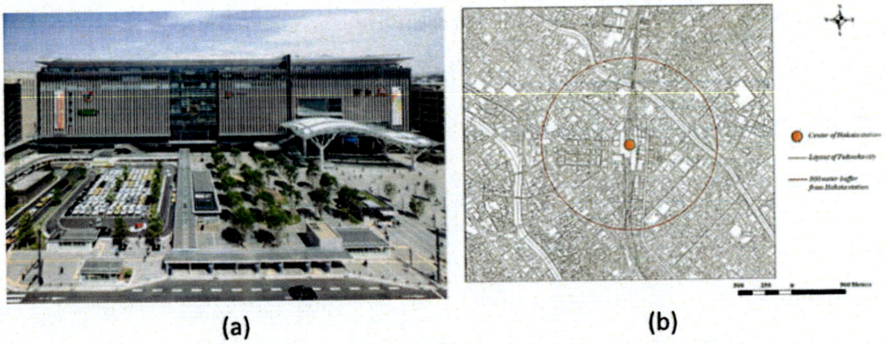

Fig. 1 Hakata station. **a** Perspective of Hakata station. **b** Layout of Fukuoka city and walking distance from the center

The retail service was planned to attached to every part of the station as shown in Fig. 2a detailed the retail operators that built upon the site of Hakata station including Amu Plaza department store which belongs to JR Kyushu (the major railway company in Kyushu region) and Hankyu department store and Fig. 2b shows proportion of transport service and commercial activity.

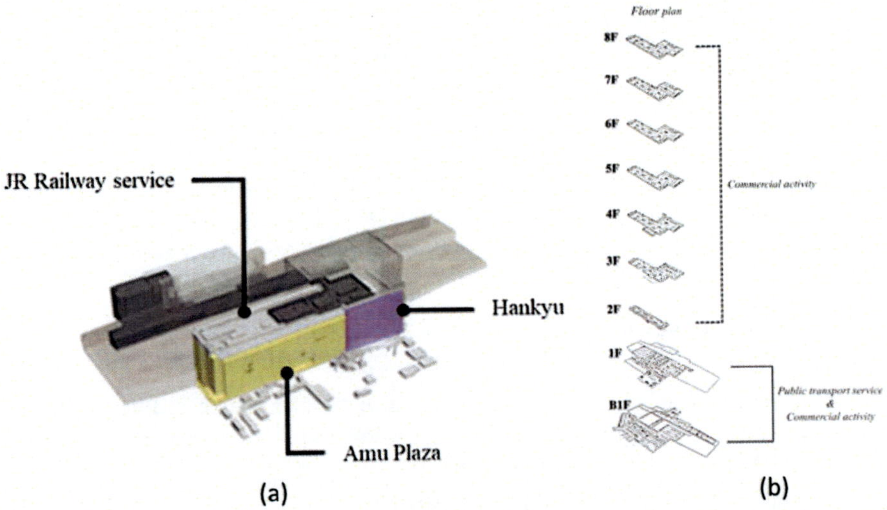

Fig. 2 Space use in Hakata station building. **a** Retail operators. **b** Layout of Hakata station building

3.2 Supply of Opportunity

The information for opportunity's supply was gathered by calculating the retail space from the master plan of Hakata station which is indicated in the square meter unit. The data collection demonstrated supply data into the point data in analysis software. Figure 3 displays how point data represented supply data that was collected from the Hakata station. The point data is essential for the competition factor accessibility measure because of the point interpreted as a reference location for the weight of supply and allows the study to observe the change of potential accessibility when the demand has changed over time.

The retail supply was simply calculated from the plot area of the subjected store. The supply of retail activity is illustrated in Fig. 4, the areas for single retail activity

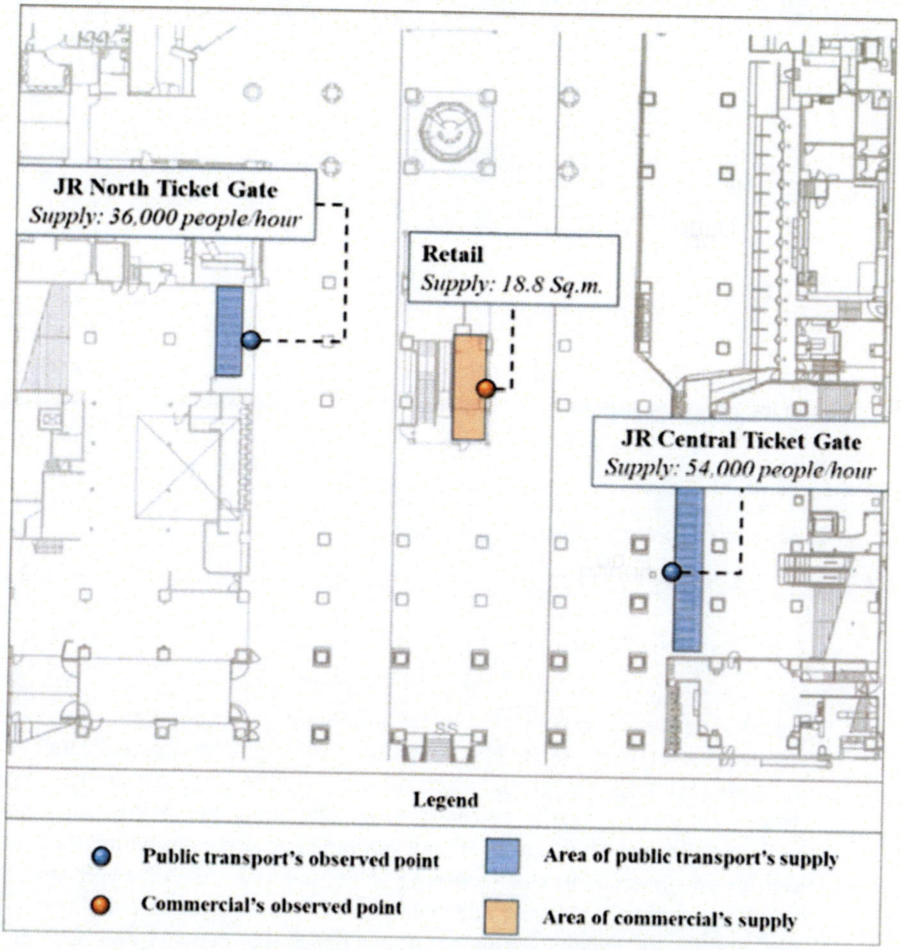

Fig. 3 Supply's node representation

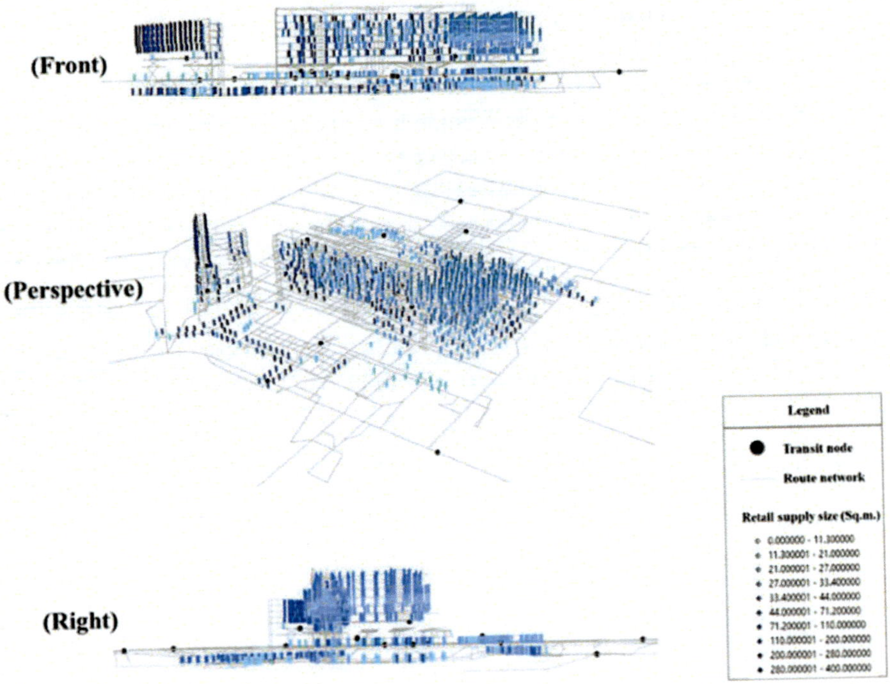

Fig. 4 The size of retail supply

vary according to its characteristics. For instant, the chain retail store or restaurant are tended to have a bigger space than local shops or less well-known stores. Therefore, this aspect is taken into account for assessing the potential accessibility through the location-based competition factors measure because the size of the shop can also mean the spatial capacity to attract people to visit and spend time and money on their shop which interpreted as attractiveness.

3.3 *Potential Demand*

Potential demand is represented by the number of pedestrians that flow through the a pedestrian network of the Hakata station building. The demand data was limited for observing only weekdays when it is easy to trace the pattern of accessibility. Pedestrian number distribution is done by two steps methods for determining the pedestrian volume throughout Hakata station. The first is to define the number and location of observer spots. According to the database, the observer spot is a location where potential demand had been observed, which presents 2576 virtual points located upon the Hakata station building's pedestrian network and node of all available activities. 40 locations from Hakata station building's circulation were chosen in order to collect the pedestrian volume throughout the day. The first 20 observer spots are located along with the main circulation and another 20 spots are on the minor areas in order to determine the most suitable search radius for simulating the most accuracy of a relative number of pedestrians through statistical analysis.

The number of observer spots and location of each observer spot were selected by consideration of equal distribution which was done using the Reach index (see Eq. 4) from Urban Network Analysis toolbox (Sevtsuk et al. 2016).

$$Reach[i]^r = \sum_{i,j \in G-\{i\}, d[i,j] \le r} W_{[j]} \quad (4)$$

where

$Reach[i]$	Is cumulative opportunity measure of location i
$W_{[j]}$	Is the total weight found in location j
$d[i, j]$	Is the shortest path between i and j that found within given travel distance r.

The Reach index has given a numeric number as 0 and 1 to define that the subjected node whether included in the weight assignment from the single nearest observed spot (pedestrian data collection spot). The Reach index can be defined as the maximum value more than 1 when that node is also included in another observed spot. Thus, the average Reach index that approaches to 1 will define the best fit distance for weight distribution assignment in order to compute potential demand distribution in a further step. As a result, shown in Fig. 5, the 50-m distance from the observed spot presented the reach index value as 0.95 which is the closest to 1 which indicated the best fit distance for weight distribution assignment in this study.

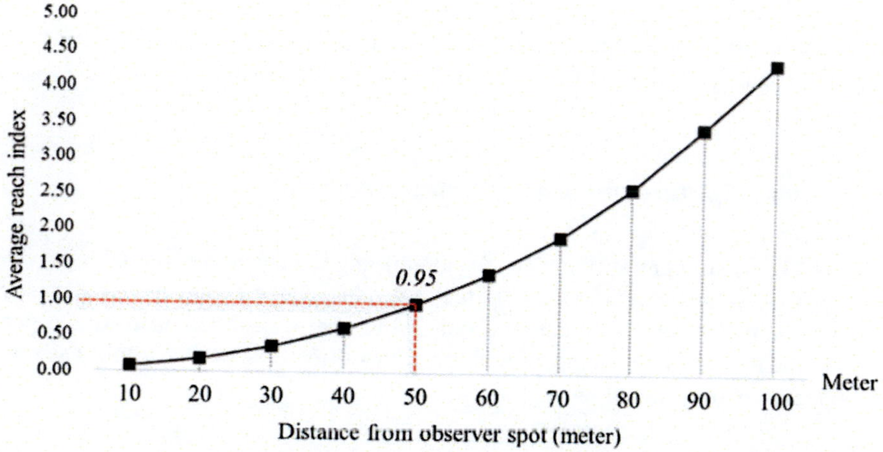

Fig. 5 Average reach index according to distance from observed spot

Then, the study estimated the number of pedestrians of each virtual node using the straightness centrality weighted by the pedestrian number from the observer spot which is assigned from the weight distribution in the previous method. The study has compared the demand potential calculated by straightness index (see Eq. 5) in each search radius and the time period in the day by Pearson correlation coefficient.

$$Straightness[i]^r = \sum_{i,j \in G-\{i\}, d[i,j] \leq r} \frac{\delta[i,j]}{d[i,j]} \tag{5}$$

where

- i Is origin
- j Is a destination
- δ_{ij} Is the Euclidean connecting route between i and j
- d_{ij} Is the shortest route between i and j
- G Is analysis network (graph)
- r Is search radius.

The pedestrian estimation result interpreted in Table 1 shows that a 50-m search radius indicated the most significant search radius distance among all the observed search radiuses. The correlation coefficient indicated the most significant as 0.711 for the first selected observed time period (T_1) 8:00–9:00 follow by 0.690 for 10:00–11:00 (T_2), 0.640 for 12:00–13:00 (T_3), 0.626 for 15:00–16:00 (T_4), and 0.607 for 18:00–19:00 (T_5).

This contribution was then processed onto the Spatio-temporal accessibility analyses throughout this study using the 50-m search radius for the demand potential simulation. As a result, Fig. 6 illustrates the volume of the pedestrian count at each observer location throughout the day and interpreted the surrounding estimated pedestrian volume using the straightness index with 50-m radius to simulate the temporal pedestrian volume and flow in Hakata station in order to represent the demand potential for the commercial competition factors accessibility computation.

3.4 *Spatio-temporal of Retail Accessibility*

The use of centrality indices for identifying the demand potential interpreted along with (Hu and Downs 2019) that introduced a computation of accessibility by comparing supply-to-demand ratio which helps scholars measure the potential of the location regards location-based competition factors accessibility as presented in Eqs. 6 and 7 as follow;

$$A_i = \sum_{i=1}^{n} R_i f\left(C_{kj,m}\right) \tag{6}$$

Table 1 Pearson correlation coefficient for the search radius of straightness index regards each observer time period

Search radius (m)	T_1 (8:00–9:00)	T_2 (10:00–11:00)	T_3 (12:00–13:00)	T_4 (15:00–16:00)	T_5 (18:00–19:00)
50	***0.711*****	***0.690*****	***0.640*****	***0.626*****	***0.607*****
100	0.487**	0.485**	0.437**	0.428**	0.417**
150	0.451**	0.462**	0.405**	0.409**	0.396**
200	0.413**	0.408**	0.377**	0.393**	0.375**
250	0.355**	0.365**	0.351**	0.360**	0.350**
300	0.286**	0.279**	0.282**	0.291**	0.286**
350	0.199**	0.200**	0.206**	0.211**	0.208**
400	0.162**	0.160**	0.166**	0.169**	0.169**

**Correlation is significant at the 0.01 level (2-tailed)

The most significant variable in the same time period is presented in Bold and Italic style

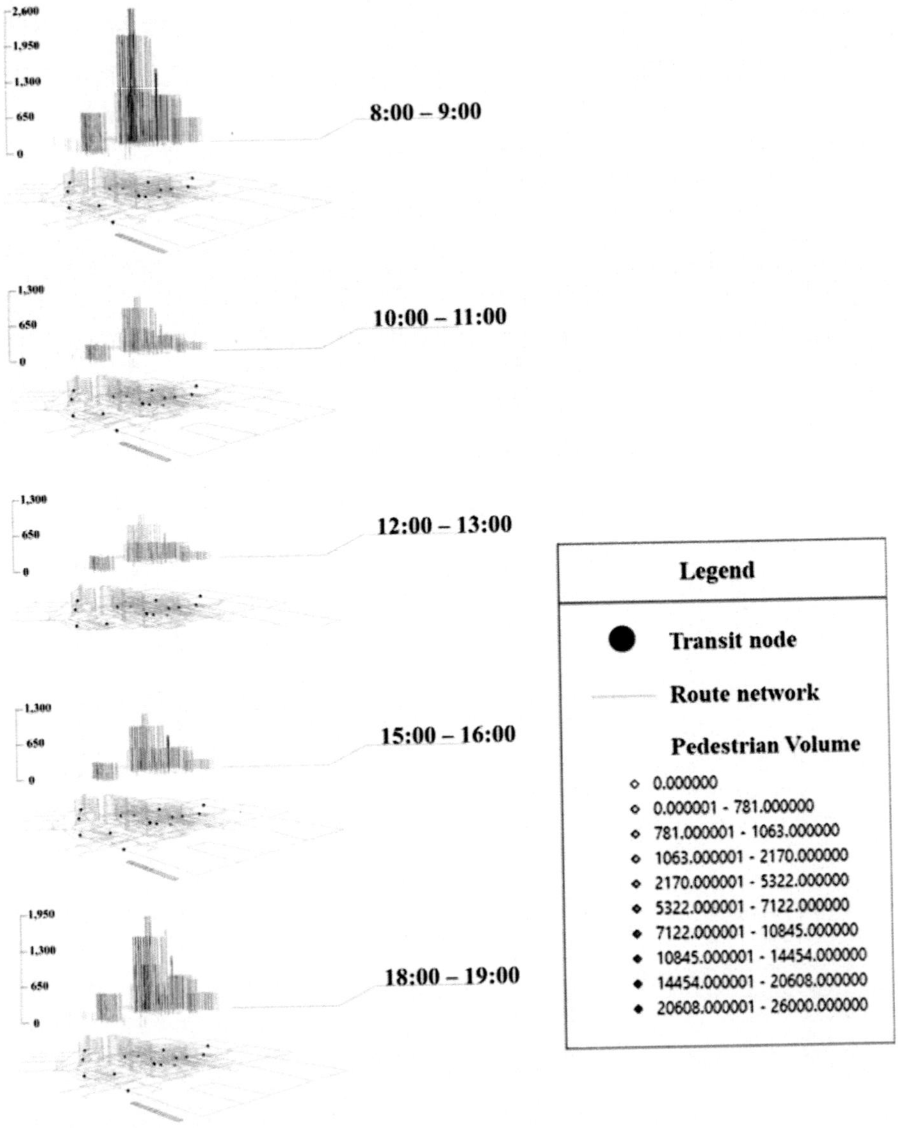

Fig. 6 Pedestrian volume in each observed time period

By

$$R_i = \frac{S_i}{\sum_{k=1}^{m} Di} \tag{7}$$

where

A_i Is an accessibility of location i;
R_i Is a supply-to-demand ratio of location i;
S_i Is a supply commercial opportunity of location i;
D_i Is a demand potential for location i.

Consequently, the study proposed the inverse proportion to supply-to-demand ratio to make it more practical as demand-to-supply for calculating the competition factors accessibility measure and be more equitable when comparing the potential of two or more locations as the proposed equation shown as Eq. 8.

$$A_i = \frac{1}{R_i} \tag{8}$$

In conclusion, the study employed the integrated competition factors using reversed proportion compared with the demand opportunity of retail activity in Hakata station area. The exponential graph displayed in Fig. 7 showed that the integrated competition factors model indicated the associated direction the accessibility value increases along with the volume of actual demand potential (people) recorded from the field survey. This indicates the law of commercial attractiveness in the spatial assessment that the attractive area should indicate the high demand in low supply.

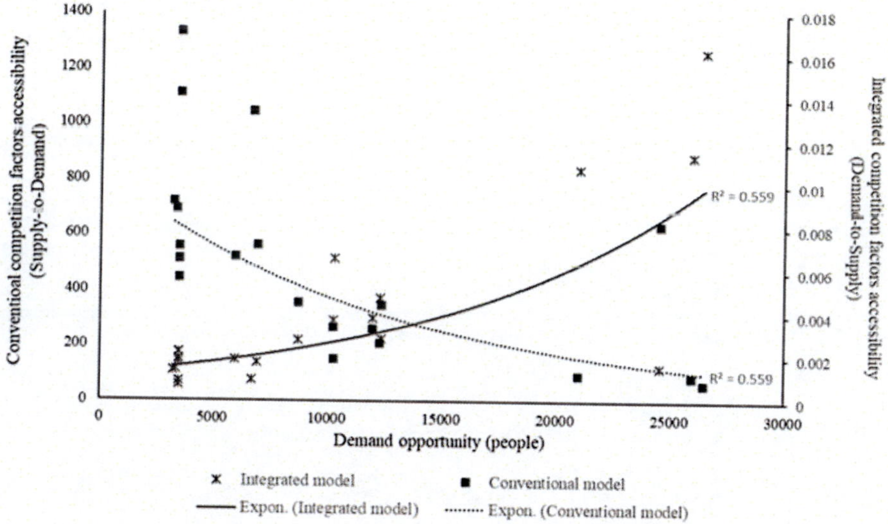

Fig. 7 The exponential graphs comparison between conventional competition factors and integrated competition factors model

4 Results

The study illustrates the Spatio-temporal analysis of competition accessibility among retail activities found in Hakata station. The central area of Hakata station, especially on the 1F floor and B1F floor, indicated the retail accessibility throughout the day, unlike the areas between 2F floor to 8F floor which are accessible during 10:00–21:00. Although there is less retailer opportunity in the central area of the station in both the 1F floor and B1F floor the competition level of retail accessibility in the central area of the 1F floor indicated the highest accessibility throughout the Hakata station area at the time of 8:00–9:00. The retail activity characteristics in this area consisted of small-plot retailers between 14 and 22 m^2 for supporting the travel trips, such as a convenience store, souvenir shops, lunchbox shops, and vending machines so that users can make quick transactions for purchasing goods. Moreover, the need for large space for flow of passengers during rush hour shaped the character of the small plot commerce in this area. To perceive the whole picture of retail potential accessibility, the study illustrates the average accessibility value classified by retail zoning regarding the characteristics of business and the geographical accessibility as displayed in Fig. 8, the characteristics of Spatio-temporal retail competition in Hakata station can be defined as morning rush hour, evening rush-hour and non-rush hour.

In the morning rush hour from 8:00–9:00, 1F floor illustrates the most significant retail accessibility by representing the accessibility value as 445.34 at Zone E followed by Zone C, B, and A as 287.65, 223.41, and 198.77 respectively. Both

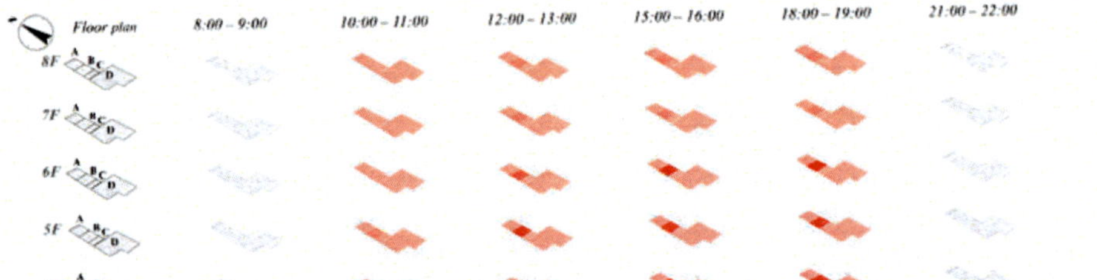

Fig. 8 Average retail accessibility divided by zone of commercial characteristic during the selected times of the day for representing the Spatio-temporal accessibility change

zones from B1F floor indicate a similar level of retail competition as 123.1 for Zone A and 121.58 for Zone B.

Non-rush hour which considered 9:01 to 18:00 determines varying characteristics of retail competitiveness. B1F floor showed a dramatic dropping of accessibility level at 10:00–11:00 as Zone A, C, and D represent 19.74, 14.3, and 15.55 respectively. The entire area of B1F floor during 12:00–16:00 is slightly increased but does not exceed 52.16. Besides, 1F floor still has the highest accessibility in non-rush hour but the accessibility level has been declined and shifts the most significant retail competitiveness to zone B. The result found that the highest accessibility value indicated as 349.24 at 12:00–13:00 time period. While the 2F floor determines the most consistent level of retail competition as the accessibility value illustrated as 17.33 to 28.94. Also, 3F floor to 8F floor indicated accessibility values between 10.84 and 49.2 except the core area which referred to Zone B of each floor from 3 to 8F floor. These areas illustrate higher retail competition than the rest of the areas and exclusively high accessibility level at 4F and 5F floor from 12:00 until 16:00 which indicated the value during 108.24–134.63.

For the evening rush hour, this peak is a little longer than morning rush hour and is statistically interpreted from 18:00 to 20:00. The retail competition factors accessibility of evening rush hour demonstrates slightly lower accessibility value than the morning rush hour. This time period performs as the combination of the high level area of morning rush hour and non-rush hour as the highest accessibility displays at the 1F floor around the central area, for which the accessibility value is indicated as 398.74 at zone E follow by zone B and D as demonstrated as 336.18 and 284.69 respectively. Besides, the core areas of 3F floor until 8F floor perform as more significant retail competition levels than other zones which share the same floor.

To demonstrate further characteristics of retail competition, Fig. 9 displays the average accessibility of retail opportunity throughout Hakata station at each time of the day, classified by type of retail including a convenience store, souvenir shop and

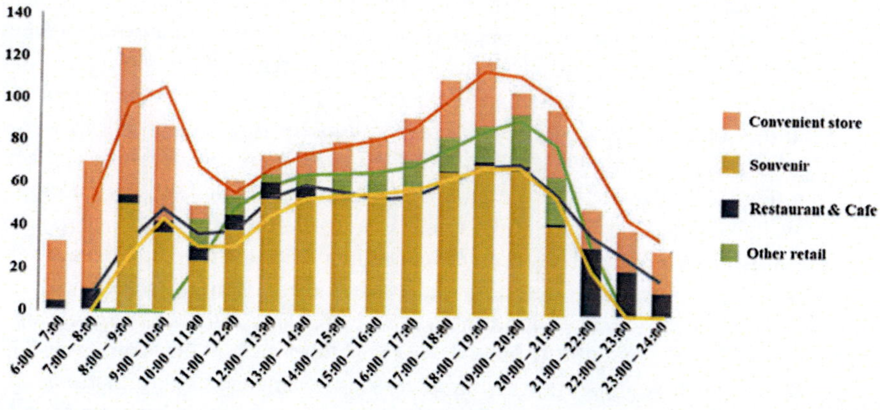

Fig. 9 Average accessibility of retail opportunity tracts by hour of the day

stand, restaurant and café, and other retailers. Results suggest that the convenience store has the highest competition factor accessibility among retailers in Hakata station due to its outstanding performance of accessibility. It shows that morning rush hour indicates the highest average accessibility as 124 at 8:00–9:00 then suddenly dropped to 86.44 and 51.74 at 9:00–10:00 and 10:00–11:00 respectively and then gradually rose to reach the second-highest at 119.8 in the evening rush hour at 18:00–19:00. Souvenir and food services have a similar trend of accessibility throughout the day. Accessibility stays between approximately 40–75 during 8:00–21:00 which indicates the highest average accessibility value as 70.17 for Souvenir and 72.27 for the restaurant and café activity.

5 Conclusions

This paper introduces an assessment methods Spatio-temporal pattern of retail potential accessibility that aimed to determine the change of competitiveness of each retail node in Hakata station. Each type of retailer service has different times and user volumes. The space–time of retail accessibility is associated with two aspects: travel patterns and operating hours of retail services. First, travel patterns referring to the concentration of passengers or users of the station that regulates the different amounts of potential demand in each particular time of the day. Second is operating hours of retail activities in the station that refers to the availability of retail services in the transport hub site. By these means, it can be concluded that there are three Spatio-temporal patterns of retail competition in Hakata station: morning rush hour, non-rush hour, and evening rush hour.

The characteristics of morning rush hour shows the strong desire in the access direction due to the time-limited and the high volume of passenger in the station which made "On-the-go foods and services" which are mostly found in convenience stores and drug stores had the most significant retail competitiveness in this time period. The time after the short-peak time of morning rush hour is classified as non rush hour, referring to "retail-oriented occupation" due to the fewer determined (time-limited) transit users so that the station's users in this time tend to have more time for enjoying retail services in the station as well as the operating hour of department stores built in the station encouraged more retail options for their customers as well. The last pattern is the evening rush hour which is the prime time of commercial activity due to both advantages from the number of transit users and the operating hour of all available retail services. Although it is a prime time for retailers, its competition value of the evening rush hour is lower than the morning peak because the get-off work times are more varied and relaxed than the morning time period.

In addition, the retail opportunities in the department store buildings significantly increased the accessibility level during the afternoon and the evening rush hour time. Although the competition factors accessibility measures the relationship between potential demand and supply of the individual node that represents any location upon the interior of the Hakata station and it is completely dependent to differences in operating times of each retail activity, but the evening rush hour is the only time that able to represent completed accessibility competitiveness due to every retail type being operated during this time period.

6 Limitations

The limitations of this study are presented in three aspects as the equal competitiveness assessment among retail activities, the data collection methods of potential demand, and data of supply as explained in the following point;

- Equal competitiveness assessment; regarding the methods presented in this paper, it is difficult to compare the accessibility competitiveness between restaurant area and convenience store in the morning rush hour as varying operating hours of retail services in Hakata station obstructs equal competitiveness assessment.
- Demand; the precision of potential demand estimation using centrality analysis seems to be more effective using the actual pedestrian counting with more frequent data collection in both time intervals of collection and the number of collection spots. Also, human detection is one of the most effective methods for potential demand data collection as well.
- Supply; According to the effects of using retail plot area for identifying the supply, it may be better to use the selling record and any further information that could better refer to retail competition than location-based accessibility.

References

Bach L (1981) The problem of aggregation and distance for analyses of accessibility and access opportunity in location-allocation models. Environ Plann A 13:955–978

Caschili S, Montis AD (2013) Accessibility and complex network analysis of the U.S. commuting system. Cities 30:4–17

Cerdá A (2009) Accessibility: a performance measure for land-use and transportation planning in the Montréal Metropolitan Region. School of Urban Planning, McGill University, SL

Dalvi MQ, Martin KM (1976) The measurement of accessibility: some preliminary results. Transportation 5:17–42

Dempsey N et al (2010) Elements of urban form. In: Dimensions of the sustainable city 2. Springer, Dordrecht, pp 21–52

Dolega L, Pavlis M, Singleton A (2016) Estimating attractiveness, hierarchy and catchment area extents for a national set of retail centre agglomerations. J Retail Consum Serv 28:78–90

Handy SL, Niemeier DA (1997) Measuring accessibility: an exploration of issues and alternatives. Environ Plann A 29:1175–1194

Handy SL, Boarnet MG, Ewing R, Killingsworth R (2002) How the built environment affects physical activity. Am J Prev Med 23(2):64–73

Hansen WG (1959) How accessibility shapes land use. J Am Inst Plann 25(2):73–76

Hillier B (1996) Space is the machine. Space Syntax, London

Hu Y, Downs J (2019) Measuring and visualizing place-based space-time job accessibility. J Transp Geogr 74:278–288
Ingram DR (1971) The concept of accessibility: a search for an operational form. Reg Stud 5:101–107
Institute for Transportation and Development Policy (ITDP) (2017) TOD standard. ITDP, New York
Jiang S, Ferreira JJ, Gonzalez MC (2012) Discovering urban spatial-temporal structure from human activity patterns. UrbComp 12:95–102
Kang C-D (2016) Spatial access to pedestrians and retail sales in Seoul, Korea. Habit Int 57:110–120
Lee K, Lee J, Kwan M-P (2017) Location-based service using ontology-based semantic queries: a study with a focus on indoor activities in a university context. Comput Environ Urban Syst 62:41–52
Porta S et al (2009) Street centrality and densities of retail and services in Bologna, Italy. Plann Des Environ Plann B 36(3):450–465
Sevtsuk A, Mekonnen M, Kalvo R (2016) Urban network analysis: toolbox for ArcGIS 10/10.1/10.2. City Form Lab, Graduate School of Design, Harvard University, Cambridge, MA
Shen Q (1998) Location characteristics of inner-city neighborhoods and employment accessibility of low-wage workers. Plann Des Environ Plann B 25(3):345–365
Tan Y, Kwan M-P, Chai Y (2017) Examining the impacts of ethnicity on space-time behavior: evidence from the city of Xining, China. Cities 64:26–36
Tribby CP, Zandbergen PA (2012) High-resolution Spatio-temporal modeling of public transit accessibility. Appl Geogr 34:345–355
Wee BV, Geurs K, Chorus C (2013) Information, communication, travel behavior and accessibility. J Transp Land Use 6(3):1–16

Land Use and Transport Integration to Promote Pedestrian Accessibility in the Proximity of Mass Transit Stations

Apinya Padon and Pawinee Iamtrakul

Abstract Bangkok and its vicinity encounter traffic problems and air pollution that has been aggravating its situation due to traffic jams and high-density buildings in urban areas. Due to the previous development plan, the area is not aligned with the needs of area development particularly the accessibility problem. Therefore, it is necessary to distinguish the area around the station of 500 m from the centroid of each station. The design factors which were considered in this study include; design, diversity, density, destination and accessibility. The classification of destination in the different stations was performed by using statistical analysis to distinguish the group of stations by using Cluster analysis. The average of all cases was compared among all pairs and analyzed from a Dendrogram. The result was classified into six groups which represented the main activity of the residential area, the mixed-use area, commercial area, recreation area, central area, and the government center. It is an approach to the development around station areas following their context to alleviate traffic congestion with reducing air pollution to achieve sustainable development and promote activities surrounding transit stations.

Keywords Transit Oriented Development: TOD · Land use · Transit stations

1 Introduction

Bangkok and its perimeter are expanding along with the main street network with inconsistent secondary road development. The congestion of traffic volume includes the problem of road safety, and the lack of public transport links, affordable housing and urban infrastructure are critical to urban public problems which Bangkok as continue to occur. The statistics of TomTom, the famous GPS manufacturers, rated the city with the most traffic jams in the world from 390 big cities in 48 countries. It was found that Bangkok represented the first ranking of congestion (BLT Bangkok, 2561) which might be due to the major cities in developed countries such as Tokyo,

A. Padon (✉) · P. Iamtrakul
Faculty of Architecture and Planning, Center of Excellence in Urban Mobility Research and Innovation, Thammasat University, Pathumthani 12121, Thailand
e-mail: thungmay37@gmail.com

W. Weerawat et al. (eds.), *Urban Rail Transit*, Lecture Notes in Mobility,
https://doi.org/10.1007/978-981-15-5979-2_10

London, Singapore, Hong Kong, Seoul and New York, using rail mass transit systems as part of traffic solutions while mitigating environmental problems. The economic growth of the community is stimulated as a result of government policy and the current national Economic and Social Development plan focuses on the development of transportation of people and goods by the rail system. Thailand's development presents lacks development of sufficient public transport such as parking, bus, and transportation centers for different communities to link with existing developments of housing, trade centers, employment areas and advanced transport planning for the future. As a developing country, infrastructure development has always been focused on an increasing of supply side, while less promote on public transportation and almost ignore for non-motorization (Iamtrakul & Zhang 2014).

For the last 15 years, since the BTS rail mass transit system was opened, Thailand was lacked plans to develop appropriate areas around the mass transit stations or Transit-Oriented Development (TOD), the design of the mass transit systems lack holistic planning which covers the development of the city to promote traveling with the public transport system. Traffic transport forecasting (transport modelling) often ignores key usage of land, such as urban forms or urban design factors contributing to all groups of users including disadvantaged people, such as children, women and people. The Non-motorized transport Mode (NMT), such as pedestrians and cyclists, has always been ignored as well as lack of capturing economic valuation, such as reduced consideration of the value of external social impacts. When designing TOD projects, a high-density building block or high-density land use will be employed in its highest utilization (Highly-dense and mixed land use) such as a commercial centre, a government institution or a residential building in the area or around the public transport stations. The density is dimmed when it is developed out of the centre. The traffic radius should be within walking distance approximately up to 800 m from the development centre or station. Rapid development in center core of urban area motivates people from rural migrates to urban for reaching better quality of life (Iamtrakul, Raungratanaamporn, and Klaylee 2018).

The area around the station poses a different context. The large area of Bangkok comprises 5 main areas which are; 1. Central business district, Bangkok city centre (CBD) 2. Urban Area 3. East Outer Ring Area (East Outer Ring Road) 4. West Outer Ring Road (Suburban) and 5. Outer of Bangkok or the perimeter (TerraBKK 2014), to provide specific development that responds to the context of the station area. This study aims to examine the physical characteristics of the station areas and employ data from the landuse to classify different groups of stations according to the features of the mass transit stations, promoting access to the TOD areas. To develop a space around the station needs planning of a spatial based area that responds to the local needs and preserve the unique development of the area to achieve sustainable mobility development.

2 Literature Review

2.1 *Transit Oriented Development: TOD*

Transit-Oriented Development (TOD) is a concept that is Important in creating the right environment around the transit station which Is the focus on system services. Public transportation coupled with areas with high population density and development areas around the station used integrated land to encourage public transportation. The area around the station, has a higher density than other areas, including a mix of residences, employment and shopping. The settlements presented by a variety of access to cities and types of locations which are easy to walking into the station (Shinkle 2000).

The community is mixed within an average pedestrian distance of 2000 ft from the bus stop and the main commercial area. TOD blends residential office shops, open spaces and public transportation in a walking environment, making it convenient for residents and staff to travel by public transport, bike and foot or car. Also, the transport hub should be located in the centre of the neighbourhood, away from the residents of 500 m or 10 min, this central location reflects the importance of transportation in the community and the overall region as in Fig. 1 (Angela 2010).

Peter Cal Thorpe introduced the TOD concept, based on the Pedestrian Pocket in The Next American Metropolis, referring to TOD as an area with mixed use. The total area is approximately 80 ha (about 12 rai), which is developed around the main transit terminal and commercial area. The TOD area is within a 10-min walking radius of approximately 600 m from the transit terminal. The next area, called the Secondary Area, consists of low-density residences, schools, parks, and commercial offices, within a 1-mile (1.6 km) radius, the TOD is a combination of residences,

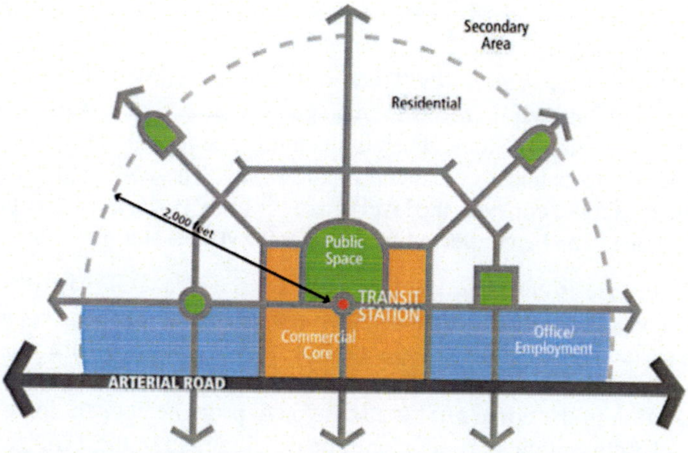

Fig. 1 Transit-oriented development. Joshi et al. (2017)

shops, offices, open spaces and public places. The concepts of both of them are similar (Calthorpe 2011).

The development of the area according to the TOD principle was successful, not just by the development in the area of TOD, but it should be developed into a unique TOD district resulting in a reduction of travel volume. Several passengers can access public transportation and people in other areas may make the trip into the TOD area with private cars rather than using public modes. It is developed according to TOD principles and creating a built environment to create a more TOD place in many areas to succeed by considering 5 dimensions of station area development as follows (Angela 2010):

- *Density to create an area of TOD*: It must have some houses with a large working-age population. There should be several tourist features in the area and they must be within walking distance. Tourist attractions play a role in trip generation to feed the BTS Skytrain because these populations are the numbers of passengers who will enter the station and result in a reduction in private car travel.
- *Diversity*: Space utilization or land utilization in areas around TOD should contain a mixture of land use in a variety of types. There are many styles of housing with different architectural styles that connect with the surrounding areas to provide access to various land use activities.
- *Design of the physical environment of the TOD area*: Facilities in the TOD area could assist commuters with walking trips or bicycle use to reduce travel by private cars and attract more people into the station area.
- *Distance to Transit*: The identification of the public commuters in access to TOD areas which represents the distance between the metro station, housing location and the employment area. To scope with travel behaviour of commuters in access to the city centre, it requires access to the various activities with fast and convenient links, which significantly reduce the time and cost of travel with the public transport system. Thus, the development of the area according to TOD takes into account the distance to the station to meet the preferences of the trip makers appropriately.
- *Destination Accessibility*: The ability to access areas of development according to the TOD principle means to manage the connection of various areas near transit stations such as shops, business centres, residential areas, and areas that attract people to connect conveniently and easily. The ability to access these areas is necessary to consider the level of accessibility of the mass transit system in which the area can be connected to the BTS (Skytrain) station.

To create an environmental dimension in the area, TOD can be applied to promote the development of successful station area planning. The concept also helps to build social interactions between people who live in the city, with a lively atmosphere, then local people will have an opportunity to know each other more. This implies a reduction in the crime rate. It also helps to promote cooperation of the community, which helps public participation. The development of the TOD area is linked to the rail system. However, a different location should be designed to respond to a main activity in the city.

2.2 Node-Place Model

The area of development that utilizes the concept around the mass transit station will focus on an interchange of transportation made and land use, thus it leads to study the differences of the area around the mass transportation. Therefore, the Node-Place Model is used in the classification of TOD by using the factors of transportation (Lyu et al. 2016). Node-Value describes the different types of transportation, and can be divided into factors as follows:

1. The number of passengers using public transport services in the station area.
2. Types of train service in stations: The station area that has various types of train or can be connected between the various railway lines will increase the value of the area with high land prices.
3. Car and bicycle parking lots: Development around the public transport station aims to reduce the number of private cars. If the public transport stations have suitable car and bicycle parking lots, it would be easier to access public transportation.

For the development factors in the land area or Place-Value, it will be mainly explained based on the land use and economy around the public transport station that can be divided into the following factors (Lyu et al. 2016);

1. The population density in the area around the public transport station;
2. The number of employees within the stage of development surrounding the public transport station;
3. Characteristics of land use.

It can be seen that using the Node-Place Model to measure the different types of bus stations will demonstrate the relationship of existing development and land usage development around the mass transit station. Lyu et al. (2016) described that if there is development in the area of transportation or Node to access easier, the land or Place Value will be more diverse (Fig. 2).

Figure 2 is a classification of dimensionals of transit station area development which consists of 5 characteristics by using the Node-Place Model. The following detail explains about its relationships (Lyu et al. 2016):

1. *Balance area* is an area with Node-Value and a moderate level of Place-Value. It can be explained that in this area, both the mass transportation system and the beneficial land usage are appropriate.
2. *Stress area* has a high Node-Value and Place-Value which can be explained that in this area, there is both mass transportation development and intense beneficial land use development which causes land prices in this area to be high, and there is less vacant space due to the significant development.
3. *Dependency area* has a Node-Value and low Place-Value. It can be explained that in this area, the mass transportation system is developed and there is less land use development which is a low-density suburban area.

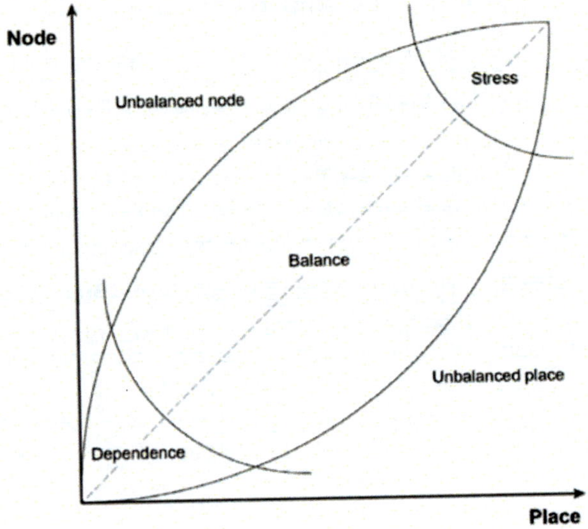

Fig. 2 The node-place model as defined. Lyu et al. (2016)

4. *The Unbalanced Node area* is an area where the Node-value is greater than the Place-Value value described in the area. It develops a high mass transit system but does not have the development of taking advantage of land to be consistent with the development.
5. *Unbalanced Place* is an area with less Node-Value than Place Value. It can be explained that this area has high land use development but the public transport system is not appropriately provided.

The Node-Place Model is used to distinguish the characteristics of each mass transit station, which will use the development of transportation and development of land, and the use of the Node-Value and Place-Value may not be sufficient to classify. Station features or even in areas where the transportation and development of taking advantage of the high level of development, but there is no access or no proper road structure. There are not enough pedestrian or bicycle paths which resulted in being not fully developed as it should be.

3 Study area

The mass transit system was operated by Bangkok Mass Transit System company since 1999. It started its operations with 17 BTS stations from Mochit to On Nut. In 2004, Bangkok mass transit system, Thailand public company Ltd. was received the concession by Bangkok Mass transit system for operated. The Metropolitan Rapid Transit (MRT) with 18 stations from Hua Lamphong to Bang Sue. In this study, the area of 500 m was considered to be the service radius in providing appropriate connections to the proximity area of transit station (Table 1). The two mass transit

Table 1 The service radius of TOD by other researchers

Service radius	Carl (2012)	Ontario (2012)	Mass Transit Administration (1988)	ITDP (2014)	Metropolitan Council (2006)
Distance (m)	600	400–800	450–500	500	500

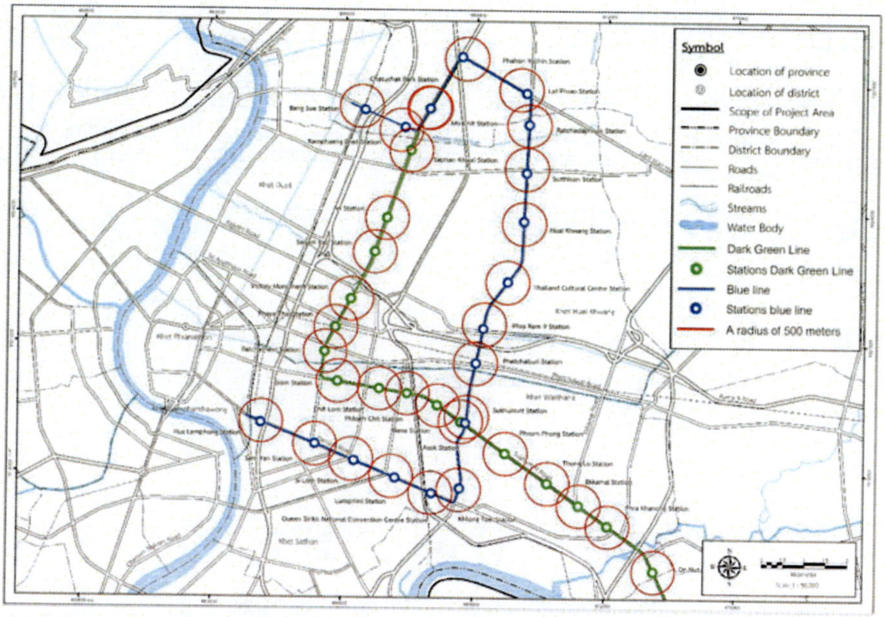

Fig. 3 Study area with radius of 500 m around BTS and MRT stations

systems of BTS and MRT focused on 35 stations which cover the major zones of trip generation in Bangkok. This study will group the typical characters of the feature of the mass transit stations while promoting walkability to the TOD area to reduce road traffic problems in Bangkok (Fig. 3).

4 Methods

In the study of the land use integration and transportation to promote pedestrian access to the mass transit system, the method can be divided into 4 main steps as explained in the following detail (Fig. 4);

1. Literature review to study factors used in the characterization of mass transit stations.

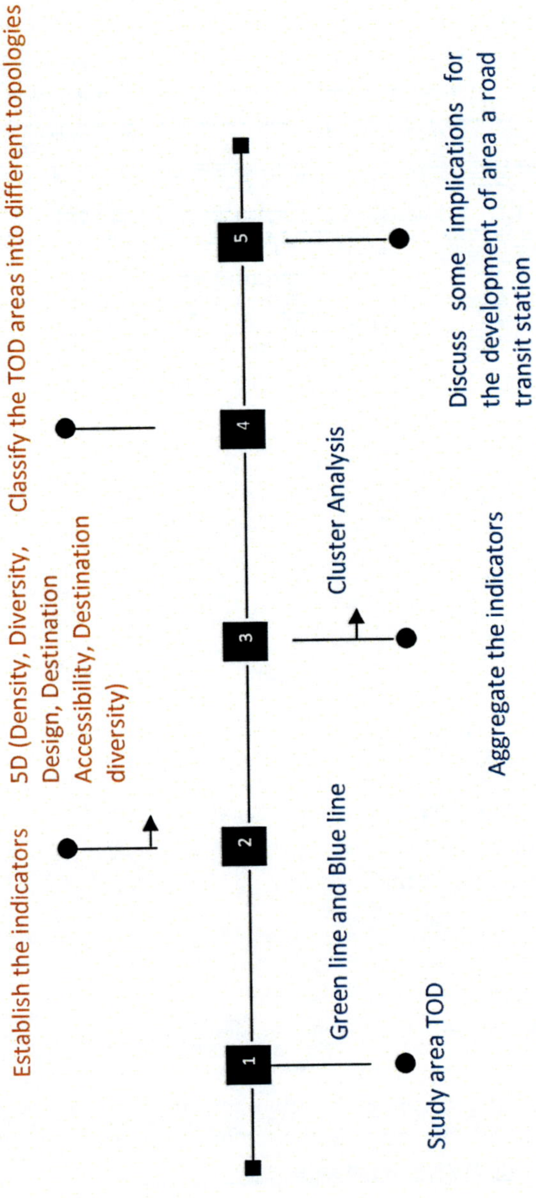

Fig. 4 Methodological framework

2. Gathering data of surrounding areas near the Mass Transit Station under 5 variables of classification: Density, Diversity, Design, Destination Accessibility, Destination diversity. There were 35 stations both BTS and MRT that were chosen for this study.
3. Classifying the characteristics of each station by using the cluster analysis technique to classify the groups of the mass transit stations with statistical analysis data.
4. Summarizing the results from the study and discussion about the classification of the 35 stations (Blue Line and Green Line) to finding the way to improve the surrounding area of TOD.

From the table of indicators, in Table 2, consisting of design diversity, density, destination accessibility and destination diversity which is the design of the physical environment of the area to be livable. There are facilities in the area that help promote travel on foot or by bicycle. The analysis used connected nodes and Ped-shed ratio. The diversity of space usage or land utilization in the area to be set to TOD should contain a mixture of land use in a variety of types, thereby using Land use mix analysis. The density will find the density of building space with the Floor Area Ratio: FAR. The destination accessibility and destination diversity analysis and the ability to access the area of the development area, according to the TOD refer to a management system that connects the major areas of activity of the surrounding area, such as shops and business centers; the residential area and the amount of space that attracts people to connect quickly and easily; the number of the utilities in the area.

5 Results of Analysis

Due to the study of influencing factors that affect TOD typology in classifying the characteristics of the mass transit stations, it was found that the factors can be divided into 5 areas according to the 5D principles of TOD: consisting of design, diversity, density, destination accessibility and destination diversity as shown in Table 2.

5.1 Indicator of analysis

5.1.1 Design

The design variable was analyzed by the connected node ratio and Ped-shed ratio using a tool of GIS to find the intersections of roads and distance from the path of 500 m from the center of the station. It was conducted by using the Network Analyst to examine the connection areas of stations and traffic routes (Figs. 5 and 6).

Table 2 The indicators for evaluating TOD in different stations

Indicator	Measures	Definition	Unit of analysis
Design	Connected node ratio	Number of road connections	Position
	Ped-shed ratio	500 m walking distance by road line	Meter
Diversity	Land use mix	Average value of land use rate at 500 m	Square kilometers
Density	Gross dwelling density	Floor area ratio: FAR	FAR 0.0–0.5 0.5–1.0 1.0–2.0 2.0–3.0 More than 3
Destination accessibility	Facilities and services	1. Count of all the pharmacies and medical facilities 2. Count of all the post offices 3. Count of all the primary and secondary schools 4. Count of all the supermarkets, shops, markets 5. Count of all the police stations 6. Count of all the cinemas, theatres, art galleries, gaming venues	Number of points
Destination diversity	Accessibility	A summary of the presence/absence of the 10 areas: 1. Childcare: Public primary and secondary schools 2. Supermarket 3. Post office 4 Parking spots for the public 5. Police 6. Hospitals 7. Medical: general practitioners, maternity health clinics, community health centers 8. Hotels 9. Convenience stores: marketing 10. Hospitals	

(continued)

Table 2 (continued)

Indicator	Measures	Definition	Unit of analysis
	Transit accessibility bus stops	Public transport point	

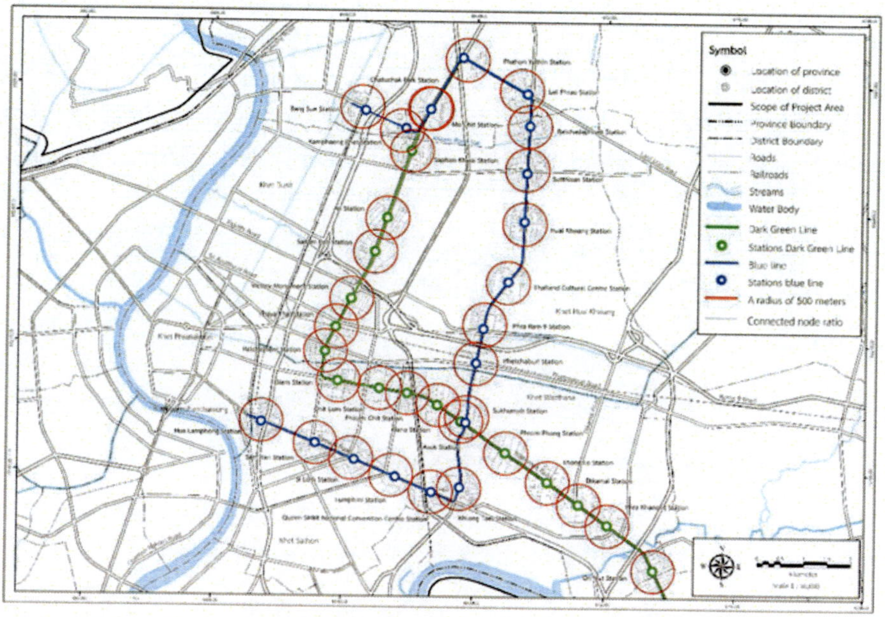

Fig. 5 Connected node ratio (intersections/all intersections)

5.1.2 Diversity

Diversity was analyzed from the average of mixed land use in a 500-m radius around the station. The area with the most mixed land use is Kamphaeng Phet Station, which was 0.79 km^2.

For the least mixed land use, there were 2 stations, which are Thong Lo Station and On Nut Station. The result showed only 0.10 km^2. Apart from that, Thonglor and On Nut have Commercial areas located on the roadside and most are residential space, actually (Fig. 7).

From the characterization of the area (Fig. 8) when combined with land usage, it is found that there is significantly demonstrated different patterns of activity. According to land usage within 500 m around the station area consisting of several types, this analysis attempted to classify the different patterns of residential, commercial, industrial, warehouse, mixed land use, public utilities, educational institutions, religious institutions, government institutions and other public utilities, agriculture, recreation, and others. To develop the area to meet the needs of development and improvement

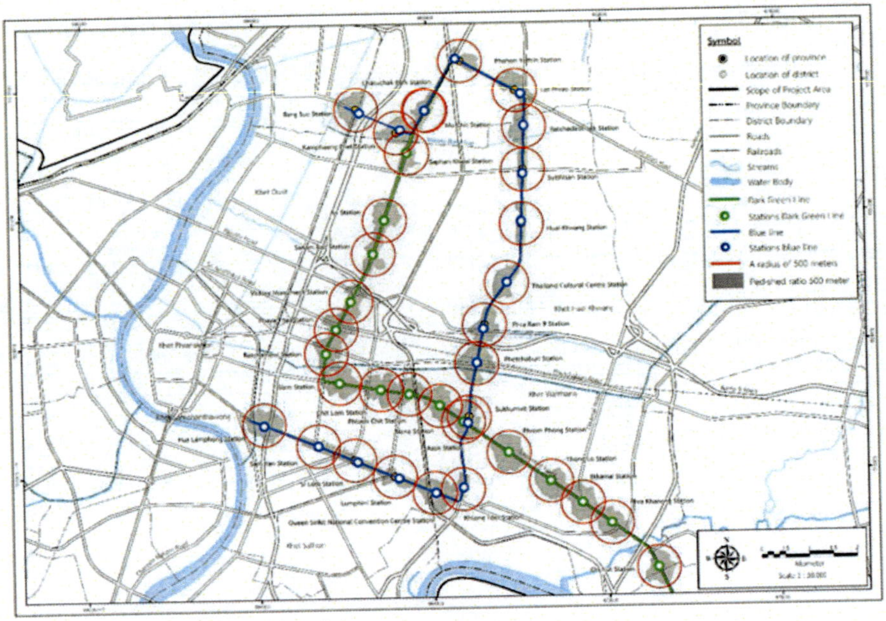

Fig. 6 Pedshed ratio of the walkability of a neighborhood

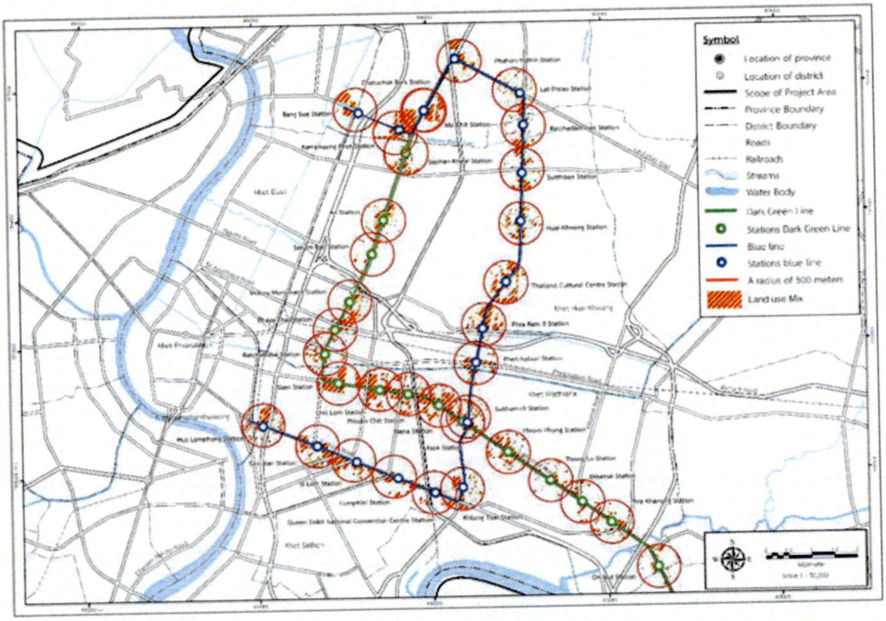

Fig. 7 Mixed use aspect

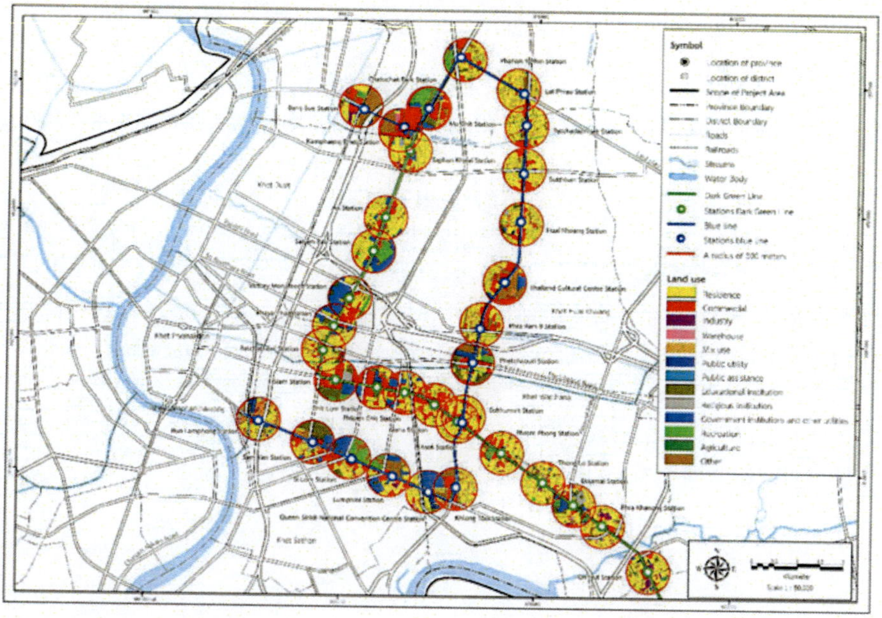

Fig. 8 The land use within 500 m

while maximizing the utilization, it is necessary to look at the main physical context of the area as well.

5.1.3 Density

Density was calculated from the floor area ratio (FAR) around the area of 500 m within the total 35 stations. It revealed that Hua Lamphong was the station with the highest FAR at 1.72 due to the high density of buildings per land plot. A lot of land use is in the major use of commercial buildings, and some are general buildings. It was found that Phetchaburi was the station with the lowest FAR at 0.59 because there was less building density and some areas are generally open space and vacant land (Figs. 9, 10 and 11).

5.1.4 Destination Accessibility

The destination accessibility analysis: the ability to access the development area, according with the TOD refers to a management system connected to the major surrounding activity areas, such as shops, business center, etc.; the residential area, the utilities and the amount of space attracts people to connect faster and easier; By

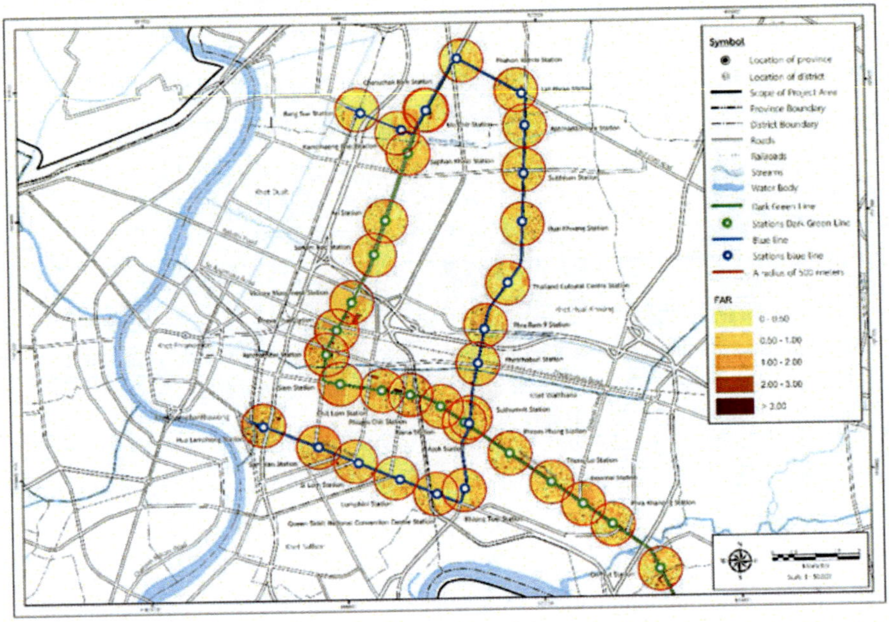

Fig. 9 Floor area ratio: FAR

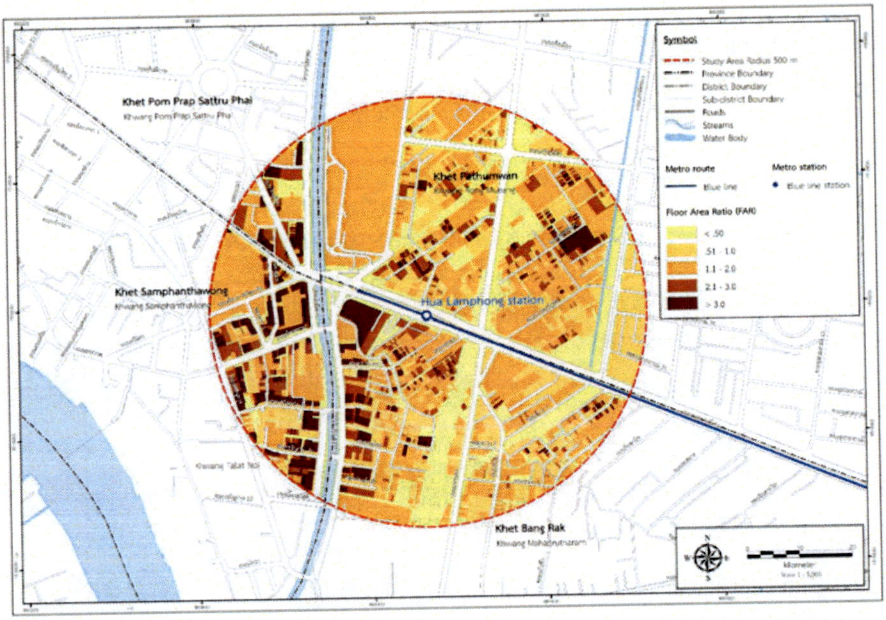

Fig. 10 FAR of Hua Lamphong station

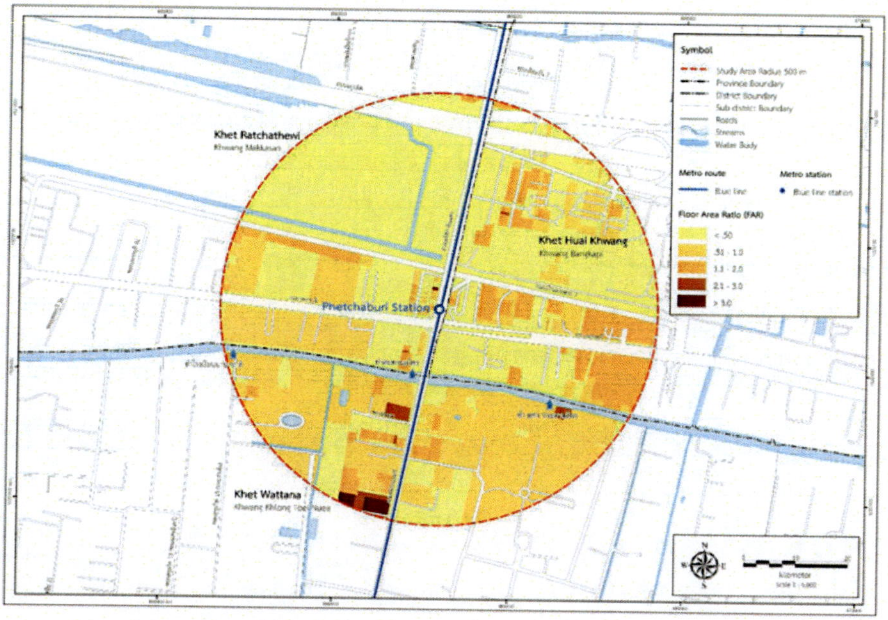

Fig. 11 FAR of Phetchaburi station

analyzing the existing locations which was retrieved by the destination points within the boundaries of each buffer (Fig. 12).

5.1.5 Destination Diversity

Using the calculation from the sum of the presence/absence of all 10 areas in the area, the result of calculation in the area of 500 m radius around the metro station can be identified from the summary of the number of each station. The summary of statistics demonstrated the computation by the number of destinations in each buffer.

5.2 Cluster Analysis

The data analysis of the area development around the mass transit stations which was divided into 5 categories: Density, Diversity, Design, Destination Accessibility, Destination diversity within BTS station area (Green line and Blur line) for a total of 35 stations. By using the statistical calculation to classify the characteristics, the Cluster Analysis technique with Hierarchical Cluster Analysis was employed for the calculation process. The method has potential, not necessary to divide as a group before. It was classified according to the development indicators to differentiate the

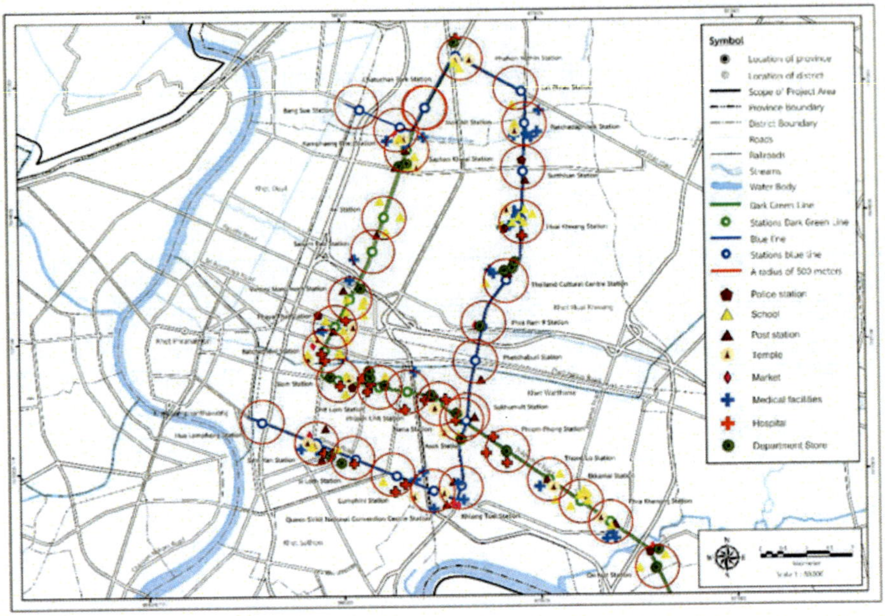

Fig. 12 Destination accessibility and destination diversity

same characteristics to be in the same group by using a between-groups linkage. The calculation of the average distance between all pairs of the case was determined. The characteristics of two BTS stations from the Dendrogram graph can be divided into 6 groups as shown in Figs. 13 and 14.

The Dendrogram graph shows the characteristics of stations which are classified based on BTS (Green Line) from On Nut Station to Mo Chit Station and MRT (Blue Line) which divided into 5 groups of classification as demonstrated by the following (shown in Fig. 15).

- *Group* 1: represents the residential areas consisting of 8 stations: Bang Sue station, Lat Phrao station, Sutthisan station, Cultural Center station, Phetchaburi Station, Ari station, Sanam Pao station and Thonglor station.
- *Group* 2: shows the Mix used areas consisting of 4 stations: Khlong Toei station, Ekamai station, Phra Khanong station and Ratchadaphisek station.
- *Group* 3: demonstrates the commercial areas consisting of 6 stations: Sukhumvit station, Ploenchit station, Nana station, Asoke station, Phrom Phong station and Chidlom station.
- *Group* 4: represents the transportation areas consisting of 4 stations: Onnut station, Hua Lamphong station, Rama 9 station and Siam station.
- *Group* 5: is the recreation area consisting of 5 stations: Chatuchak Park station, Lumpini station, Silom station, Phahonyothin station and Mo Chit station.
- *Group* 6: illustrates the government centre office area consisting of 8 stations: Huay Kwang station, Queen Sirikit National Convention Center station, Sam Yan

Agglomeration Schedule

Stage	Cluster Combined		Coefficients	Stage Cluster First Appears		Next Stage
	Cluster 1	Cluster 2		Cluster 1	Cluster 2	
1	3	5	.478	0	0	8
2	11	26	1.315	0	0	7
3	4	35	1.721	0	0	26
4	13	30	2.766	0	0	13
5	19	28	2.990	0	0	6
6	18	19	3.620	0	5	8
7	11	25	4.546	2	0	11
8	3	18	5.275	1	6	14
9	16	17	6.095	0	0	25
10	27	31	6.353	0	0	15
11	11	24	7.994	7	0	15
12	14	15	8.665	0	0	18
13	13	34	9.155	4	0	21
14	3	10	9.494	8	0	19
15	11	27	9.631	11	10	28
16	7	21	10.803	0	0	33
17	1	9	11.168	0	0	24
18	14	33	11.275	12	0	26
19	3	6	12.138	14	0	20
20	3	8	13.638	19	0	21
21	3	13	14.788	20	13	23
22	12	22	15.119	0	0	25
23	3	29	15.898	21	0	28
24	1	23	16.431	17	0	27
25	12	16	16.853	22	9	31
26	4	14	17.150	3	18	30
27	1	2	18.519	24	0	29
28	3	11	20.113	23	15	29
29	1	3	21.550	27	28	30
30	1	4	23.246	29	26	32
31	12	32	24.396	25	0	32
32	1	12	29.450	30	31	33
33	1	7	29.979	32	16	34
34	1	20	43.224	33	0	0

Fig. 13 Agglomeration schedule

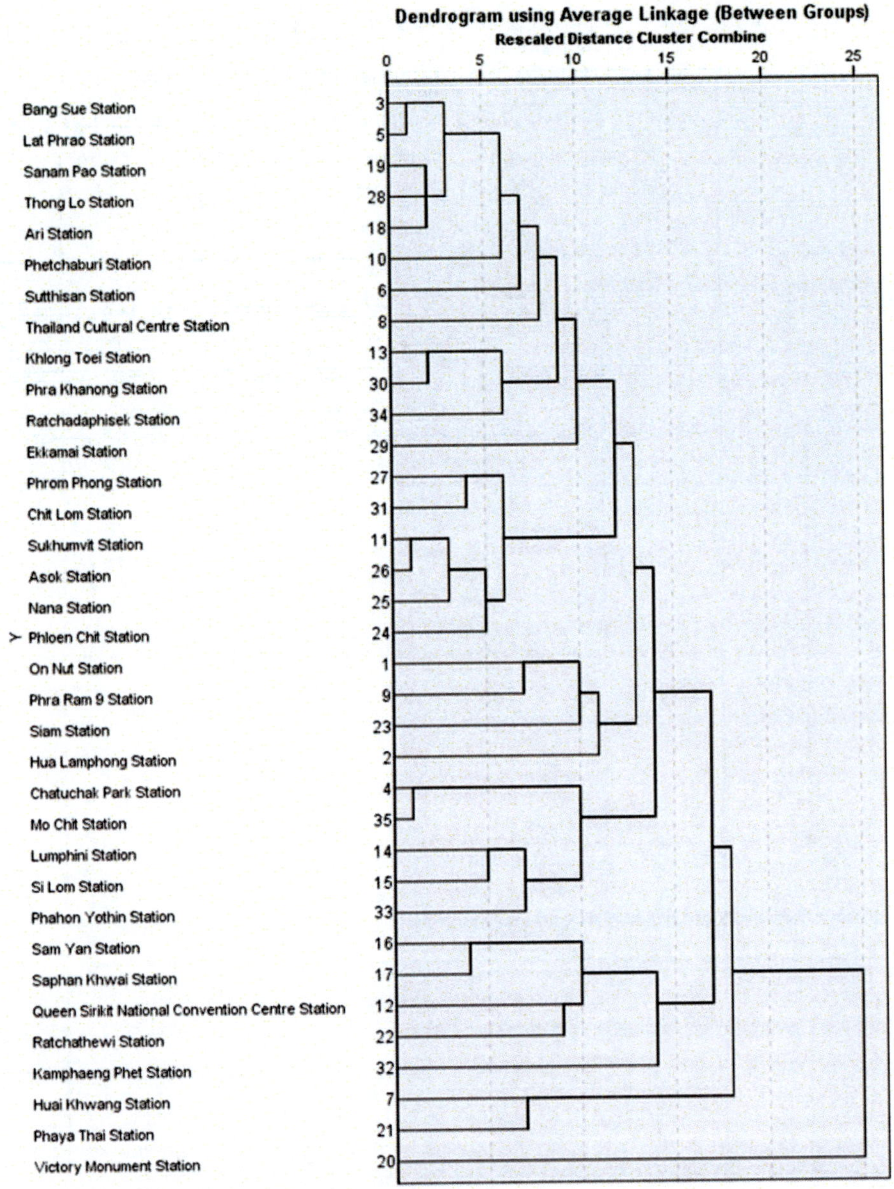

Fig. 14 Graph classification of station

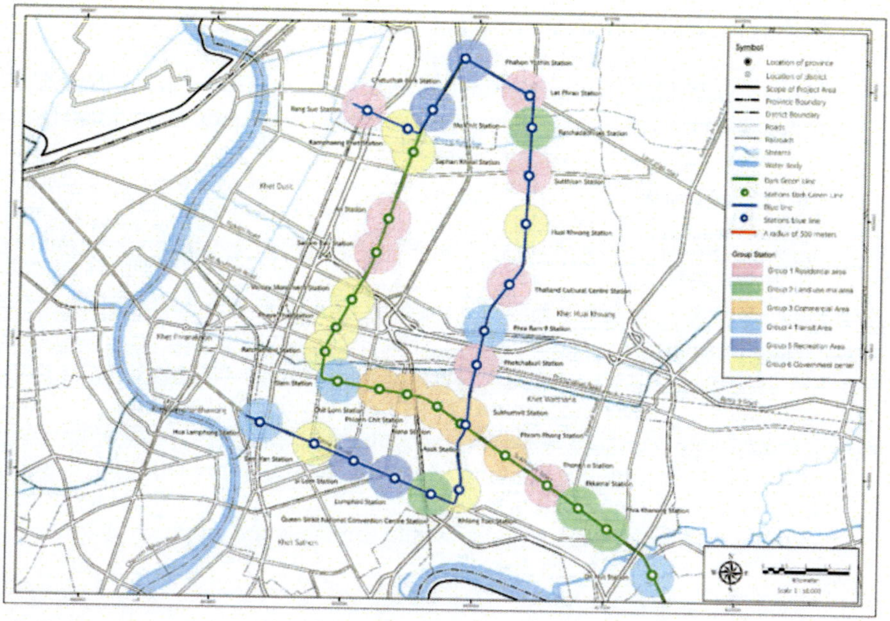

Fig. 15 Feature classification of BTS and MRT stations

station, Saphan Khwai station, Victory Monument station, Phaya Thai station, Ratchathewi station and Kamphaeng Phet station.

Group 1: Residential Area

The residential area has 8 stations: Bang Sue station, Lad Phrao station, Sutthisan station, Cultural Center station, Phetchaburi station, Ari station, Sanam Pao station and Thong Lor station. These areas are located on large and expensive real estate zones. Lad Phrao station, especially is in a dense residential area and nearby there are many educational institutions. Moreover, it is also close to government offices, especially. The Department of the Court of Justice as well. Apart from that, Lad Phrao is located adjacent to Lad Phrao Road which is an important station for connecting from Lad Phrao Road to other places, such as Chok Chai Si, Wang Thonglang and Bang Kapi. The area around Sutthisan Station is residential zones with a variety of accommodation in the heart of the city. The Sutthisan Winitchai Road is the main route that connects Vibhavadi Rangsit Road in Din Daeng area and connects to Soi Lat Phrao 64 (Soi Ketunuti) to various communities that are scattered along are Soi Lat Phrao Road. Huai Khwang, Wang Thonglang, and Thailand Cultural Center station located in the business property, shopping centre, and residential area and near significant arts and cultural sites such as the Thailand Cultural Center. Additionally, it can connect from Huay Kwang station at ground level and allow passengers to enter the Huai Khwang BTS Maintenance Center by connecting both the airport link to Suvarnabhumi Airport at Makkasan Station and Asoke Pier. The surrounding area is a mixed activity, both job sites, important educational institutions and large department stores. Besides that, it also connects as a centre of the station in the direction from Huay Kwang station as ground level to Huai Khwang Electric Railway Maintenance Center. Phetchaburi station connects Suvarnabhumi Airport, Makkasan station and Asoke Pier. The surrounded areas are mix-use space, workplace, educational institutions and department stores. This is the same as Ari Station, Sanam Pao and Thong Lo which represent a mixed-use land area surrounded by government offices, companies, stores, workplaces, of famous street food, restaurants, community malls, etc.

Group 2: Land Use Mix

The Mixed-use areas: Khlong Toei station, Ekamai station, Phra Khanong station and Ratchadaphisek station: the areas have a variety of activities and choices of markets, retail stores, bus stations, schools and educational institutions. With more options of activities, it is worthwhile commuting to the area. It is a must to promote the community to be a part of the economic district. Thus, it can be more self-reliant in times of crisis and manage the land to provide more jobs and activities.

Group 3: Commercial Area

The area of this station is used for the city's infrastructures such as hospitals, universities, with mixed land use. There is a variety of land use which includes the residential type and building density as well. Being in the centre of the city, there are many high buildings and offices in this area. Commercial land use in the initial stages of establishing the city is usually located in the centre, along with important roads, while small shops are often mixed with the residences. An essential factor in selecting a shopping is location is the convenience of travelling by customers. Therefore, shopping is often located at road intersections, like in the past there were large settlements along canals or rivers. In the commercial area, there are not only shops that sell goods or products but also various service centres such as barbers, the beauty zones, hospitals, movies, etc.

Group 4: Transit Area

The result discovered 4 stations that are Onnut station, Hua Lamphong station, Rama 9 station and Siam station. All these places are stations that convince people to travel by public transport and by car because of the potential of connection points to change the route to the sky train station or freeway in the same area. Rama 9 and Siam Station in rush hour is faced with traffic jams sometimes. The transport part which is roads, parking and transportation stations are dispersed differently in both cities and urban areas. The current network functions like veins of the body to nourish the parts of the city. Managing land in different types of building allows users with a good plan to be effective continuously if traffic is not congested. Nevertheless, if the transportation system has failed, activities in the city such as trading operations will be disrupted.

Group 5: Recreation Area

There are 5 stations in this group which are Chatuchak Station, Lumpini Station, Silom Station, Phahon Yothin Station and Mo Chit Station which is the majority of land consisting of the main public parks of the city. It is found that Chatuchak Station and Mo Chit Station are within a radius of 500 m of the Chatuchak Park, Wachirabenchat Park and Queen Sirikit's Garden. Lumpini Station has of Lumpini Park. Therefore, it is an area that is suitable for recreation activities. If this area is developed, the investment will not be higher in creativity. Only promoting policy guidelines is needed to encourage people to use this type of land or activity. Most of them belong to the government, including parks, playgrounds, youth centres, sports stadiums; and some of them are occupied or owned privately, such as amusement parks, swimming pools, etc. Using this type of land, every city should well managed for all users and should be an economical service or free, as it helps to reduce the stress of daily life.

Group 6: Government Center

The central government office area consists of 8 stations, namely Huay Kwang Station, Queen Sirikit National Convention Center Station, Sam Yan Station, Saphan Khwai Station, Victory Monument Station, Phaya Thai Station, Ratchathewi Station and Kamphaeng Phet Station. Mentioning about the area that the Government agencies and government offices are located, some agencies require large land areas such as hospitals and educational institutions. Some agencies require small areas such as police stations, fire stations, etc. Government offices that have to provide services related to the daily life of people are usually located near residential or downtown areas. Currently, some government buildings are expanding to the suburbs and many cities include government offices.

6 Conclusions

From the analysis and classification of 22 Bangkok Transit System Skytrain (BTS) and 18 Metropolitan Rapid Transit stations (MRT), the total of 35 stations from the classification of the different groups considers all 5 factors or 5D of the TOD, consisting of Design, Diversity, Density, Destination accessibility, and Destination diversity. This study applied statistical calculations to classify the characteristics by using Hierarchical Cluster Analysis technique of the cluster analysis. With this method, it is not necessary to know how many groups are divided in the first place. Classifying groups is according to the development indicators around the mass transit station by focusing on the same characteristics to be in the same group by using the between-groups linkage method. It will calculate the average distance of all pairs of cases. The characteristics of the two mass transit lines from the Dendrogram graph can be divided into 6 groups of stations. All these groups of stations can be classified, and searching for appropriate planning and policies that can help area development or any design recommendation for typical TOD in accordance to its context.

Acknowledgements The researcher was supported by the research fund of the graduate development plan from the National Research Council of Thailand (2020).

References

Angela B (2010) Transit-oriented development [Online]. Available at: https://tram.mcgill.ca/Teaching/srp/documents/Brinklow_SRP.pdf. Accessed 30 June 2019

BLT Bangkok (2012) Bangkok is the most traffic champion in Asia [Online]. Available at: https://www.bltbangkok.com/CoverStory/BangkokisthemosttrafficchampioninAsia. Accessed 15 June 2019

Calthorpe Associates (2011) Transit-oriented development design guidelines [Online]. Available at: https://planning.saccounty.net/PlansandProjectsIn-Progress/Documents/General%20Plan%202030/GP%20Elements/TOD%20Guidelines.pdf. Accessed 4 Nov 2019

Carl F (2012) Walkability of transit-oriented development: evaluating the pedestrian environment of Metro Vancouver's regional city centres [Online]. Available at: https://qspace.library.queensu.ca/bitstream/handle/1974/7212/Funk_Carl_R_201205_MPL.pdf?sequence=1&isAllowed=y. Accessed 4 Nov 2019

Iamtrakul P, Zhang J (2014) Measuring pedestrians' satisfaction of urban environment under transit oriented development (TOD): a case study of Bangkok metropolitan, Thailand. Lowland Technol Int 16(2), 125–134

Iamtrakul P, Raungratanaamporn I, and Klaylee J (2018) The impact of urban development on social capital in urban fringe area of Bangkok, Thailand. Lowland Technol Int 20(3):331–340

ITDP (2014) TOD Standard [Online]. Available at: https://www.itdp.in/wp-content/uploads/2014/04/01.TOD-Standard_Final.pdf. Accessed 4 Nov 2019

Joshi R, Yogi J, Kavina P, Vishal D (2017) Transit-oriented development: lessons from Indian experiences [pdf]. Available at: https://cept.ac.in/UserFiles/File/CUE/Working. Accessed 10 July 2019

Lyu G, Bertolini L, Pfeffer K (2016) Developing a TOD typology for Beijing metro station areas. J Transp Geogr 55:40–50

Mass Transit Administration (1988) Baltimore central light rail transit—environmental effects report [Online]. Available at: https://204.52.129.214/repositories/2/archival_objects/23825. Accessed 10 July 2019

Metropolitan Council (2006) Pedestrian-oriented features guide for transit-oriented development [Online]. Available at: https://metrocouncil.org/Communities/Services/Livable-Communities-Grants/Maps,-forms-misc/Metropolitan-Council-TOD-Guide-Pedestrian-Oriented.aspx. Accessed 4 Nov 2019

Ontario (2012) Transit-supportive guidelines [Online]. Available at: http://www.mto.gov.on.ca/english/transit/pdfs/transit-supportive-guidelines.pdf. Accessed 4 Nov 2019

Shinkle D (2000) Transit-oriented development in the states [Online]. Available at: http://www.ncsl.org/documents/transportation/TOD_final.pdf. Accessed 4 Nov 2019

TerraBKK (2014) Details of area division [Online]. Available at: https://www.terrabkk.com/news/10657/Detailsofareadivision. Accessed 15 June 2019

Measuring the Changes of Subway Accessibility Through the Service Area Territories: A Case Study of Fukuoka Subway Network

Somsiri Siewwuttanagul, Takuro Inohae, and Nobuo Mishima

Abstract The expansion of subway networks can stimulate changes of accessibility of surrounding communities. The plan of future a network needs to be evaluated to examine the effect on existing networks and the accessibility of surrounding communities in order to prepare them for the changes of spatial physical dynamics. This research focused on the plan of a new subway station in the Japanese city of Fukuoka. The new station will be located in a prime site of the city and aims to link the old and new CBDs of Fukuoka city (Tenjin station district and Hakata station district). This methodology was introduced for measuring the changes of accessibility in order to indicate how much such plans could potentially benefit surrounding buildings. In this study, 12,237 buildings in the subject area including four major usages of Public Service, Mixed Use, Residential and Commercial were considered in the analysis, which uses a Reach Index to simulate the changes of accessibility between before and after operation of new subway stations. The results can be used to compare how expanding subway accessibility is associated with surrounding buildings. The changing of service area territories indicates the influence that a new subway station has on transportation networks.

Keywords Service area territory · Accessibility · Public transportation · Land use

1 Introduction

The changing of public transport networks will always have an effect on physical urban forms and also the overall dynamic of transport networks in a given city (Miller 2004; Broitman and Koomen 2015; Dubé et al. 2013; Loo et al. 2010), especially in metropolitan areas that contain a high density of population which implies a high number of users of public transportation as well. Therefore, new

S. Siewwuttanagul (✉)
Faculty of Engineering, Mahidol University, Salaya, Thailand
e-mail: jamessomsiri@gmail.com

T. Inohae · N. Mishima
Department of Civil Engineering and Architecture, Saga University, Saga, Japan

W. Weerawat et al. (eds.), *Urban Rail Transit*, Lecture Notes in Mobility,
https://doi.org/10.1007/978-981-15-5979-2_11

construction projects or expansion projects of public transportation networks need to be evaluated in terms of their effect on existing transit networks and the accessibility of surrounding communities in order to prepare for the changes of spatial physical dynamics (Esch et al. 2014; Dempsey et al. 2010) which leads to changes in user's behavior (Ceder et al. 2013; Golias 2002; Guo and Wilson 2011; Hernandez and Monzon 2016). According to the current project that is focused on here, the study aimed to investigate the effects of a new subway station construction which will create a link between two highly populated (or 'centers') of Fukuoka city, Japan. The project proposed to provide better accessibility by linking two main interchange stations, one located in the Tenjin area (including interchanges linking the subway, bus center and Fukuoka city's local train system called Nishitetsu), and the other, located in the city's Hakata area (which hosts the biggest train station in the Kyushu region and an inter-region bus center). The project included a plan to build a new subway station that would link those two city centers in a very popular area of Fukuoka city that includes the largest shopping center and is also close to Nakatsu-Kawabata, which has the highest density of buildings in the city. The resulting changes of completion and subsequent operation of the new facility became a focal point of this research. The comparison of the change of accessibility due to the new subway station is subjected to investigation by the application of the specific methods of this research.

The interpretation of this research was inspired by Golias (2002) who examined the impacts on travelers' behavior and transport modes of choice following the construction of a new subway system. However, this study has applied a different approach for defining the results by using more physically spatial analysis method and using a computer-aided-analysis technique on route analysis (Wu and Chen 2012; Chen et al. 2014; Litman 2006, 2012) in order to determine the impacts of a future public transportation network. The relationship of connecting distance and public transportation service area territories was interpreted by implementing a methodology to investigate the change of a future public transportation network. Also, this framework has been developed to investigate urban development dynamics by considering physical urban configurations that work with the interaction between urban spatial construction and transportation networks (Papa et al. 2013; Papa and Bertolini 2015).

The recent effective tools that are able to assist planners to examine the changes of accessibility between origin and destination were introduced by the contours catchment of Spatial network analysis for multi-modal urban transport systems (SNAMUTS) (Curtis et al. 2012) and Reach index of Urban network analysis (UNA) (Sevtsuk and Mekonnen 2012; Sevtsuk and Kalvo 2015). These tools operate the network measurement that allows planners to investigate the interaction between places and urban network within the assigned distance from one node to another which is measured using the shortest path route (Lee et al. 2017). These indicators support (Fransen et al. 2015) who introduced a function that determines the number of access options at any given time point that could be used to reach a particular destination from each community. It is also related to the use of node-place dynamics for achieving the successful development of a station area, which helps to better understand the complexity of urban systems through gathering spatial information (Bertolini 2008). This method assisted the examination of transit accessibility, with measurements taken from any activities nearby transit nodes (Lowry and Lowry 2014), such as commercial, residential, and mixed-use buildings, as well as public services and places of work including offices. To contribute to forming a new perspective on the expansion of transit networks, we have developed a methodology for determining the change of accessibility of future public transportation networks through the construction of new transport facilities. We chose Fukuoka city in Japan for this case study which involved a real-time project that was predicted to have a high impact regarding the future connection of two urban cores both considered as areas having a very high potential in the city.

2 Methodology

2.1 Overview of Data and Study Area

The procedure proposed for measuring the changes of accessibility on the existing subway's network compared with the future subway's network, which includes one new station linking the two urban cores and has been proposed as the completion of the public transportation network in Fukuoka city, Japan. Fukuoka city already provides modes of public transportation including an urban bus network and a subway network which are both operated by the same company (Nishitetsu). The network covers the area of Fukuoka city as shown in Fig. 1.

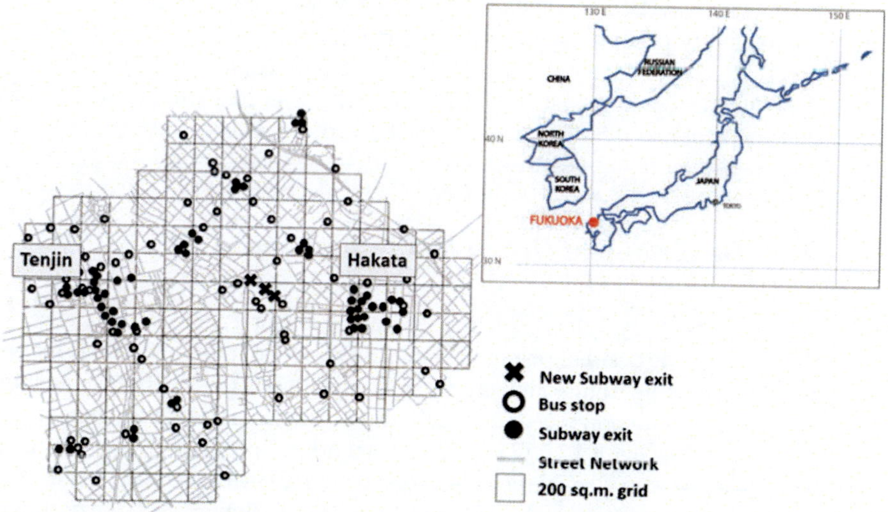

Fig. 1 The study area (Fukuoka city, Japan)

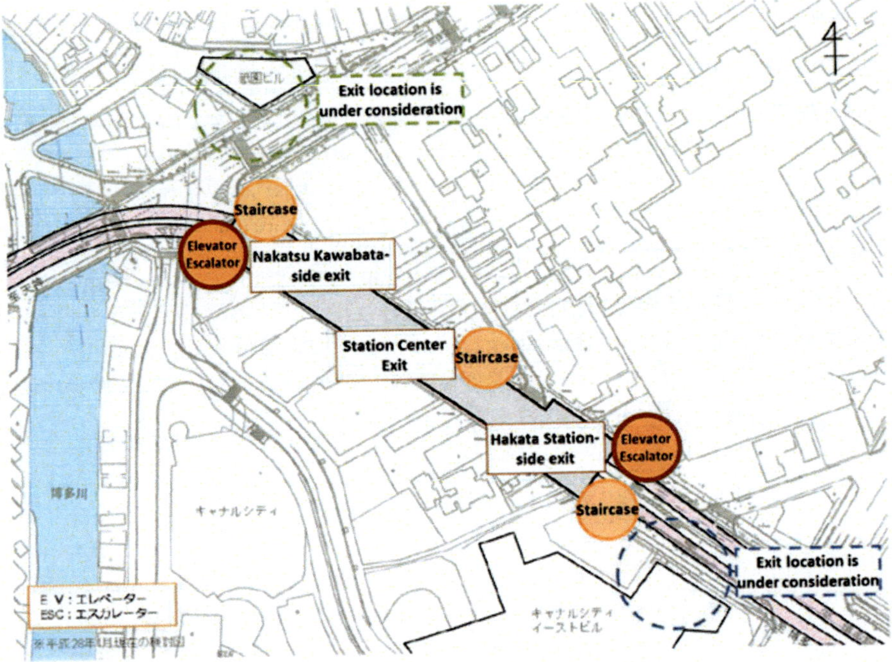

Fig. 2 The master plan of new subway exits (Chuukan station) (Fukuoka City 2016)

According to the new plan, Fukuoka city decided to construct a future subway network which will have one new station called Chuukan Station along the Nanakuma line, which will be extended by 1.4 km and aims to link the old and new CBDs of Fukuoka city (Tenjin station district and Hakata station district). The station was proposed to include five exit gates (see Fig. 2) and would be connected to the largest department store in Fukuoka named Canal City. The area also includes a convention hall and a high-end hotel located in the heart of Fukuoka's main business area.

2.2 *The Setting of Origin and Destination*

The scope of analysis has been divided into (1) the origin point, which uses building data; and (2) the destination point, which uses nodes of public transport for observing the impacts of the future subway network that affects the level of accessibility with regards to both qualitative and quantitative elements. The total subject area covered is 6,240,000 m^2 with 12,237 buildings categorized into four groups of major usage, namely Public Service, Mixed Use, Residential and Commercial; these are used as the origin points to access the destination (nodes). Following this, the destinations are divided into two parts. The first part uses only on the subway's gates to compare the accessibility of the existing and future networks. The second part uses the subway's gates as well as bus stops in the area; this is then analyzed using the same methods as the first part.

2.3 *The Details of Index*

According to a review of accessibility conducted by Geurs and Wee (2004), the relationships between components of accessibility illustrated the association of land-use and transport component which relied on the location and characteristics of origin and destination that could refer to transit nodes and building locations in this study. This study classified infrastructure-based measurements on transport component perspectives in order to investigate the changes of accessibility due to the promoted plan of the new subway station in the prime site of Fukuoka city. To compare the effect of the new subway station, potential measure (Bhat et al. 2002; Hansen 1959; Vickerman 1974) in order to measure the changes of accessibility of surrounding buildings to the subway station. In a direct relationship to (Bhat et al. 2002), this study operated centrality tool assay, using the Reach Index (Eq. 1) in Urban Network Analysis (UNA) toolbox in order to simulate the impacts on accessibility due to the operation of a new subway station in Fukuoka city. The Reach Index was introduced by Sevtsuk and Mekonnen (2012) and Sevtsuk and Kalvo (2015) in their centrality analysis technique of urban network analysis (UNA). The Reach Index measured the number of accessible nodes of public transport within the designated search radius. This study set every building contained within the study area as the Origin point (i) and public transport nodes, such as a subway gate or a bus stop as a Destination point (j) in order to examine the number of nodes that each building has access to, as defined within the study.

$$Reach[i]^r = \sum_{J \in G-\{i\}, d[i,j] \leq r^{W[j]}} \qquad (1)$$

Variations in the length of street segments were difficult to define in terms of the territories of zoning; accordingly, to solve this issue, the study designed a set of cell grids (Nedovic-Budic et al. 2016), each covering an area of 200 m^2, in order to identify the subjected observation areas (see Fig. 1). Also, the variations of the accessible distances and street networks in each subject area that could be affected by the expansion of the public transport service area, as well as the facilitation of the public transport network in Fukuoka city, were all monitored, too.

2.4 The Setting of Boundary

The study measures infrastructure-based accessibility by observing the changes of accessibility of each building along the subway network. The buildings subjected to consideration in this study are all within the public transit node's pedestrian catchment areas (PCAs) according to Jun et al. (2015). Not only the walking distances were considered but also the urban development density, urban form, and interstation length which caused PCAs shorter than 800 m. The study applied Reach index to investigate the buildings that access the subway station with standard catchment area as 800 m (Jun et al. 2015; O'Sullivan and Morrell 1996; Sherwin and Parkhurst 2010) to observe the boundary of the subway station's service area of Fukuoka city. This initial analysis shows that the urban development agglomeration of Fukuoka is clustered within 500 m from the subway station (Fig. 3).

According to initial analysis and the review of Jun et al. (2015), this study proposed the boundary for observing changes of accessibility which were presented 500 m in radius from each related subway station and 200 m in radius from each bus stop which connects the two central business districts (CBDs), namely Tenjin area and Hakata.

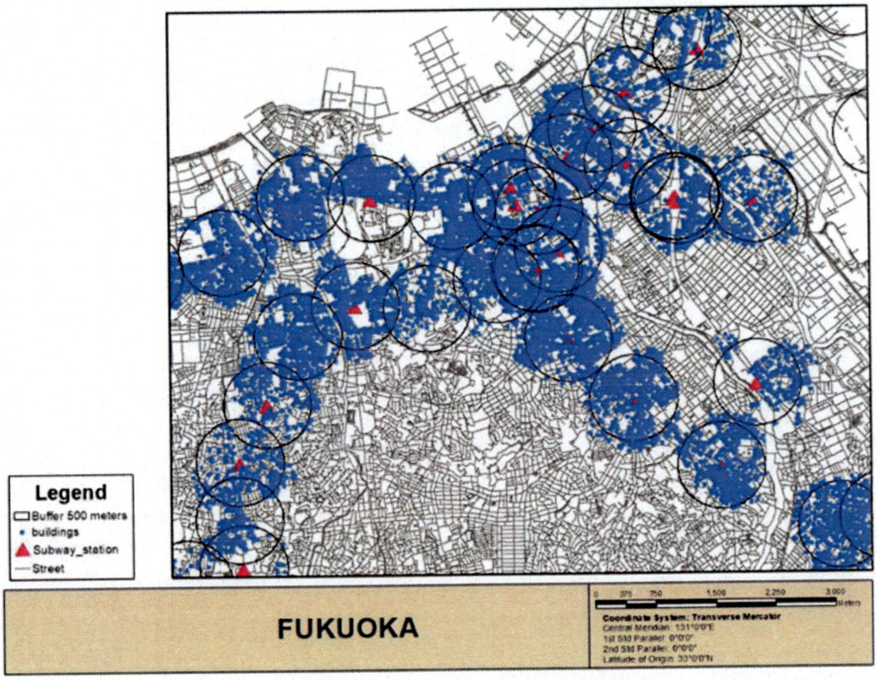

Fig. 3 Building accessibility to subway station within 800 m radius

2.5 *Definition of Terms*

This study has investigated the changes of transit node accessibility on the expansion of a public transport network, comparing the existing network and the future network of Fukuoka, which had one new subway station constructed in a prime area of the city. Then, the study divided the dataset into four parts in order to create a comparison between the subway network and the whole public transportation system with regards to accessibility change. Therefore, this part gives the relevant definitions which have been used in this study as:

SE A Subway's exit in the existing network; the number of nearby buildings able to access subway's exit within the study area (9 stations in total) at a 100, 200, 300, 400 and 500 m radius from each subway exit.

SF A Subway's exit in the future network; the number of nearby buildings able to access subway's exit within the study area (10 stations in total) at a 100, 200, 300, 400 and 500 m radius from each subway gate.

3 The Changes of Accessibility

The results illustrated the spatial accessible area of the existing subway network (SE) and the future subway network (SF) which has one new station with 5 exits. The spatial map shows the new station helps filling the transit desert of the study area and made surrounding buildings more highly accessible for public transport networks, by connecting adjacent public transportation nodes along with the public transport network. Future network accessibility was investigated in terms of the most accessible area in a 100 m radius with 10 access choices to the public transport network and the highest accessible area is within a 500 m radius of which there were indicated 59 access choices to the public transport's nodes.

The impact of the new subway network has increased the accessibility in the area surrounding the subway station. The future subway network (SF) exhibited a shift in access potential, in particular the area which is illustrated in Fig. 4 by two large gaps between the existing and future subway from the access point possibility of 6 choices to 9 choices. The study shows that the access points of 6 and 7 in the existing subway network in a 100 m radius from new subway station have shifted to 7, 8, 9, 10, 11, and 12 choices in the future subway network as indicated by the distribution rates of 39.6%, 5.3%, 23.7%, 7.7%, 16.1% and 7.6% respectively.

In addition, the existing subway's network provided a 3.9% subway accessible score for the total area when considered within a 100 m radius from each subway's gates. The pedestrian catchment area has expanded according to the service radius from subway's station as follows: the accessible area in a 200, 300, and 400 m radius is generated as 13.9%, 29.2%, and 47.4% respectively. In a 500 m access radius, an outstanding accessibility ratio is represented as 66.4% of total area in terms of subway accessibility.

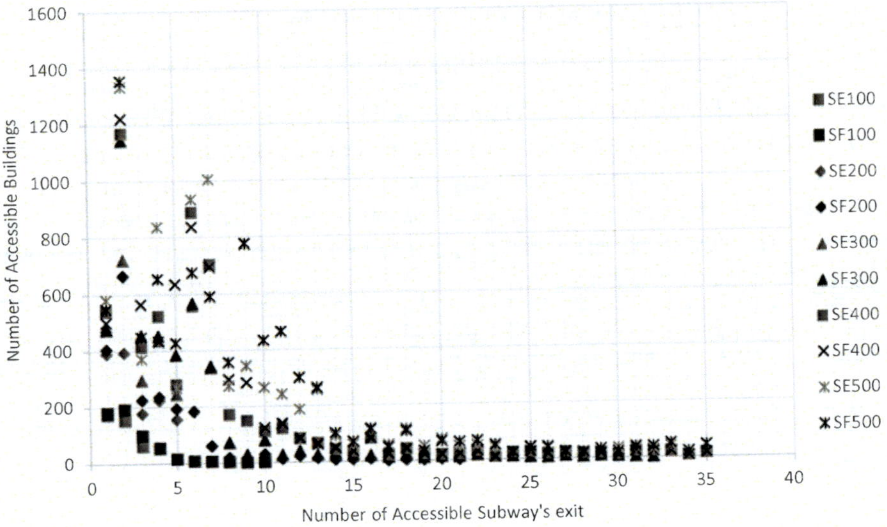

Fig. 4 The changes in accessibility between existing network and future network

Figure 5 illustrates that a 300 m radius access distance shows the highest expansion of subway accessibility as a + 4.29% increase from the existing network. This is followed by a + 3.84% increase at 400 m, a + 2.88% increase at 500 m and a + 2.687% increase at 200 m. Within a 100 m search radius, the new subway station influences the subway network with an increased expansion at only + 0.8% (as shown in Fig. 6).

To consider the distance of the subway's accessibility around the Chuukan station area, Table 1 shows that the new subway network can shorten the subway's exit accessibility distance by 80% in a 100 m radius, from 500 m (in the existing network) to less than 200 m (in the new network) and also shorten the accessibility distance by 25% of buildings which are located within 300 m from new station, since at a 400 m distance from the new station the accessibility was found to be at the same level as existing network due to the intersection of the surrounding subway's service area. In other words, if we consider access distance as the time spent to reach the network measured at a walking speed of 4 km per h, the new network can save 1.5–6 min in terms of the access time within a 300 m area surrounding the new station.

4 Conclusion and Discussion

The overall purpose of the construction of Chuukan station is mainly for the expansion of the subway's network. The difference between the pair of relationships (SE&SF) indicates that the new subway station is affected in terms of its expansion of accessibility, which provided a greater area and more potential for users to engage with

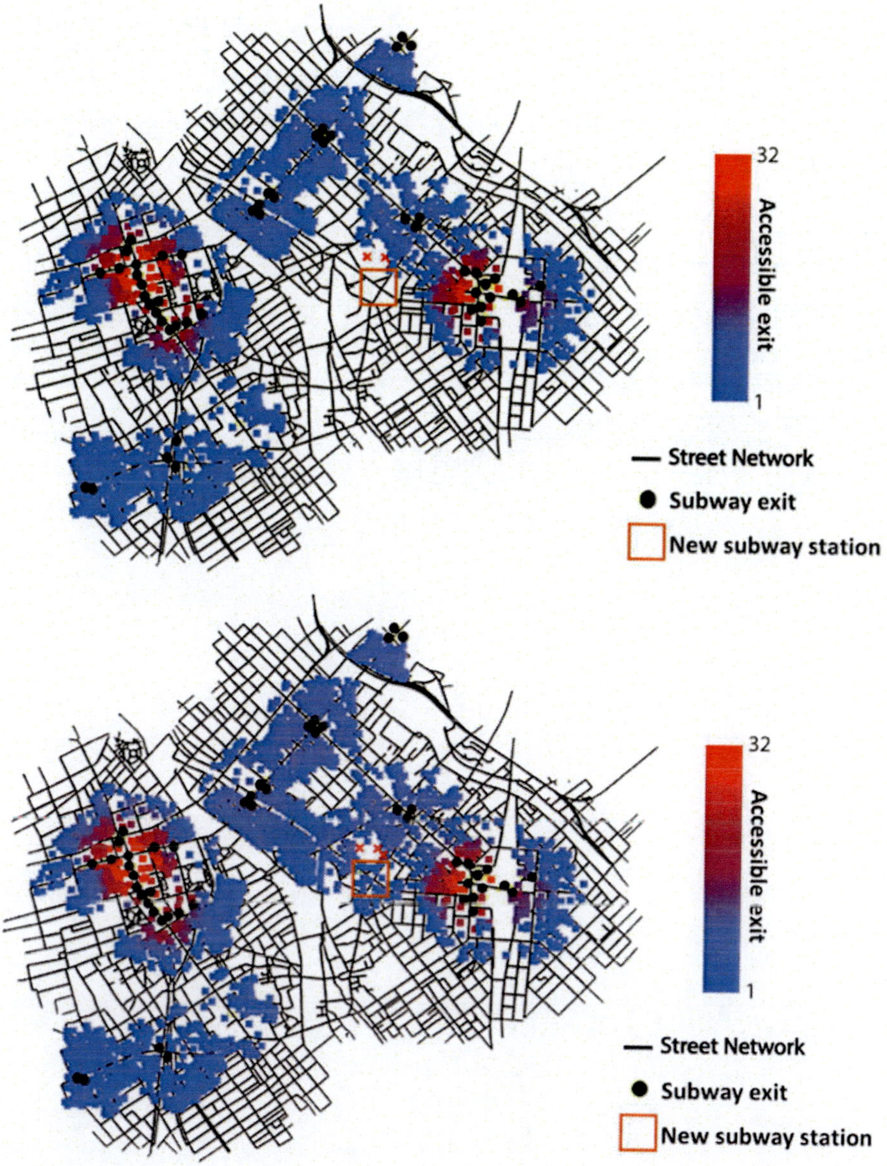

Fig. 5 The subway accessibility maps between the existing subway network (upper) and future network (lower) in a 300 m search radius

Table 1 Average number of accessible subway exits

Distance from each radius (m)	Average number of possibly accessible subway exit									
	Buildings at 100 m radius		Buildings at 200 m radius		Buildings at 300 m radius		Buildings at 400 m radius		Buildings at 500 m radius	
	SE	SF	SE	SF	SE	SF	SE	SF	SE	SF
100	0.0	0.9	0.0	2.6	0.0	4.0	0.0	5.0	1.0	6.0
200	0.0	0.2	0.0	1.5	0.0	2.9	0.2	4.5	2.8	7.8
300	0.0	0.1	0.0	0.2	0.3	1.9	2.9	5.7	5.2	9.7
400	0.0	0.0	0.3	0.4	2.8	3.1	4.2	5.7	6.0	9.0
500	0.2	0.2	1.2	1.2	2.4	2.4	4.4	4.9	8.1	9.7

the subway network. Chuukan station clearly fills the subway's desert of the prime area of Fukuoka city due to the location of the new subway station located between two central business districts (CBDs) of Fukuoka city which were never linked by a railway system. Even though the new subway station improves the accessibility of the subway network, it does not significantly affect the whole public transport network (including impacts on bus stops and subway gates) and also does not affect the existing highest accessible area. This suggests that the area in which the new subway station is located was already providing a full service area within a 200 m radius. Besides, the impacts on travel demand apparently show the same tendency as the impacts on node accessibility, which can be seen to identify the clear territory of the new subway station effects on the urban form of the chosen study area.

This study shown the access distance scale of the analysis for introducing the methods in order to measure the impacts of the construction of new transport facilities. The methodology of this study was used to detect the effective distances to access transport facilities by comparing the changes of subway accessibility conditions within 500 m radius areas which were represented by the average value taken at every possible choice of accessing each particular node of subway's exit. Following this step, the change of the accessible building was investigated and the number of accessible choices which related to the distance that was required to reach any particular node of public transport was interpreted. The methodology used in this study proposes the measurement of accessibility between land-use and transportation. The study selected Reach index which measures the accessibility from the origin location to the destination location along with the street network. Reach index measures the urban network in a direct way of field surveying which measures the actual distance between two subject locations within a given search radius but it has an advantage in saving more time than manually recording data. This tool allows urban planners and related researchers to measure a larger scale within a shorter time. This research discusses the subway catchment that was observed from 100 to 500 m radius with 100 m interval. If we re-scaled the simulation with more frequencies of distance, then we could find the movement of effect in any access distance. Notwithstanding, in this discussion we have not been able to illustrate the change after applying a smaller scale of distances within this analysis; however, we still insist that the results at a 500 m distance will be the same, even though, for example, the results of accessible buildings at a 500 m radius compared between before and after operating the Chuukan station probably show some things that are slightly different due to the resolution of details. However, the actual walking distance and speed are dependent on various factors such as the length of walking distance which could be influenced by such factors as weather, physical condition, or attractiveness of places nearby, etc. Subway's exits take passengers in different directions depending on the street segments they were located in. Regardless of this statement, the analysis result can be seen as being more reliable due to the fact that no matter which exit a user takes to access the subway system, they will eventually reach the same platform and thus are able to choose the direction of transit which is connected inside the underground station where one does not need to take a detour to reach the platform. Moreover, any transfer to other routes or modes of mass transit, rather than using the subway itself, is already considered as part of the same public transport system. For the future research agenda, this research leads to the development of the evaluation methods on spatial accessibility which interacted between public transport network and surrounding land-use in urban area. The development of accessibility indicator use which assists planners assesses the effect of changes of accessibility around relative transit nodes and communities. Moreover, the assessment of the other relative variables that could affect the transit station accessibility such as climate condition, infrastructure's level of service, and the attractiveness of the surrounding attractions could also help to develop its methods according to the proposed methodology of this study.

References

Bertolini L (2008) Station areas as nodes and places in urban networks: an analytical tool and alternative development strategies. In: Bruinsma F, Pels E, Priemus H, Rietveld P, Wee BV (eds) Railway development impacts on urban dynamics. Physica-Verlag, pp 35–58

Bhat C, Handy S, Kockelman K, Mahmassani H, Gopal A, Srour I, Weston L (2002) Development of an urban accessibility index: formulations, aggregation, and application. Center for Transportation Research, The University of Texas at Austin, Austin

Broitman D, Koomen E (2015) Residential density change: densification and urban expansion. Comput Environ Urban Syst 54:32–46

Ceder A, Chowdhury S, Taghipouran N, Olsen J (2013) Modelling public-transport users' behaviour at connection point. Transp Policy 27:112–122

Chen L-C, Wu C-H, Shen T-S, Chou C-C (2014) The application of geometric network models and building information models in geospatial environments for fire-fighting simulations. Comput Environ Urban Syst 45:1–12

Curtis C, Scheurer J, Mellor R (2012) Spatial network analysis for multimodal urban transport systems (SNAMUTS). Accessibility Instruments for Planning Practice. COST Office, pp 49–55

Dempsey N, Brown C, Raman S, Porta S, Jones C, Bramley G, Jenks M (2010) Elements of urban form. In: Jenks M, Jones C (eds) Dimensions of the sustainable city, vol 2. Springer, pp 21–52

Dubé J, Thériault M, Des Rosiers F (2013) Commuter rail accessibility and house values: the case. Transp Res Part A 54:49–66

Esch T, Marconcini M, Marmanis D, Zeidler J, Elsayed S, Metz A, Dech S (2014) Dimensioning urbanization: an advanced procedure for characterizing human settlement properties and patterns using spatial network analysis. Appl Geogr 55:212–228

Fransen K, Neutens T, Farber S, De Maeyer P, Deruyter G, Witlox F (2015) Identifying public transport gaps using time-dependent accessibility levels. J Transp Geogr 48:176–187

Fukuoka City (2016) The plan of high speed rail number 3, 19 Jan 2016. Retrieved from Fukuoka City: https://subway.city.fukuoka.lg.jp/nanakumaline_extension/data/pdf/pdf_h29-01.pdf

Geurs K, Wee BV (2004) Accessibility evaluation of land-use and transport strategies: review and research directions. J Transp Geogr 12:127–140

Golias J (2002) Analysis of traffic corridor impacts from the introduction of the new Athens Metro system. J Transp Geogr 10(2):91–97

Guo Z, Wilson NH (2011) Assessing the cost of transfer inconvenience in public transport systems: a case study of the London underground. Transp Res Part A 45(2):91–104

Hansen W (1959) How accessibility shapes land use. J Am Inst Plann 25(2):73–76

Hernandez S, Monzon A (2016) Key factors for defining an efficient urban transport interchange: users' perceptions. Cities 50:158–167

Jun M-J, Choi K, Jeong J-E, Kwon K-H, Kim H-J (2015) Land use characteristics of subway catchment areas and their influence on subway ridership in Seoul. J Transp Geograph 48:30–40

Lee K, Lee J, Kwan M-P (2017) Location-based service using ontology-based semantic queries: a study with a focus on indoor activities in a university context. Comput Environ Urban Syst 62:41–52

Litman TA (2006) Urban transportation management. In: Munier N (ed) Handbook on urban sustainability. Springer, pp 353–385

Litman TA (2012) Evaluating accessibility for transportation planning: measuring people's ability to reach desired goods and activities. Victoria Transport Policy Institute

Loo BP, Chen C, Chen ET (2010) Rail-based transit-oriented development: lessons from New York City and Hong Kong. Landsc Urban Plann 97(3):202–212

Lowry J, Lowry M (2014) Comparing spatial metrics that quantify urban form. Comput Environ Urban Syst 44:59–67

Miller E (2004) Integrated land use/transport model requirements. In: Hensher D, Button K, Haynes K, Stopher P (eds) Handbook of transport geography and spatial systems. Elsevier Ltd., pp 147–165

Nedovic-Budic Z, Knaap GJ, Shahumyan H, Williams B, Slaev A (2016) Measuring urban form at community scale: case study of Dublin, Ireland. Cities, pp 148–164

O'Sullivan S, Morrell J (1996) Walking distances to and from light-rail transit stations. Transp Res Rec 1538:19–26

Papa E, Bertolini L (2015) Accessibility and transit-oriented development in European metropolitan areas. J Transp Geogr 47:70–83

Papa E, Moccia FD, Angiello G, Inglese P (2013) An accessibility planning tool for network transit oriented development. Planum J Urban 2(27):1–9

Sevtsuk A, Kalvo R (2015) Urban network analysis toolbox for rhinoceros 3D

Sevtsuk A, Mekonnen M (2012) Urban network analysis—a new toolbox for ArcGIS. Rev Int Géomat 2:287–305

Sherwin H, Parkhurst G (2010) The promotion of bicycle access to the rail network as a way of making better use of the existing network and reducing car dependence. Centre for Transport and Society, Department of Planning and Architecture, University of the West of England, Bristol

Vickerman R (1974) Accessibility, attraction, and potential: a review of some concepts and their use in determining mobility. Environ Plann A 6(6):675–691

Wu C-H, Chen L-C (2012) 3D spatial information for fire-fighting search and rescue route analysis. Fire Saf J 48:21–29

Train Approach Information Platform and Service System

Lalita Narupiyakul and Waressara Weerawat

Abstract Important information for passengers using mass transportation in a metropolitan city is the public transport schedule. People can plan their trips and estimate the travel time to their work place or to meeting locations. Currently, people in Bangkok can know the train arrival time when they reach the train platform. This is not convenient for passengers. The train platforms are crowded since passengers do not know the train arrival time of each platform. The objectives of this research are to (i) design the architecture to develop the system for exchanging information between the third-party users and the train telegram system, (ii) design the data standard for exchanging the train approach information among train telegram systems, (iii) implement the system as API web services that can serve third-party users to gain information about trains approaching each platform in real time.

Keywords Train approach services · Train approach API · Data exchange · Arrival time · Departure time · Train telegram

1 Introduction

To commute in Bangkok by the metro train system, a passenger must estimate or plan the time or duration on their own. They need to consider whether it is a busy time or how long to wait for the next train. This is not convenient for most passengers. With the current technology, each rail operator has their own mobile application. The information from these applications cannot be shares. People cannot search, estimate

L. Narupiyakul (✉)
Faculty of Engineering, Mahidol University, 25/25 Salaya, Phutthamonthon, Nakhon Pathom 73170, Thailand
e-mail: lalita.nar@mahidol.ac.th

W. Weerawat
Faculty of Engineering, Mahidol University, 25/25 Salaya, Phutthamonthon, Nakhon Pathom 73170, Thailand

L. Narupiyakul
Integrative Computational BioScience Center, Mahidol University, 999 Salaya, Phutthamonthon, Nakhon Pathom 73170, Thailand

W. Weerawat et al. (eds.), *Urban Rail Transit*, Lecture Notes in Mobility,
https://doi.org/10.1007/978-981-15-5979-2_12

times or calculate fees when they have to transit to different routes. However, some mobile applications such as BKK Rail (2019) can direct the passenger and provide information about transit between routes, estimate the traveling time and calculate the total fee for a trip. Currently, all mobile applications from any rail operators provide only an approximate time schedule and no application can provide real-time data for the train approach time for each platform. This paper describes the implement of a system that the train operators can use to share data with the same schema or the same data model via web services. The idea of this work is to have a data service center that can gather the train approach information from different rail operators, manage these data and provide the data to third-party users/applications, such as organization units, or developers, as a service. The advantage of this system is to help the rail operators provide train approach information and to contribute a set of API services for any third- party users or developers so they can develop new knowledge or new applications which can contribute directly to the rail passengers.

In this paper, the introduction is described in Sect. 1 and reviewing related works are explained in Sect. 2. Section 3 introduces the system architecture, data exchanging model, and the system design and development. The experimental results are reported in Sect. 4 while the conclusions are summarized in Sect. 5.

2 Backgrounds

Presently, rail transportation is the main public transportation in many countries. In some countries, companies or the government have open data policies and regulations on how to provide information in a standard format. Examples of data formats are CSV, KML, SHP, JSON, XML, GTFS and so on.

In Singapore, the MRT rail transportation data center provides a set of offline data such as the train station codes and their location (Govtech 2019). The data center provides static data about the train transportation information represented as a CSV file for data, SHP file for map, and KML file for location.

For public transportation in each state of Canada, the government also has open data policies. The rail transportation information in each state is published on the government's website (Open Government Canada 2019). Examples of rail transportation are lists of station names in English and French and their codes, the express and regular train schedules, and the location of train stations. This information is provided in JSON, CSV, XLS and KML file formats.

To review real-time data on the mobile applications in the different countries, most applications can show the information of map and train schedules. In many countries, there are mobile applications that can provide real-time data such as arrival time, departure time and the current train's position. For example, the MTR application in Hong Kong (MTR Cooperation Limited 2019) can show the current train's positions and also show the real-time arrival time at each platform for each station. This real-time data can assist passengers to calculate the time to reach the platform.

In addition to General Transit Feed Specification (GTFS), the GTFS Realtime extension (GTFS-realtime) (GTFS 2019) which is the feed specification that can be used to provide userd with transit data updates in real time. GTFS-realtime provides up-to-date information about current arrival and departure times for the passengers. The scope of GTFS-realtime information includes (i) trip updates i.e. delays, cancellations and changed routes, (ii) service alerts such as unforeseen events affecting a station and (iii) vehicle positions—information about the vehicles including location and congestion level. One example of the research has compared the GTFS static scheme and GTFS-realtime scheme. The GTFS static data has been used and implemented in many applications and the uses of GTFS are reported in Antrim and Barbeau (2013). It can be used to report station stop locations with high accuracy, but the travel time must be based on the schedule instead of the actual time. On the other hand, the GTFS-realtime data can be used to report the actual time of approach at each station (Yuan and Frey 2018). In conclusion, the GTFS-realtime data may need to be combined with other data sources to complete a real time train approach information system.

For an alternative data standard, RailML (2019) is a standard interface for railway data exchange in XML format. The RAILML contains 4 main data areas including (i) timetable, (ii) infrastructure for the description of tracks and signaling equipment, rolling stock for the description of vehicles and (iv) interlocking for the description of signaling routes (Nash et.al. 2004). Using the XML format, the railway data can be exchanged based on the data exchanging platform as a web service provider. However, our train approach information which receives the data from the train telegram system can be partially implemented following the RAILML data standard. There is some real-time information that is defined by the ARL system.

3 Train Approach Information Service System

In this section, detail about the design of train approach service system is explained based on exchange of data between servers. The next subsection introduces a data model representing the train approach information for each train at each platform from each station. In our system, the information is requested from the server to track the arrival time at each platform. In the last subsection, the system processes are described according to 2 main modules: API service management (ASM) module and API services module.

3.1 Design of Train Approach Information Platform

This section is composed of 2 main subsections: (i) system overview and (ii) system architecture.

3.1.1 System Overview

To exchange train approach information between the rail operators and the client's application, the most important concern is about keeping the security standard of the rail operating system without interfered from public networks or unknown networks. Therefore, if the train telegram system provides the train approach information to the client, it is necessary that this information must be transferred to a middleware system before sending the information to the third-party users/applications.

The middleware in this system is the API Service Management (ASM) server that was two main features: API service management system and API services. This ASM server works between the ACLARE server from Airport Rail Link (ARL) which is one of Thai rail operators and the client applications accessed from the public network. The ACLARE is named for the server that has been used for data exchanging under the collaboration between the ARL and the Cluster of Logistics and Rail Engineering, Mahidol University. The ASM server is authorized to pull the data from the ACLARE server and the data exchanging between theses server can be secured by using SSL technology and also using firewall filters only the authenticated IP address to access this ACLARE server. The API service management system can assign the privilege of data access to the users, generate a token or API key for a service request, and set the authentication for user and API key to access the train approach data according to a period of time. The ASM system is a service provider that can return the information according to the third-party user's request. The API services allow only authenticated applications to request the information by verifying the API key before returning the information.

The ACLARE server is located on the ARL network that allows only the ASM middleware by filtering only a specific IP address to request the train approach information in the form of a log file. The ACLARE server can provide the information about the train approach information, such as the arrival times of three consecutive trains for each platform at each station, train direction information and station codes.

The last part is the third-party application or client application that allows the API key's owner to request the information for their own application from the API services. Connecting to the API service in the ASM serve is very convenient in terms of managing the users who can request the data, and is also made secure by using the https with SSL which are security standards for the application level and for data interchanging. The overview of this system is illustrated in Fig. 1.

3.1.2 System Architecture

After overviewing the system, it is designed based on how the data can be exchanged between the rail operating system and the third-party applications. The system architecture is composed of 3 main parts: ACLARE server, ASM server and the third-party applications as illustrated in Fig. 2. In the first part, the ACLARE server contains two private API services that can retrieve the train approach information contained in the trace.log file from the train telegram system in the private network of the ARL

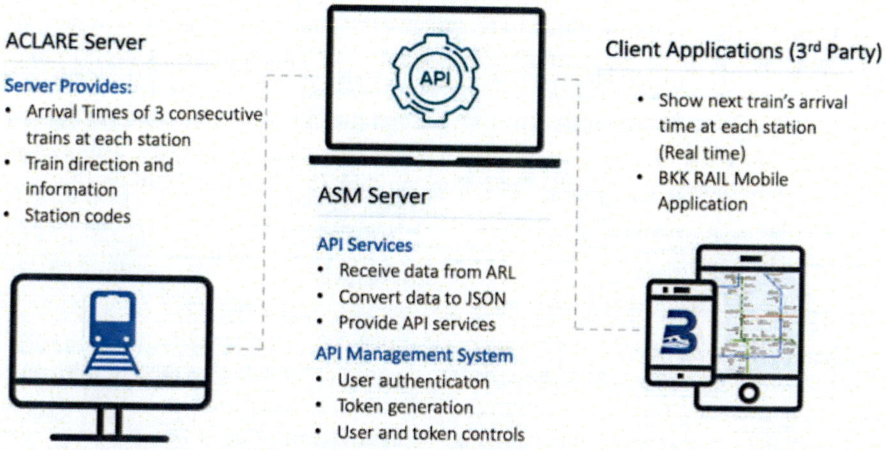

Fig. 1 Overview of train approach information system

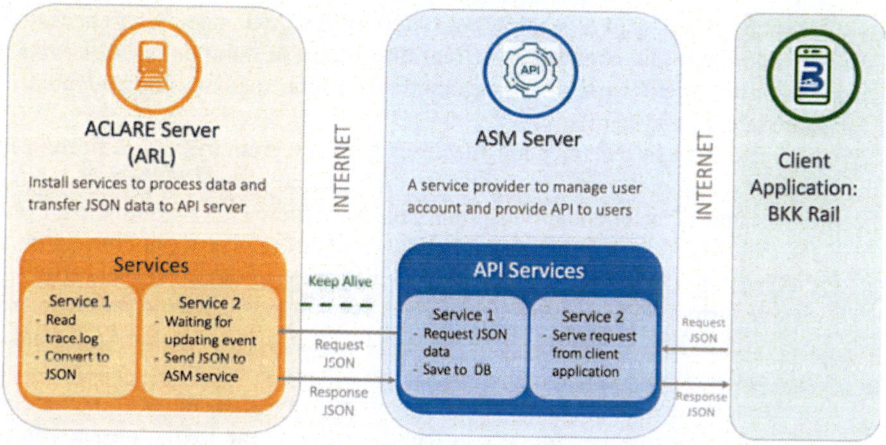

Fig. 2 System architecture

system. This conversion service converts the train approach data from trace.log file to a JavaScript Object Notation (JSON) file. Another service monitors the updating event of the trace.log file from the system. When the updating event is sent from the system, this service can trigger the conversion service to convert data to a JSON file and send the JSON file to the ASM API services.

For the ASM server, there are two main services operating on this server. The first service is to request the JSON data and save the data into the database system. Another service can serve a request from the client applications. Any applications that have a valid API key can request the train approach information from the API services. In this system the API key is issued to the registered users verified by the rail operator's staff. In this ASM server, the API service management system is operated to provide the data service to the client applications. The client can receive an API key when the rail operator's staff authenticates the registered users. The staff can define the duration of data using service and also can monitor the service usage of each client. Between the ASM server and the client applications, the information is transferred by using the security standard with SSL technology. This can guarantee that all data will be encrypted before sending to client applications.

3.2 Data Model

According to the train approach information received from the rail telegram system, there are 19 fields in a total of 44 characters, i.e. sequence number, station code, platform number, arrival time and departure time. These 19 fields were defined by the ARL operator for the purpose of providing the duration of train arrival at each station. The signaling system sends the train location to each sensor and then the rail telegram system calculates the data and interprets the duration of train arrival at each station. Our system is permitted by the ARL operator to access only the trace.log data that containing the real-time report of duration of train arrival at each station. Figure 3 represents an example data in a trace.log file generated from the train telegram system.

Since data in the trace.log file from train telegram system is in the form of a character string, it is not easy to retrieve the information such as time, and station code. Furthermore, the client applications cannot use these data without the knowledge of the data structure defined for this sequence of string. Thus, our conversion services on the ACLARE server must convert the sequence of data string and structure these data in form of JSON format as shown in Fig. 4. Using JSON to represent these data can be easy and compatible with many types of applications that access data via the API and web services.

Discussing the data from other train telegram systems, the same technique can be applied to convert the sequence of the data string to the JSON format. The structure of a character string is required from each train telegram system and all the rail operators can share the same JSON format as shown in Fig. 5. The client applications then can request the information in JSON format via the API from the ASM server.

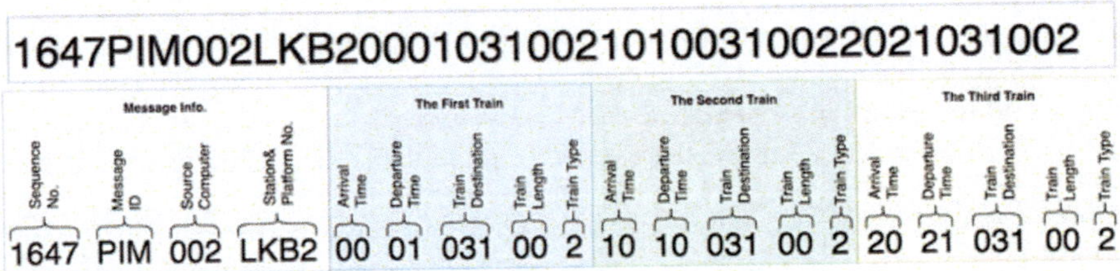

Fig. 3 Example of data and the data structure derived from trace.log file

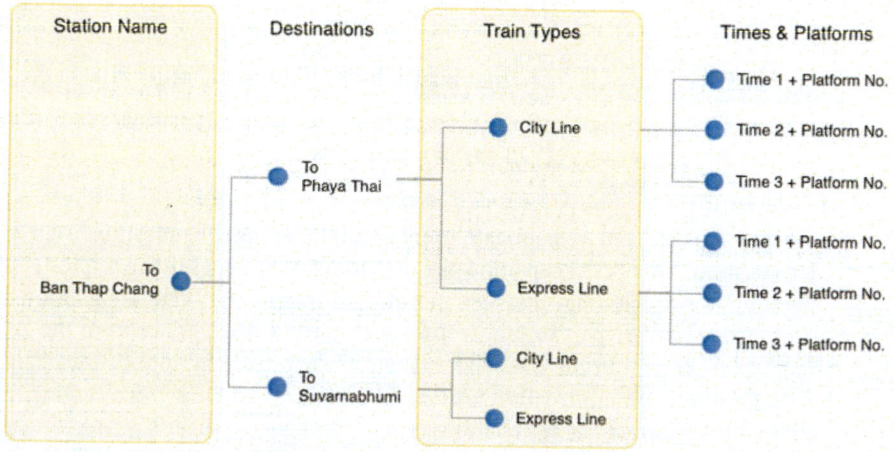

Fig. 4 Example of JSON structure

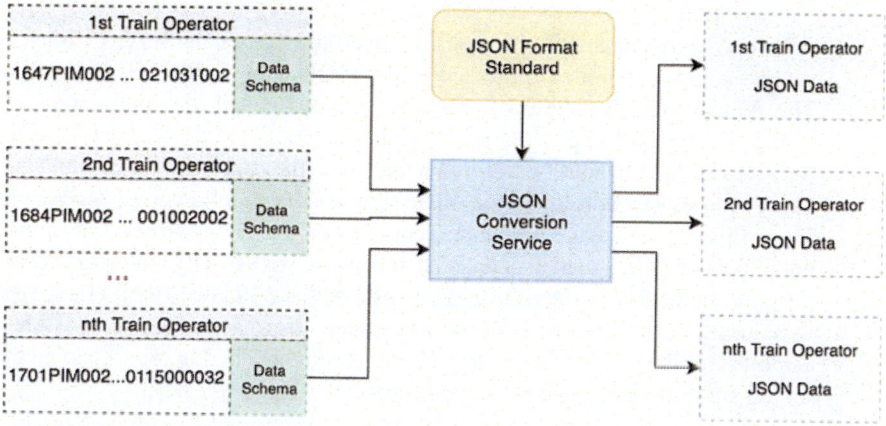

Fig. 5 JSON conversion service

The data standard for train approach information can be assigned to serve different rail operators and the rail operators can share the same set of information. The main advantage of this sharing is that the client application can utilize these data to develop many applications that can assist passengers.

3.3 System Processes

This section is divided into 2 main parts: the API service management processes and API services. In the first part, the API service management is the part that can

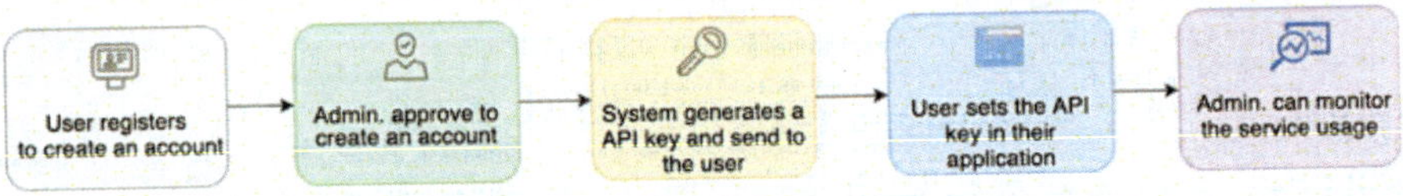

Fig. 6 The system process of API service management

manage the data and authenticate users or applications to request the train approach information. This ASM can allow the user to register for a service request, define the duration of service usage and also monitor and report the client application usage of each user. Another part in this section is to receive JSON data from the ACLARE server and keep it in the database system while the client applications can request the train approach information via this API service and the service can respond with JSON data to the client applications.

3.3.1 API Service Management Processes

API service management processes are used to manage the services for a user and set up the privilege for a user to access the JSON data via the API and web services. The API service management processes are composed of 5 main steps as shown in Fig. 6. The first step is the user registration to create an account for a user. The next step is that the administrator must approve the user registration and activate the user account. The administrator can assign the duration to access the API services. The API service management then generates the API key and sends it to the user by email. This API key can be updated according to the security and user access control purpose. In the fourth step, the user can embed the API key in their client application to request the JSON data of the train approach information. Finally, the ASM server can monitor the service usage of each user account. The log report can be generated by the administrator to summarize the service usage regularly.

3.3.2 API Services

The API services can serve any client applications that request the train approach information and respond with JSON data sent to the client applications. An example of JSON data is represented in Fig. 7. This JSON data contains the current station name and destination in the top level of JSON structure. The sublevel of JSON structure contains the train type, platform number, the arrival time, departure time, timestamp of 3 consecutive approaching trains to each platform at each station. Providing train approach information as a service, can assist the developers to develop various applications that can facilitate many passengers.

```
"_id" : "BTC",
"destination" : {
    "PTH" : {
        "2" : [
            {
                "platform" : "2",
                "time_to_arrival" : 3,
                "time_to_arrival_format" : ISODate("2019-08-13T11:18:30.030+07:00"),
                "time_to_departure" : 4,
                "time_to_departure_format" : ISODate("2019-08-13T11:19:30.030+07:00"),
                "train_length" : 0,
                "train_type" : "2",
                "updated_time" : ISODate("2019-08-13T11:15:30.030+07:00")
            },
            {
                "platform" : "2",
                "time_to_arrival" : 9,
                "time_to_arrival_format" : ISODate("2019-08-13T11:24:30.030+07:00"),
                "time_to_departure" : 10,
                "time_to_departure_format" : ISODate("2019-08-13T11:25:30.030+07:00"),
                "train_length" : 0,
                "train_type" : "2",
                "updated_time" : ISODate("2019-08-13T11:15:30.030+07:00")
            },
            {
                "platform" : "2",
                "time_to_arrival" : 22,
                "time_to_arrival_format" : ISODate("2019-08-13T11:37:30.030+07:00"),
                "time_to_departure" : 22,
                "time_to_departure_format" : ISODate("2019-08-13T11:37:30.030+07:00"),
                "train_length" : 0,
                "train_type" : "2",
                "updated_time" : ISODate("2019-08-13T11:15:30.030+07:00")
            }
```


Ban Thup Chang Station
To Phaya Thai
Updated time + Train duration
Platform No.2
Train type 2 City Line

Fig. 7 Example of the JSON data for the train approach information

4 System Evaluation

According to the ACLARE service system, the train approach information has been sent to the ASM server and collects them in the database. For the ARL system, there are 8 stations including (i) Phaya Thai station, (ii) Ratchaprarop station, (iii) Makkasan station, (iv) Ramkamhaeng station, (v) Hua Mak station, (vi) Ban Thap Chang station, (vii) Lat Krabang station, and (viii) Suvarnnabhumi station. The ACLARE service was monitored for several weeks and the system operated regularly without internet disconnection or missing data. In this experiment, the ACLARE service can filter the destination codes that are necessary for the client applications such as the codes to Phaya Thai station: 024, 031 and the codes to Suvarnabhumi station: 009, 010. The different codes can represent the destination with the train direction and train track together. For the few weeks of train approach data monitoring, there are totally 18 destination codes that the system can filter. These codes are defined by the ARL system and they have been used for the internal operation. Four of them are filtered and kept in the database. These four codes can be used to inform the client applications about the train approach information such as arrival time and departure time. The other codes were used for the train telegram system and are not included in this project.

Another system evaluation is to test the connection between a client application and the API services. In this testing, the BKK Rail mobile application was implemented to connect the API service and integrated with the BKK Rail user interface to

Table 1 Comparing the departure time between the schedule time and the time derived from API services

Station	Weekend (14:00–16:00): Phaya Thai to Suvarnabhumi		
	No. of timeslot	No. of Train arrivals on schedule within 1 min (times)	No. of Train arrivals within 2 min (times)
1. Phaya Thai	12	11	1
2. Ratchaprarop	12	11	1
3. Makkasan	12	11	1
4. Ramkhamhaeng	12	11	1
5. Hua Mak	12	11	1
6. Ban Thap Chang	12	11	1
7. Lat Krabang	12	11	1
8. Suvarnabhumi	12	10	2
	96	87 (~90%)	9

inform the passengers of the departure time of 3 consecutive trains to each platform at each station. The time is collected from the API services and compared to the time schedule of each station. This test is conducted during the weekend to avoid delays to trains during busy times. The results are reported in Table 1. Approximately 90% of the trains were on schedule (including ±1 min). The reason that our test accepts ±1 min from the schedule due to the BKK Rail application show time in the format of hour:minute as shown in Fig. 8. It did not show seconds on the interface. The application rounds the time up or down before showing the time on the interface. Therefore, it is possible that the time from the API may be different from the scheduled time. In the future, the departure time derived from API services will be compared with the real-time approach time when the train departs from the platform to measure the precision of departure time that will be shown on the client applications.

5 Conclusions

In Bangkok, there were more than 900,000 passengers per day commuting by using rail systems in 2018 reported by Bangkok Expressway and Metro Public Company Limited (BEM). There are three rail operators: BTS, ARL and MRT providing rail transportation services in Bangkok. For each day, each station's platform is crowded with passengers waiting for the next train arrival during the rush hours. The passengers can estimate the approaching time of the train based on the train schedule. One of the problems for Bangkok's passengers is that there is no way to check the real-time of the train approaching each station. The passengers also do not know whether there are the additional trains approaching each platform during the rush hours. The passengers cannot know when the next train will arrive in real-time. Previously, there was

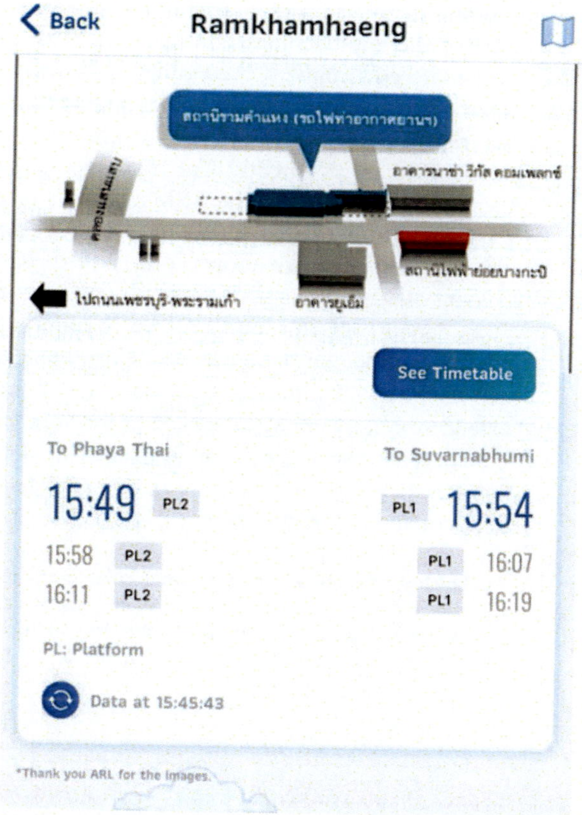

Fig. 8 Departure time shown on the third-party application (BKK Rail)

no web or mobile application that could provide the real time of train approaching to each platform of each station. Therefore, the passengers cannot manage their time to reach the platform. For this paper, the train approach information platform and service system are developed to demonstrate that the train operator can provide the real-time of the train approach information as a service for third-party developers to implement web or mobile applications that can be an advantage to serve passengers.

In this paper, the train approach information platform and service system is separated into two main parts. The first part is the ACLARE services that can retrieve the train approach information from the train telegram system and then convert to the JSON file every time when the data has been updated. This JSON file is sent to the ASM server via the web services. Another part is the ASM service management and API services. The system manages and authenticates the users to access the API services. Furthermore, the system can receive JSON files from the ACLARE server and keep them in the database system. The API services can serve the client application to request and respond to train approach information in the form of the JSON format called by the API services with the validated API key.

Our system can present useful information such as the arrival time and departure time of the approaching train that can be shared to any client applications under the secure environment. According to our design system, our API services do not interfere or interrupt the train telegram system. Only the ASM server can retrieve the train approach information. Then requests from the client applications are sent to the ASM server directly. The train operators do not to worry about open data on the public applications. The ASM system is designed to manage users who want to request the train approach information. Only approved API keys from applications can request this information via the API services. The ASM system also monitors amount of service usage from each user's account.

In conclusion, this system is an example of a part of system development for other train telegram systems. It can demonstrate data sharing between an operator and client applications. In the future, this concept of system development can be extended to implement a data center for train telegram systems and the API service system can serve as a data center, gathering data from many train telegram systems and can share information to any client applications with a standard data exchanging protocol and security standard.

Acknowledgements The authors would like to express our gratitude to Airport Rail Link, SRT Electrified Train Co., Ltd. that collaborates, supports and provides us with useful information. This research is partially supported by the Newton Fund-Industry Academic Partnership Program-Thailand under the SmartMetro Project (Grant No. 18-19/100) and Crown Property Bureau Funding, Thailand.

References

Antrim A, Barbeau SJ (2013) The many uses of GTFS data—opening the door to transit and multimodal applications. In: Proceedings of the ITS America's 23rd annual meeting and exposition

BKK Rail (2019) [Online]. Available at: https://play.google.com/store/apps/details?id=th.ac.mahidol.clare.gorail&hl=en. Accessed 1 Aug 2019

Google Inc (2019) GTFS-realtime specification. Available at: https://github.com/google/transit/tree/master/gtfs-realtime/spec/en. Accessed 1 Aug 2019

Govtech (2019) Government of Singapore [Online]. Available at: https://data.gov.sg/dataset?q=mrt. Accessed 1 Aug 2019

MTR Cooperation Limited (2019) [Online]. Available at: https://www.mtr.com.hk/en/customer/services/mtr_mobile.html. Accessed 1 Aug 2019

Nash A, Huerlimann D, Joerg S, Paul KV (2004) RailML—a standard data interface for railroad applications. In: Proceeding of the 9th international conference on computer in railways (Comprail IX), Dresden. WIT Press, Southampton, United Kingdom

Open Government Canada (2019) Railways [Online]. Available at: https://open.canada.ca/data/en/dataset/a0acfce7-52ba-599a-9870-1cfccd9dae03. Accessed 1 Aug 2019

RAILML (2019) RAILML.org. Available at: https://www.railml.org. Accessed 1 Aug 2019

Yuan W, Frey H (2018) Evaluation of general transit feed specification data for route characteristics and travel activity data of an electric train system

Short-Turning Management During Railway Network Disruptions

Ratthaphong Meesit and John Andrews

Abstract Railway network operations are vulnerable to unplanned events such as failures of network components and natural disasters. These events normally cause timetable disruption, which might lead to serious impacts on both passengers and freights in the network. This paper presents a railway disruption management model that can be applied to assess and compare the efficiency of short-turning strategies implemented to solve an unplanned-track blockage situation. The model is constructed based on a stochastic-discrete event simulation concept. The application of the proposed model is demonstrated using the urban railway network in the UK. The results of the computational experiments illustrate that the proposed model is a useful tool to support a decision-making process to solve a track blockage situation in the real world.

Keywords Railway disruption management · Short-turning operation · Resilience model

1 Introduction

Unexpected disruptions are a significant problem affecting the reliability in the railway network operation (Schmöcker et al. 2005). Once these events occur, train services might need to be delayed and cancelled, which causes thousands of passengers to be impacted. Thus, it is essential to manage and deal with this problem efficiently in order to ensure the best possible services to passengers during disruptions.

Railway disruption management is the process of dealing with the occurrence of a disruption on railway networks (Jespersen-Groth et al. 2009). This process is set

R. Meesit (✉)
Department of Civil Engineering, Faculty of Engineering, Burapha University, Mueang Chonburi 20131, Thailand
e-mail: ratthaphong.me@buu.ac.th

R. Meesit · J. Andrews
Resilience Engineering Research Group, Faculty of Engineering, University of Nottingham, Nottingham NG7 2RD, UK

W. Weerawat et al. (eds.), *Urban Rail Transit*, Lecture Notes in Mobility,
https://doi.org/10.1007/978-981-15-5979-2_13

up by infrastructure managers, traffic controllers and train operators. The mitigation strategies applied involve the modification of service timetables and rolling stock and crew schedules during and after the disruption. The implementation of a mitigation strategy basically depends on the severity of the disruption: low and high level impact disruptions (Cacchiani et al. 2014).

Low impact disruptions refer to short duration incidents that are often caused by passengers (e.g. congestion at stations) or small defects in rolling stock (e.g. door failures) (Schmöcker et al. 2005). The possible strategies that can be applied to solve these events are called "*dispatching rules*". This type of strategy is related to an adaptation of service timetables in order to avoid or absorb the delays, and it requires simple control actions from traffic controllers without the difficulty of managing the availability of resources such as trains and crews (Blenkers 2015). The common dispatching rules are: using buffers in a timetable, overtaking, changing the stations stop pattern, inserting on-time trains, turning the affected train early, and cancelling train services (Hofman et al. 2006; Landex 2008; Jespersen-Groth et al. 2009).

High impact disruptions, on the other hand, are incidents which have relatively large magnitudes and durations (e.g. bridge collapse or the occurrence of a natural disaster). This type of disruption can cause a temporary blockage of a railway line, leading to a significant reduction in the railway service performance. In this case, traffic controllers and train operators need to deal with the limited availability of resources such as trains and crews, along with the modification of service timetables (Cacchiani et al. 2014). Therefore, a set of predefined strategies, such as rerouting, short-turning and providing replacement services can be used in order to help traffic controllers manage the situation in a timely manner (Ghaemi et al. 2017).

In recent years, a considerable amount of literature has focused on the development of disruption management models for railway networks. However, according to Cacchiani et al. (2014) and Ghaemi et al. (2017), only limited research has been conducted to support the management of a railway network during high impact disruptions. The effect of complete track blockages, where some parts of the network need to be closed for several hours, on train services and passengers is barely studied, and a model that can be used to evaluate the mitigation strategies for this situation is still lacking. This study introduces a model that can be used to simulate and manage rail traffic during track blockage situations. The model is developed based on the stochastic-railway network simulation model by Meesit and Andrews (2018) and it focuses on a short-turning operation strategy. These capabilities allow traffic controllers and train operators to evaluate and compare the efficiency of different short-turning solutions before selecting the proper one to mitigate a track blockage situation.

The remainder of the paper is presented as follows. Section 2 explains the framework of the proposed model. Section 3 demonstrates an application of the proposed model using the urban railway network case study. Finally, Sect. 4 concludes the paper.

2 Short-Turning Operation Modelling

Short-turning operation is a strategy to operate train services on a part or parts of the disrupted routes. For example, once a complete-track blockage situation occurs, the services on the disrupted routes might need to be delayed and cancelled. Traffic controllers can thus apply this strategy to allow the trains to run to the closest stations to a disruption and turn around to provide services in the opposite direction of their routes. In this way, the services on the non-disrupted parts of the routes will be maintained and the impact on passengers will be reduced.

To simulate this strategy, the study applies the stochastic-railway network simulation model by Meesit and Andrews (2018) as a basis of the new model. This simulation model was developed using a discrete event simulation concept, and it can be used to predict the system performance of a railway network during disruptions, both in terms of train services and passengers. Thus, the model proposed in this study will be built by introducing new events to this simulation model to imitate the short-turning operation. These new events can be modelled by considering the phase changes of the timetable during the disruptions as can be seen in Fig. 1.

2.1 *Phase 1*

In the normal operation, the performance of the network is assumed to be at P_0. When a blockage occurs, the performance of the network will dramatically decrease over time until it has met with the robustness of the system (P_d). If there is no mitigation strategy applied, the performance will be at this point until the disruption is recovered. However, if there is a mitigation strategy applied (e.g. short-turning

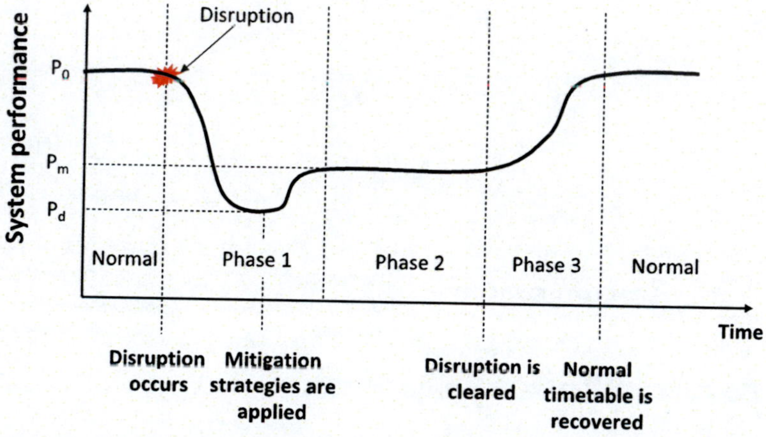

Fig. 1 Railway network performance graph during a disruption

operations), it still takes some time for the performance to increase to a certain level (at P_m). This is because, during the situation, some trains on the disrupted route may be at unplanned locations (e.g. facing the disruption or at the middle of the route). Therefore, to implement the short-turning services, the traffic on the disrupted route needs to be managed. A decision on the trains on the network need to be made in order to solve the situation efficiently. This circumstance can be modelled into three conditions based on the location of the trains on the disrupted route as follows.

- If a train is at a terminal station, the model investigates a short-turning station on the route based on the train direction. If there is a short-turning station ahead, the model allows the train to continue its service as planned until the short-turning station. Otherwise, the train is required to wait at the terminal station until the disruption is recovered.
- If a train is at an intermediate station (Fig. 2), the model checks whether the current station is the short-turning station. If the condition is true (e.g. T1), the short-turning event is created by considering the next departure time (T_D) at this station in the opposite direction. However, if the train is delayed, the allowable delay at the turning station (t_{ad}) will be taken into account for the occurrence time of this event. Nevertheless, if the condition is false, the train can continue running based on the normal procedure of the simulation, unless it is at the station nearest to the disruption (e.g. T2), the next condition can be applied.
- If a train is facing the disruption (e.g. Fig. 2, T2 and T3), traffic controllers can decide whether the train needs to wait, run back to the previous station and wait or run back to the short-turning station and join the short-turning services. These options can be set in the model by updating variable F_{rb} to 0, 1 and 2, respectively.

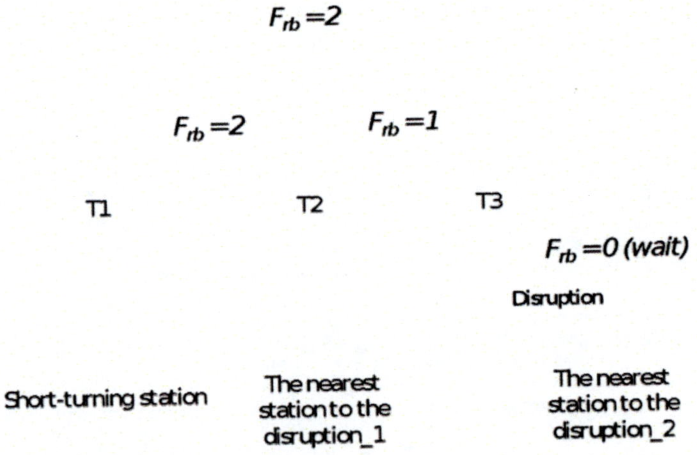

Fig. 2 Example of rail traffic management options at phase 1

2.2 *Phase 2*

In this phase, the trains on the disrupted route are operated in the short-turning mode. Thus, the performance of the network will be constant at P_m until the disruption is cleared. The way to simulate this phase is the same as explained in the second condition of phase 1 (trains are turned around at the short-turning station).

2.3 *Phase 3*

Phase 3 is related to the change of the timetable when the disruption is recovered. The trains that are operating in the short-turning mode or waiting at stations need to return to the normal operation. Therefore, traffic controllers are required to make a decision again in this phase in order to bring the original timetable back to passengers. This phase is also modelled based on the position of the trains on the disrupted route as described below.

- If a train is at a terminal station, the model calculates the departure time for the train based on the original timetable. This is done by considering the allowable delay at this terminal station (t_{ad}). If the simulation clock (the current time in the simulation, T_c) is less than the considering departure time (T_D), the train can begin the service at T_D. However, if T_c is more than T_D but still less than T_D + t_{ad}, the train can begin the service at T_c (with delay). Otherwise, the next T_D will be considered and the previous T_D, that the train could not serve, are counted as service cancellations (see Fig. 3).
- If a train is at an intermediate station, the model checks whether the current station is a short-turning station. If the answer is yes (see Fig. 3), traffic controllers can decide whether the train needs to wait and continue its service based on the original route or make a short-turn one more time to balance the number of trains in each direction on the route. This can be modelled by changing the turning factor (F_{tr}) to 0 and 1, respectively (e.g. in Fig. 3, $F_{tr} = 0$). Otherwise, the train is authorised to proceed its service based on the normal procedure of the simulation.
- Finally, if a train is facing the disruption (i.e. waiting trains, e.g. T2 and T3 in Fig. 2 when $F_{rb} = 0$), the train can continue its service immediately once the disruption is cleared.

3 Model Application

3.1 *Case Study*

This study uses the Liverpool railway network in the UK to demonstrate the model application (see Fig. 4). The network has 67 stations, with the length of 120 km

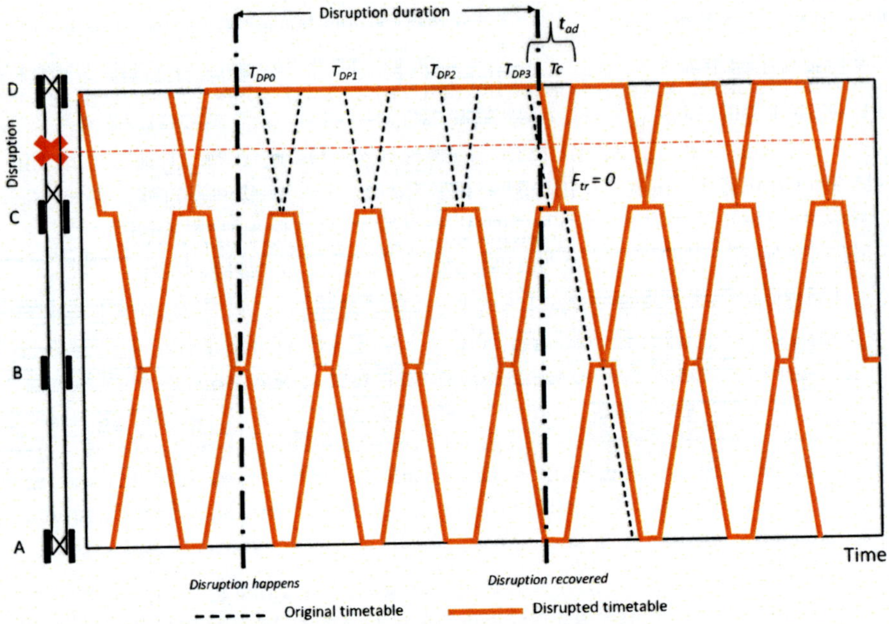

Fig. 3 Example of rail traffic management options at phase 3

(double track). Seven service routes are operated daily from 6:00 to 24:00. These include Southport to Hunts Cross (R0), Ormskirk to Liverpool Central (R1) and Kirkby to Liverpool Central (R2), and four loop routes from four terminal stations: Ellesmere Port (R3), Chester (R4), West Kirkby (R5) and New Brighton (R6), via the Liverpool Central station. The trains on each route are the British Rail class 507/508 (3 coaches). The timetable and passenger data applied in the simulation were acquired from the study of Meesit and Andrews (2019).

It is noted that the proposed model was built in C++11 environment with Microsoft Visual Studio 2015. Then, a personal computer, with a dual core Intel i3 processor, CPU 3.50 GHz, 8 GB of RAM running on Window 7 (64-bit), was used to conduct all computational experiments. According to the stochastic behaviour of the model, 1000 simulations were carried out to ensure that the results obtained are converged. The model took approximately 5 min to run this number of simulations.

3.2 *Computational Experiment*

To demonstrate the capability of the model, it was assumed that there is a blockage disruption between Blundellsands and Crosby (ID12) and Waterloo station (ID65) (Route R0 in Fig. 4 which is normally operated with 4 trains per hour). This disruption occurred at 10:55 and last for about 2 h until the recovery actions were completed.

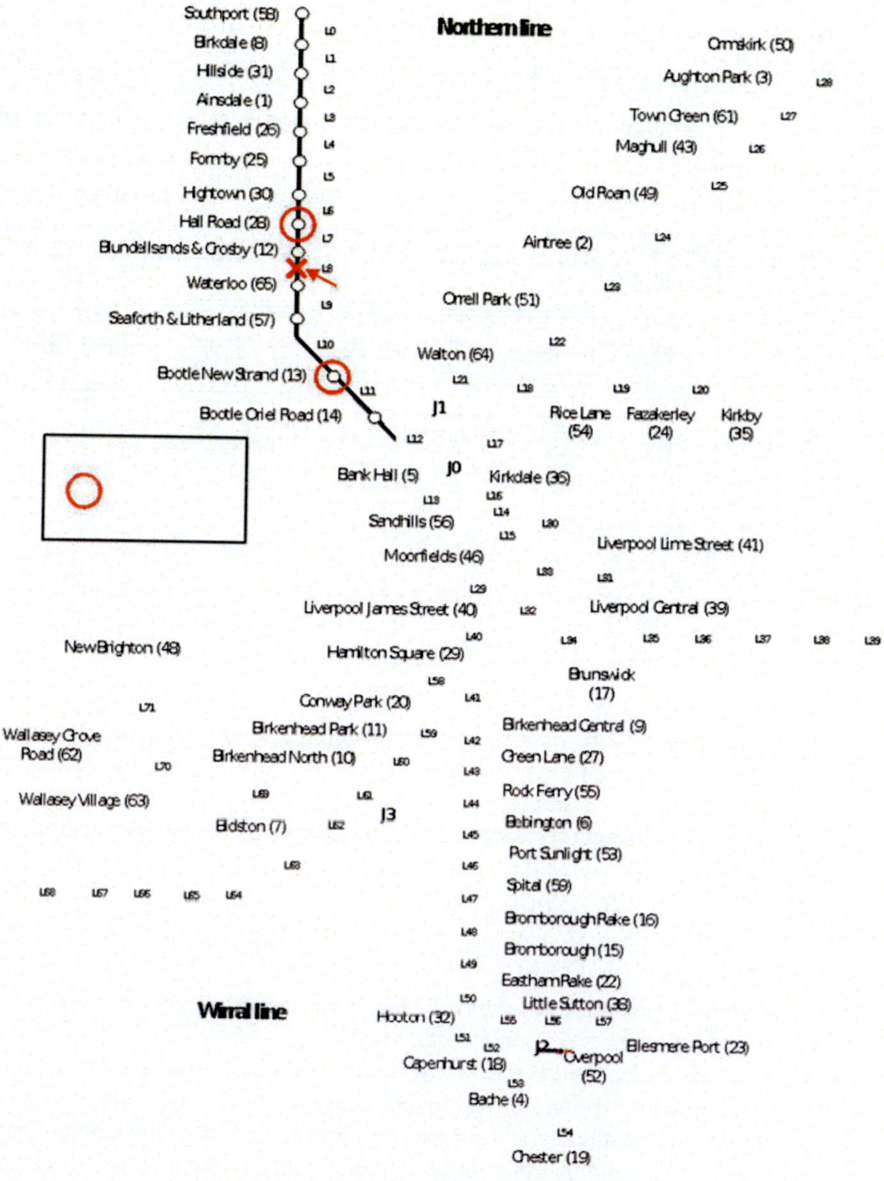

Fig. 4 Liverpool railway network in the UK (Meesit and Andrews 2019)

In detail, the recovery actions include six main steps: responding to the disruption, preparing the site, investigating the problem, waiting for equipment and machinery, solving the problem and re-opening the line (RSSB 2016), and the time to complete each step is assumed to follow a Uniform (U) and Log-normal (LN) distribution (see Table 1).

Table 1 Recovery time prediction of the example case

Recovery time	Responding time (s)	Preparing time (s)	Investigating time (s)	Waiting time (s)	Resolving time (s)	Reopening time (s)
	U (1140, 1260)	U (570, 630)	U (1140, 1260)	0	LN (8.1, 0.1)	U (570, 630)

Table 2 Short-turning solutions considered to solve the disruption

Solution no	Model parameters			
	F_{rb} for each turning station (0, 1 or 2)	F_{tr} for each turning station (0, 1)	Allowable delay at each turning station (ID28, ID13) (min)	Allowable delay at each terminal station (ID58, ID34) (min)
0	Without short-turning operation			
1	(0, 0)	(0, 0)	(0, 0)	(0, 0)
2	(0, 0)	(1, 1)	(0, 0)	(0, 0)
3	(1, 1)	(0, 0)	(0, 0)	(0, 0)
4	(2, 2)	(0, 0)	(0, 0)	(0, 0)
5	(2, 2)	(1, 1)	(0, 0)	(0, 0)
6	(0, 0)	(1, 1)	(10, 10)	(10, 10)

To mitigate this disruption, the short-turning operation was implemented. Hall road (ID28) and Bootle New Strand station (ID13) were considered as the short-turning stations. The minimum turnaround time at these stations was set to 3 min in the model. Six different configurations of the short-turning operation were considered and compared. The detail of each solution is explained below and depicted in Table 2.

- Solution 1 focuses on phase 2. Trains facing the disruption need to wait, and the others can be short-turned at the short-turning stations until the disruption is recovered;
- Solution 2 is similar to solution 1, but at phase 3 the trains are required to make a short turn even if the route is cleared;
- Solutions 3 deals with the first phase of the disruption. The trains facing the disruption in both directions are authorised to run backward to the previous stations (if they are not already at a station) and wait there for the next authorisation;
- Solution 4 is similar to solution 3, but the trains are shunted to the previous short-turning stations and join the short-turning service;
- Solution 5 is the combination of solutions 2 and 4. Thus, both phases 1 and 3 are managed;
- Solution 6 is also similar to solution 2, however, the allowable delay of 10 min is applied at both short-turning and terminal stations on the disrupted route.

3.3 Results

In this study, four key performance indicators (KPIs) were predicted in order to compare the efficiency of each solution. These KPIs consist of total train service delays, the number of departure service cancellations, total passenger delays and the number of passenger journey cancellations. Passenger behaviour during the disruption was modelled based on the change of the timetable. Before the disruption, all passengers travel using a normal timetable. When the disruption occurs, passengers use the disrupted timetable which is either real-time information announced at stations or displayed on smartphones. Existing passengers on the network then reconsider their routes according to the new information. If the expected delay of the new route is longer than their acceptable delay, passengers cancel their journeys. The passenger acceptable delay parameter was set to follow a Normal distribution with the mean of 3600 and 300 s standard deviation. Passengers are then back to using the new updated timetable after the disruption is cleared (after phase 2).

The results of the simulations are presented in Fig. 5. It is obvious that solution 0 (do nothing) is the worst solution in this example. The best possible solutions seem to be solutions 4 and 5. This is because these solutions generated low impact on both train services and passengers. However, to decide which solution should be implemented, the overall impact on either train services or passengers ($I_{T/P}$) can be used. This can be obtained using a weight summation of the train delays or the passenger delays ($D_{T/P}$) and the number of train service cancellations or passenger

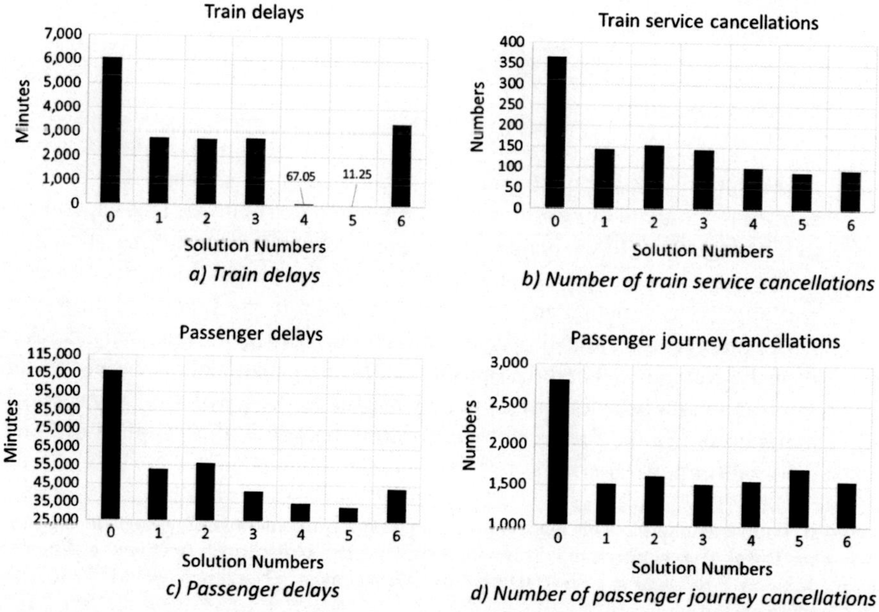

Fig. 5 KPI results of all short-turning configurations

journey cancellations ($C_{T/P}$) as presented in Eq. (1). The weighting factors ($w_{1(T/P)}$ and $w_{2(T/P)}$) are the penalties for delays and cancellations of either train services or passengers, respectively. Since the main goal of implementing a mitigation strategy is to reduce the impact of the disruption on passengers, only I_P was thus considered. Let set $w_{1(P)}$ and $w_{2(P)}$ to 1 and 50 cost unit per delay and cancellation. Thus, I_P of solutions 4 and 5 were equal to 113,567 and 119,015 cost unit. This means solution 4 should be selected for this situation.

$$I_{T/P} = D_{T/P} \cdot w_{1(T/P)} + C_{T/P} \cdot w_{2(T/P)} \tag{1}$$

where *T*/*P* refers to the impact on train services or passengers. For example, if the passenger perspective is focused, only *P* will be in the equation.

Other interesting results can be seen by comparing solutions 1 and 3, and 2 and 6. For the first case, the KPIs: train delay, service cancellations and passenger journey cancellations obtained from both solutions were quite similar. However, solution 3 tends to perform better on reducing the passenger delay. This is because although running to the previous station and waiting for an authorisation could not make much difference in the time-distance graph, passengers at the station were still able to board the delayed train once the disruption was recovered, which means they do not have to wait for the next trains to get to their destinations. For the second case, it is found that setting the allowable delay to 10 min at both short-turning and terminal stations as in solution 6 led to an increase in train delay by 25% compared to solution 2. However, it can help reduce the number of service cancellations and passenger impacts in the network significantly as can be seen in Fig. 5b–d.

4 Conclusion

A disruption management model is presented in this paper. The model can be applied to simulate rail traffic during unplanned disruptions. This capability enables traffic controllers to evaluate and compare different short-turning strategies before acting to solve a complete-track blockage situation. In addition, the proposed model was applied to simulate and solve the disruption in the Liverpool railway network. Different potential short-turning solutions were considered and analysed. As a result, it can be concluded that the proposed models are capable of providing the significant information for operating trains to increase the resilience of a railway network during a disruption. In the future, an optimisation method will be applied to find the best solution for passengers.

Acknowledgements This work was developed at the University of Nottingham and supported by the Royal Thai Government. Professor John Andrews is the Royal Academy of Engineering and Network Rail Chair in Infrastructure Asset Management. The authors would like to acknowledge these organisations.

References

Blenkers LL (2015) Railway disruption management. Delft University of Technology

Cacchiani V, Huisman D, Kidd M, Kroon L, Toth P, Veelenturf L, Wagenaar J (2014) An overview of recovery models and algorithms for real-time railway rescheduling. Transp Res Part B Methodol 63:15–37

Ghaemi N, Cats O, Goverde RMP (2017) Railway disruption management challenges and possible solution directions. Public Transp 9(1–2):343–364

Hofman M, Madsen L, Groth JJ, Clausen J, Larsen J (2006) Robustness and recovery in train scheduling—a simulation study from DSB S-tog a/s. In: 6th workshop on algorithmic methods and models for optimization of railways (ATMOS), pp 1–22

Jespersen-Groth J, Potthoff D, Clausen J, Huisman D, Kroon L, Maroti G, Nielsen M (2009) Disruption management in passenger railway transportation. In: Robust and online large-scale optimization. Springer, Berlin Heidelberg, pp 399–421

Landex A (2008) Methods to estimate railway capacity and passenger delays. Technical University of Denmark

Meesit R, Andrews J (2018) Vulnerability assessment modelling for railway networks. In: 10th IMA international conference on modelling in industrial maintenance and reliability (MIMAR 2018), Manchester, 13–15 June 2018, pp 1–6

Meesit R, Andrews J (2019) Optimising rail-replacement bus services during infrastructure possessions. Infrastruct Asset Manag 1–18

RSSB (2016) Incident response planning & management. London

Schmöcker J-D, Cooper S, Adeney W (2005) Metro service delay recovery: comparison of strategies and constraints across systems. Transp Res Rec 1930(1):30–37

A Train Rescheduling System Using Timed Coloured Petri Nets

Somsak Vanit-Anunchai

Abstract A train rescheduling system using Timed Coloured Petri Nets is proposed. Traffic in a single track area named, "Chumphorn" is modelled and analysed. This area is selected because of its high traffic density. Even with a small traffic disruption in this area, the consecutive delays can be severely propagated. Although Timed Coloured Petri Net (TCPN) formalism is a natural choice to model a train scheduling problem, we rarely see TCPN used in real world practice. We envisage that there are two problems. First, interacting with the TCPN model requires programming skills. Second, analysis of the traffic in a large railway network often encounters state explosion. This paper suggests the solutions to these problems.

Keywords Block section · Dispatcher · Timetable · Train graph · Single-track

1 Introduction

Railways usually manage their traffic through an off-line timetable which is carefully designed for several months in advance. This planned timetable defines routes, orders and timing of all trains. Designing a planned timetable is such a complex process that it requires sophisticated simulation tools such as OpenTrack and RailSys. Traffic controllers or dispatchers, who supervise all train traffic in their control area, normally attempt to maintain the traffic according to this planned timetable. When any disruptive events happen, the traffic shall be recovered back to the original timetable as quickly as possible. Based on real-time monitoring of track occupation and clearance, as well as train priorities, the traffic controller may reschedule delayed trains from the next station onwards.

Currently Thailand has been investing enormous amounts of budget in many rail projects ranging from light-rail systems to high-speed trains. Among these projects, the State Railway of Thailand (SRT) is undertaking several projects that will transform the existing single-track into double-track rail networks. SRT plans to install

S. Vanit-Anunchai (✉)
Suranaree University of Technology, 111 University Ave., Muang, Nakhon Ratchasima 30000, Thailand
e-mail: somsav@sut.ac.th

W. Weerawat et al. (eds.), *Urban Rail Transit*, Lecture Notes in Mobility,
https://doi.org/10.1007/978-981-15-5979-2_14

new signalling systems using Computer Based Interlocking (CBI) together with Centralized Traffic Control (CTC). Nowadays SRT traffic controllers reschedule trains manually. As far as we are aware, SRT has no plan to introduce computerized real time rescheduling systems. This paper proposes a computer-based real-time rescheduling system using Timed Coloured Petri Net (TCPN) models. Given the position and delay of each running train, every possible sequence of the train movement or "state space" can be generated from the TCPN model. In other words, the state space already includes all feasible timetables from which the dispatcher can select one.

It has been long recognized that Timed Coloured Petri Net (TCPN) formalism is a natural choice to model a train scheduling problem but we hardly ever see TCPN used in real world practice. According to our experience, interacting with the formal model requires programming skills which the users normally do not have. To conceal the programming code from the user, programs are generated via MATLAB GUI. These codes are used to initialize and analyse the model.

This paper is organised as follows. Section 2 describes the literature review. Section 3 explains design rationale and assumptions of the model. Section 4 describes a generic TCPN model of Train Traffic. Analysis of the model is discussed in Sect. 5. Section 6 explains how MATLAB GUI interacts with our model in CPN Tools. Section 7 presents conclusions and future work.

2 Literature Review

According to Fokkink and Hollingshead (1998) the railway signaling system is classified into three layers: infrastructure, interlocking and logistics layers. The infrastructure layer is related to equipment along the track and in the yard. The interlocking layer is the interface between infrastructure and logistics layers. The interlocking controls all equipment in the infrastructure layer such that the railway operation is safe from accidents caused by human errors or equipment failure. The logistics layer mainly involves train scheduling which requires efficiency and absence of deadlocks. It is related to the train operation of the large scale railway network (Hagalisletto et al. 2007; Janczura 1998) so that the state explosion is encountered.

Regarding safety of the interlocking layer, we modeled and analysed the interlocking tables in Vanit-Anunchai (2009, 2010, 2012, 2016, 2018) using Coloured Petri Nets. Petri Nets have been extensively used to model train scheduling. Van der Aalst and Odijk (1995) modeled train movement through railway stations and analysed throughput and waiting times of trains using Interval Timed Coloured Petri Nets. Hagalisletto et al. (2007) modeled the Oslo subway and analysed the train schedule using Coloured Petri Nets (CPN). Due to the state explosion, Hagalisletto et al. (2007) transformed the CPN model to Maude (Clavel et al. 2003) and simulated the Maude model.

3 Design Rationale and Assumptions

3.1 Absolute Block

Railway companies normally divide their railway lines into ***sections***. In one ***section*** only one train is allowed at a time in order to avoid collisions. The moving authorization can be issued if there is no object blocking the passage of the train. After the train receives the moving authorization, the train can enter the ***section***. Usually the ***section*** is classified into two types: within the ***station*** area and between two stations called "***block section***". When there is one section in the ***block section*** and the trains cross each other at the stations, it is called "***Absolute block system***".

3.2 Single Track Model

Following an incremental approach, we start with an abstract model of an ***absolute block system*** on a single track area controlled by "Chumphorn" dispatcher center. This center is responsible for 33 stations, running 30 trains each day and covering 258.5 km. Our TCPN model was created and maintained using CPN Tools (Jensen et al. 2007; Jensen and Kristensen 2009). Station and block sections are modelled by Places. Trains are represented using tokens. The State Railway of Thailand (SRT) intends to model not only the existing whole network but also their future network. It is mandatory that this experimental model shall be generic so that the CPN net structure can be easily reused for other dispatching areas by only instantiating initial tokens.

3.3 Assumptions

We assume that all required information affecting the train speed is already in the planned timetables so that our model does not include the speed profile, track curvature and terrain. The dwell time and running time of each train can be derived from the planned train schedules so that the full detailed description of the track infrastructure and rolling stock can be abstracted. Trains can be delayed and cannot leave the station before the schedule. The speed of each train in each block section is fixed and cannot be faster. This "fixed speed" assumption has two advantages. Firstly, it guarantees safety because the train always runs at the safe speed. Secondly, during the state space generation the train delays in the child states are always greater (or equal to) than in their parent states. If any states have train delays exceeding the target values, we can stop generating their descendants. Thus, the state space analysis of the model can be simplified. Later, this "fixed speed" assumption can be relaxed. The

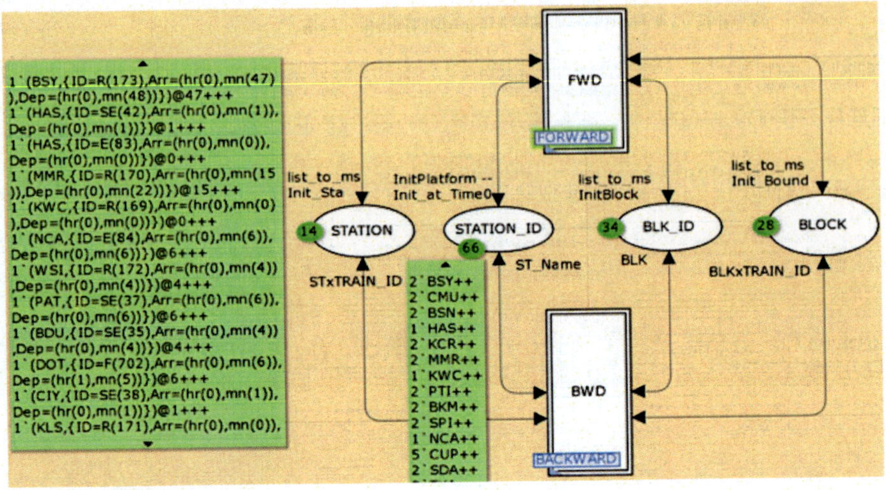

Fig. 1 The top level CPN page

train in the model can run faster by reducing the arrival times in the planned train schedule.

4 A Generic CPN Model of Train Traffic

4.1 Overview

Our CPN model comprises two hierarchical levels, 4 places, 2 substitution transitions and 4 executable transitions. The top-level page of the model is illustrated in Fig. 1. Two substitution transitions (FWD and BWD) represent the train movement outward and toward Bangkok respectively. We view a train as a token moving from a station platform (Place STATION) into a block section (Place BLOCK) and vice versa. All stations are folded into Place STATION and all block sections are folded into Place BLOCK so that we add the train's location and time into the train token.

4.2 Declarations

Figure 2 defines the data structure of Place STATION (typed by STxTRAIN_ID, line 12) and Place BLOCK (typed by BLKxTRAIN_ID, line 14). They are the products of train location (Station or Block section) and TRAIN_ID with time stamp. TRAIN_ID (Line 10) is defined as a record of Train Identification Number (TID), Arrival time and Departure time. Train Identification Number (TID, line 6) is defined as a union of

```
1: colset ST_Name = with North | BSY | CMU | BSN | HAS | KCR | MMR
2:                  | KWC | PTI | BKM | SPI | NCA | CUP | SDA | TKA | WSI | SWI
3:                  | KSU | PAT | KHM | LSU | KKN | LMA | BDU | KTL | DOT
4:                  | TCN | KPN | CIY | TAC | KLS | MLU | TPJ | SRN | South;
5: var pre, post, sta:ST_Name;
6: colset TID = union SE:INT + E:INT + R:INT + O:INT + F:INT + Mix:INT;
7: colset HR = union hr:INT;
8: colset MN = union mn:INT;
9: colset HrxMn = product HR * MN;
10: colset TRAIN_ID = record ID:TID * Arr:HrxMn * Dep:HrxMn;
11: var train_id:TRAIN_ID;
12: colset STxTRAIN_ID = product ST_Name * TRAIN_ID timed;
13: colset BLK = product ST_Name * ST_Name;
14:colset BLKxTRAIN_ID = product BLK * TRAIN_ID timed;
```

Fig. 2 The definition of TRAIN_ID, STxTRAIN_ID and BLKxTRAIN_ID

integers associated with the train's class: Super Express (SE); Express (E); Rapid (R); Ordinary (O); Freight (F); and Mix trains. Places STATION_ID (typed by ST_Name, line 1) and BLK_ID (typed by BLK, line 13) store the platform identifications and the block section identifications that are still unoccupied. ST_Name is defined as a multiset of the abbreviation of the station names. BLK is defined as a product of ST_Name.

4.3 BACKWARD Page

Substitution transitions FWD and BWD link to the second level pages, FORWARD and BACKWARD respectively. Both pages are very similar so that only BACKWARD page is illustrated in Fig. 3. Transition STA2BLK models the train movement from a station platform into a block section while transition BLK2STA models the train movement from a block section into a station platform. Function delayDown() computes the running time in the block section according to the planned schedule. Then it calculates the actual arrival time entering the next station. Similarly, function wait2() calculates the dwell time according to the planned schedule. Then the actual departure time is computed. The conditions on the outgoing arcs in Fig. 3 are used to handle the model's behaviour when the trains are passing the boundary stations (BSY and SRN). The guard function allows only the train returning toward Bangkok to be executed in this page.

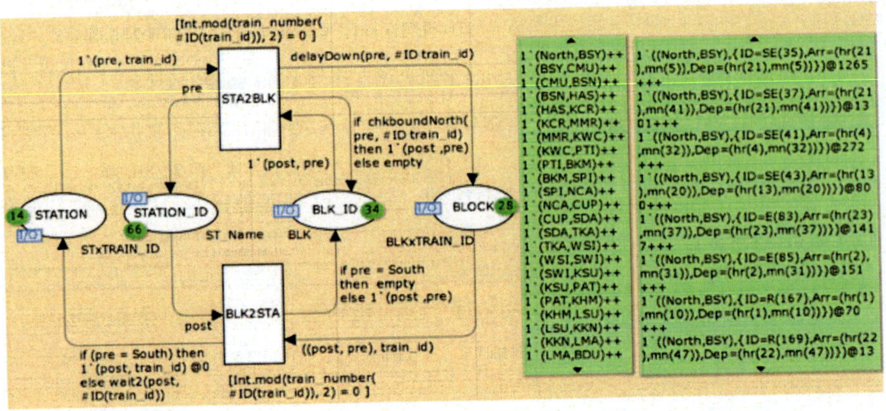

Fig. 3 BACKWARD page

5 Analysis

5.1 Initial Markings

Because a lot of initial data is required before starting simulation or analysis, we suggest entering this initial data via Excel. We then use MATLAB, transforming this initial data to the program files (*.sml) which create the initial markings in Fig. 1. Tables 1 and 2 illustrates the initial markings in Excel format; e.g. the number of loop lines transformed to "Init_Sta.sml"; the initial positions of trains transformed to "Init_at_Time0.sml"; the arrival and departure times of the trains when entering the boundary stations transformed to "Init_Bound.sml".

5.2 Experimental Results

With the particular initial markings shown in Tables 1 and 2, every train runs according to the plan and there is no delayed train. Starting at 0.00 h, the trains are located at the stations according to Table 1. We keep feeding the trains passing two boundary stations according to their schedule, Table 2, until 24.00 h. After no new train enters into the area, we let the model run until no train is left in the area. If there are any trains left, it is an undesired deadlock. Using the full state space generation, the analysis of "Chumphorn" model takes only 9 s to generate the total state space of 8048 nodes, 14,999 arcs and 2 desired terminal markings.

However if we assume that only train (R172) is delayed for 10 min, the analysis takes 41 s to generate the total state space of 28,780 nodes, 53,069 arcs and 8 terminal markings. The state space in both cases, 8048 nodes and 28,780 nodes, have no undesired deadlock where the trains cannot move further.

Table 1 Initial markings in excel format. The number of loop lines and the initial train positions

Station	No of loop lines	Train no	Arr (hr)	Arr (min)	Dep (hr)	Dep (min)
BSY	3	R 173	0	47	0	48
CMU	2					
BSN	2					
HAS	3	E 83	0	0	0	0
HAS		SE 42	0	1	0	1
KCR	2					
MMR	3	R 170	0	15	0	22
KWC	2	R 169	0	0	0	0
PTI	2					
BKM	2					
SPI	2					
NCA	2	E 84	0	6	0	6
CUP	5					
SDA	2					
TKA	3					
WSI	2	R 172	0	4	0	4
SWI	2					
KSU	2					
PAT	2	SE 37	0	6	0	6
KHM	2					
LSU	3					
KKN	2					
LMA	2					
BDU	2	SE 35	0	4	0	4
KTL	2					
DOT	2	F 702	0	6	1	5
TCN	3					
KPN	2					
CIY	2	SE 38	0	1	0	1
TAC	2					
KLS	2	R 171	0	0	0	0
MLU	2					
TPJ	4	F 777	0	13	0	13
TPJ		SE 36	0	2	0	2
SRN	5					

Table 2 Initial markings in excel format. The arrival and departure times of the trains when entering the boundary stations

Boundary-station	Train no	Arr (hr)	Arr (min)	Dep (hr)	Dep (min)
(North, BSY)	R 167	1	10	1	10
(North, BSY)	R 177	1	46	1	46
(North, BSY)	E 85	2	31	2	31
(North, BSY)	SE 41	4	32	4	32
(North, CUP)	O 445	6	29	6	29
(North, BSY)	F 701	8	36	8	36
(North, BSY)	F 721	11	38	11	38
(North, BSY)	SE 43	13	20	13	20
(North, BSY)	O 225	15	4	15	4
(North, BSY)	R 171	19	44	19	44
(North, BSY)	SE 35	21	5	21	5
(North, BSY)	SE 37	21	41	21	41
(North, BSY)	R 169	22	47	22	47
(North, BSY)	E 83	23	37	23	37
(SRN, South)	SE 36	23	53	23	53
(SRN, South)	SE 38	23	24	23	24
(SRN, South)	F 702	21	59	21	59
(SRN, South)	R 172	21	19	21	19
(SRN, South)	E 84	21	0	21	0
(SRN, South)	SE 42	20	30	20	30
(SRN, South)	R 170	20	10	20	10
(SRN, South)	E 86	18	22	18	22
(SRN, South)	R 168	17	32	17	32
(SRN, South)	R 174	16	36	16	36
(SRN, South)	O 446	13	21	13	21
(SRN, South)	SE 40	10	39	10	39
(LSU, South)	O 254	5	44	5	44
(LSU, South)	R 178	20	19	20	19

5.3 *An Example of the Branch Conditions*

Many researchers often transform the Petri Net models into other formalisms that can exploit better analysis tools and methodologies. However, we suggest that in our case the explicit state space exploration and its state space reduction techniques are still appropriate. To avoid the state explosion we have two suggestions. Firstly, because of the "fixed speed" assumption, during the state space generation we can use "branch option" in CPN Tools to rule out the inferior scenarios. Table 3 shows an

Table 3 Branch conditions

Train	Lower_Delay (min)	Upper_Delay (min)
SE 40	0	12
R 172	0	15
SE 37	0	3
SE 43	0	10

example of the branch conditions in Excel. The states with the train delays outside the specified intervals will not generate their child states. Again we use MATLAB to transform the branch conditions to "branch_cond.sml" which shall be executed before the state space generation. Secondly, one of the natural progress measures for the sweep-line analysis (Jensen et al. 2012) is "time". We use the sweep-line to analyse the "Chumphorn" model using the System Time as a progress measure. Our preliminary result reveals that the sweep-line method can reduce the memory usage to about 5%. If we could combine the branch conditions together with the sweep-line, the reduction ratio could be improved.

5.4 *An Example of the Query Conditions*

The dispatchers are normally concerned with the trains' delays when their trains are leaving the boundary stations. Table 4 shows an example of the query conditions which are the range of acceptable delays. After the dispatcher prescribes the ranges, our analysis can project sequences of events that lead to the desired states. Using MATLAB the query conditions are transformed into "Init_Query.sml" which shall be executed after the state space is generated. After selecting a terminal state that satisfies the dispatcher, he can export a sequence of events, which leads to that terminal state, into MATLAB. Then he can use MATLAB to create a suggested timetable and train graph accordingly.

Table 4 Query conditions

Block		Train	Lower_Delay (min)	Upper_Delay (min)
North	BSY	SE 40	0	5
North	BSY	E 84	0	5
SRN	South	SE 37	0	5
SRN	South	SE 43	0	5

6 Graphic User Interface

Normally exploring or searching the state space would require some expertise in SML programming language. To assist the train dispatcher using our train rescheduling system, we conceal the SML code behind MATLAB Graphic User Interface. Figure 4 illustrates the interaction between MATLAB Graphic User Interface and CPN Tools. After collecting initial data such as the planned timetables, track layouts, trains' locations, branch and query conditions using Microsoft Excel, the user opens the template model and conducts the following steps.

1. Instantiate the initial data and the branch conditions into the template model (using GUI).
2. Generate the (partial) state space (using CPN Tools).
3. Instantiate the query conditions (using GUI).

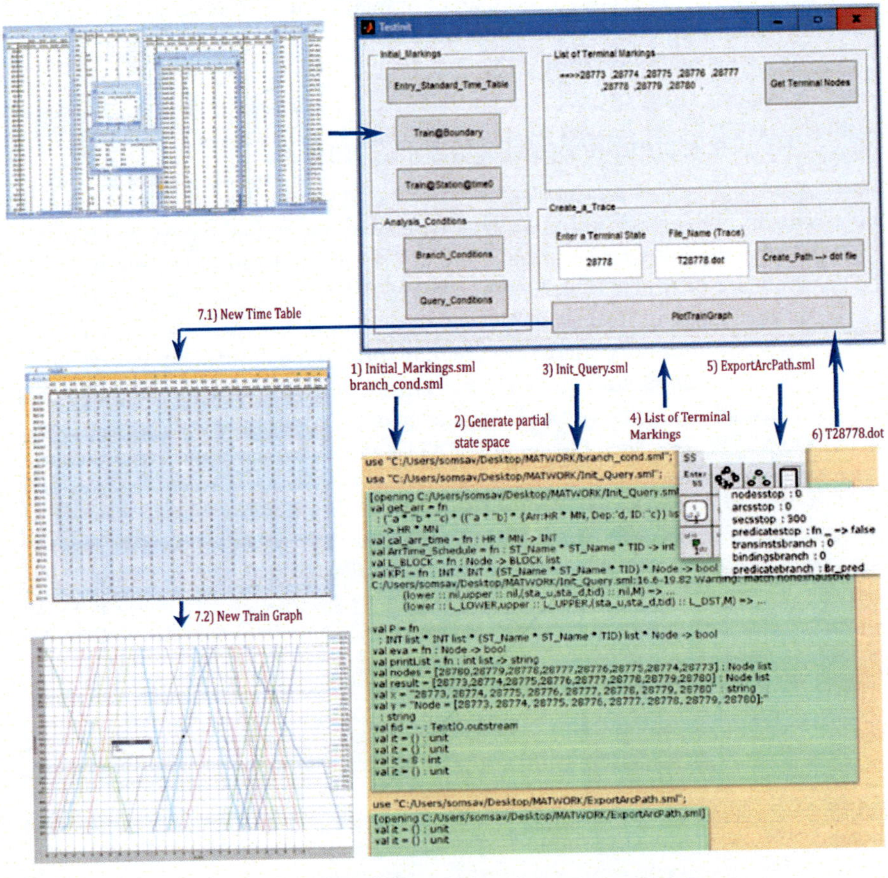

Fig. 4 The interaction between MATLAB graphic user interface and CPN tools

4. Search for the desired terminal markings and send them to MATLAB (using CPN Tools).
5. Choose a terminal marking and a file name to create a trace (using GUI).
6. Export the trace into a dot file (using CPN Tools).
7. Generate the new timetable and plot the new train graph from the dot file (using GUI).

7 Conclusion and Future Work

This paper proposes the train rescheduling system that comprises three applications: Microsoft Excel, CPN Tools and MATLAB GUI. Microsoft Excel is mainly used for entering the planned timetables, initial markings, current train locations, branch and query conditions in tabular form. The center of the system is the generic TCPN model in CPN Tools. Given train positions as an initial marking, CPN Tools is used to generate every possible scenario of train traffic. Each possible scenario is a possible timetable or a possible train graph. The train dispatcher gives the query conditions to select the solutions from all possible solutions. MATLAB GUI is originally used to plot the train graph. Later we extend its usage to conceal the SML code from the user so that he can use all three applications without program coding. We also suggest exploiting the fixed speed assumption to reduce the size of the state space using "branch option" in CPN Tools.

We canvas the future research theme in two directions. First, refining the model includes more details such as train length, platform length, varieties of network topology and signalling systems. We also wish to apply our work to the whole SRT network. Second, to further alleviate the state explosion, we would like to investigate combining the branch option with the sweep-line method.

Acknowledgements The author is thankful to his colleagues from the State Railway of Thailand, Patama Sridaranop, Anan Phonimdang, and Sommart Klinsukol.

References

Clavel M, Durán F, Eker S, Lincoln P, Martí-Oliet N, Meseguer J, Talcott C (2003) The Maude 2.0 system. In: Proceedings of rewriting techniques and applications, June 2003. LNCS, vol 2706. Springer, Heidelberg, pp 76–87

Fokkink WJ, Hollingshead PR (1998) Verification of interlockings: from control tables to ladder logic diagrams. In: Proceedings of 3rd workshop on formal methods for industrial critical systems (FMICS'98), Amsterdam, May 1998. Stichting Mathematisch Centrum, pp 171–185

Hagalisletto AM, Bjørk J, Yu IC, Enger P (2007) Constructing and refining large-scale railway models represented by petri nets. IEEE Trans Syst Man Cybern Part C 37(4):444–460

Janczura CW (1998) Modelling and analysis of railway network control logic using coloured petri nets. PhD thesis, School of Mathematics and Institute for Telecommunications Research, University of South Australia, Adelaide

Jensen K, Kristensen LM (2009) Coloured petri nets: modelling and validation of concurrent systems. Springer, Heidelberg

Jensen K, Kristensen LM, Wells L (2007) Coloured petri nets and CPN TOOLS for modelling and validation of concurrent systems. Int J Softw Tools Technol Transfer 9(3–4):213–254

Jensen K, Kristensen LM, Mailund T (2012) The sweep-line state space exploration method. Theoret Comput Sci 429:169–179

Van der Aalst WMP, Odijk MA (1995) Analysis of railway stations by means of interval timed coloured petri nets. Real-Time Syst 9(3):241–263

Vanit-Anunchai S (2009) Verification of railway interlocking tables using coloured petri nets. In: Proceedings of the 10th workshop on practical use of coloured petri nets and CPN tools, Aarhus, 19–21 Oct 2009

Vanit-Anunchai S (2010) Modelling railway interlocking tables using coloured petri nets. In: Clarke D, Agha G (eds) COORDINATION 2010. Lecture notes in computer science (LNCS), vol 6116. Springer, Heidelberg, pp 137–151

Vanit-Anunchai S (2012) Verifying flank protection: reaching the unreachable. In: Proceedings of the 3rd international conference of information and communication technology for embedded systems, Bangkok, 22–24 Mar 2012

Vanit-Anunchai S (2016) Application of coloured petri nets in modelling and simulating a railway signalling system. In: Beek MH, Gnesi S, Knapp A (eds) FMICS-AVOCS 2016. Lecture notes in computer science (LNCS), vol 9933. Springer, Heidelberg, pp 214–230

Vanit-Anunchai S (2018) Modelling and simulating a Thai railway signalling system using coloured petri nets. Int J Softw Tools Technol Transfer 20(3):243–262

Train Movement Under the Virtual Coupling System

Naphat Ketphat, Anthony Whiteing, and Ronghui Liu

Abstract The train virtual coupling system can be considered as an alternative control system introduced for increasing rail line capacity. The separation distance between consecutive trains is theoretically less compared to the other controls. However, the velocity of a following train may fluctuate significantly due to frequent changes of acceleration rate. Moreover, the separation distance between trains may possibly be lower than the minimum safe distance resulting in unsafe situations. In this paper, approaches for controlling the movement of following trains are introduced. The simulated results demonstrate that trains can proceed safely in that the actual separation distance between trains is not less than the critical safe distance. The trains have also to be driven smoothly, in that the following train is catching up the leading train with a constant velocity and has been platooned with the leading train where the velocity and the distance separating it from the leading train has been maintained. The capacity can be controlled by the range of safe distance, which ensures that the gap between trains is in the range of safe distance. In addition, the acceleration and deceleration rates of the train are realistic. The rates are limited to realistic acceleration and braking capabilities.

Keywords Virtual coupling · Separation distance · Safe distance · Diverging junction · Following train

1 Introduction

As rail passenger demand increases, there is a need for train services to be increased to support this growth. However, in some operating lines, the capacity utilization is already close to or at the maximum allowable capacity. Thus, adding new services, additional trains, into the system might not be possible. Construction of new lines

N. Ketphat (✉)
University of Leeds, Leeds LS2 9JT, UK
e-mail: tsnk@leeds.ac.uk

A. Whiteing · R. Liu
Institute for Transport Studies, University of Leeds, Leeds LS2 9JT, UK

W. Weerawat et al. (eds.), *Urban Rail Transit*, Lecture Notes in Mobility,
https://doi.org/10.1007/978-981-15-5979-2_15

is one solution used for supporting the increase in passenger demand. However, there are many problems including investment funds and availability of land for construction. Another possible solution is to manage trains running on the existing line with a closer headway. Modern signalling systems have been developed mainly to decrease the separation distance between trains. In order to increase line capacity, the signalling control called Moving Block Signalling (MBS) such as communication-based train control (CBTC) and European Train Control System (ETCS) level 3 has been developed as an effective and economical solution to reduce the train headway. Based on the MBS, the current position and velocity of each train are detected and sent to the radio block control centre (RBC) in order to determine the movement authority of each train within the control area. The braking curve (allowable braking distance vs. velocity) is calculated in real-time by the on-board computer installed in the train. It is, at least, equal to the full braking distance of the following train, which depends on its velocity, braking ability and safety margin caused by any error such as detection and communication delay time. Compared to more traditional fixed block signalling systems (FBS), it can be demonstrated that the MBS can increase line capacity and reduce train energy consumption (Takeuchi et al. 2003; Zhao et al. 2016). The MBS also has higher reliability than the existing control and offers more flexible and automated operation in e.g. train timetable recovery and junction management. Moreover, the trackside signal and detection equipment such as track circuits can be eliminated, reducing the cost of system construction and maintenance. MBS can be considered as a safe controlling system that relies on the safe separation distance between trains. This means that the following train can achieve a full stop without collision with the leading train even if the latter suddenly stops. However, such safety distance is longer than required considering that when the leading train brakes, it will not stop instantly but continue to run forward due to its movement inertia. As a result, the required safety distance to stop the train without collision is shorter than the full braking distance (minimum safe distance under MBS) (Rothery 1992), and thus the trains can run closer to each other. To use the benefit from such a shorter required safe distance, new signalling control methods called the Virtual Coupling System (VCS) have been introduced and are being developed for use in the real network.

2 Train Virtual Coupling System

As stated above, rail line capacity can be increased by applying VCS for train movement control. This can be confirmed by previous studies (CDP 2016; Mitchell 2016; Zhao et al. 2016; Flammini et al. 2019). The concept of the train virtual coupling system is to group trains as a convoy which can be merged, platooned, and split during transit (CDP 2016). Hence VCS is a new way of controlling trains by creating multiple convoys of trains (Flammini et al. 2018). Based on this system, the current data of the leading train is sent frequently to the train behind to calculate movement authority. A train can run closer following its leader proceeding on the same track. Two successive trains must however be separated by a sufficient space in order to ensure that the following train has a sufficient braking distance and can stop before reaching the leading train after the leading train applies its brakes. Before joining a train in a convoy, the convoy proposal must be accepted (Flammini et al. 2019). After accepting the convoy proposal, the leading train's data is sent frequently to the train driving behind for calculation of the optimal velocity. In the merged state, the following train proceeds with a higher velocity in order to reduce the gap from the front train. It will decelerate when the actual distance from its front train equal is to or slightly less than the minimum safe distance. This will transfer the two trains to the platooning state where the velocity of both trains is the same. There are essentially two sub-controls included in the system (Fig. 1). In the first control, the non-convoy state, the trains proceeding on the same track move independently and are controlled based on MBS. Real-time velocity and location of each train are sent frequently to the RBC. Then, data on all trains within the control area will be used to calculate the signal status sent back to each train (Harriss 2016). The separation distance between two trains must be not less than the absolute braking distance to ensure that the following train can brake on time in the case that the leading train applies its brakes or is suddenly stopped.

The second control will be used when the trains are in a convoy state. The following train will merge itself to a train or a group of trains in front by sending the convoy proposal to its leader. The leading train can either accept or reject the proposal. The first train in the convoy is still controlled under MBS but the allowable braking distance of the following trains will be calculated by those trains using the data sent from the convoy leader. The safe space between two trains must be not less than the relative braking distance which depends on the different velocities of the trains, the following train's braking ability, and buffer distance provided for avoiding any error due to communication and detection delay time. The safety of VCS relies on the safety integrity level of the communication between trains. In the case that communication between trains is lost, the train operation must be switched back to non-convoy state. The following train will react to the train in front by adjusting the distance separated from the front train. Generally, the following train will accelerate if the space between trains is longer than the minimum safe distance and decelerate if the space apart from the leading train is less than the safe distance (Henke et al. 2008; Li and Gao 2011).

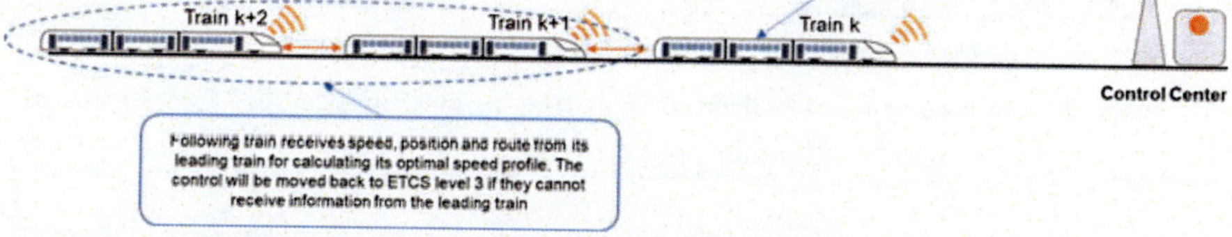

Fig. 1 Train virtual coupling system. Modified from Mitchell (2016)

2.1 Minimum Safe Distance

The required minimum separation distance between trains depends on the types of control and the location that the trains have passed; plain line, junction, station. Along the plain line, the minimum safe distance of different controls is shown in Table 1.

When the trains under VCS come to a diverging junction, the minimum separation distance is not the same as the distance required along plain line. That is due to the point operation time. If two successive trains must continue onto different lines after passing such a junction, the following train must wait until the whole length of the leading train has passed the junction. According to the recommendation by McNaughton (2011), when trains are passing diverging at such a junction, they should be separated by more than the minimum distance depending on the factors shown in Fig. 2.

Table 1 Minimum safe separation distance along the plain line. Modified from Zhao et al. (2016)

Control	Min. safe distance	Comment
MBS	$\Delta x_{k+1}^{MBS} = \frac{v_{max}^2}{2b_{k+1}} + SM$	The worst-case braking distance which depends on the maximum velocity allowed along the line (v_{max}) and the practical braking rate (b_{k+1}) of the following train
	$\Delta x_{k+1}^{MTB} = \frac{v_{max} v_{k+1}(t)}{2b_{k+1}} + SM$	This is calculated by using the allowed maximum velocity (v_{max}) and the following train's velocity ($v_{k+1}(t)$)
	$\Delta x_{k+1}^{PMB} = \frac{v_{k+1}^2(t)}{2b_{k+1}} + SM$	The safe distance depends on the current velocity of the following train itself $v_{k+1}(t)$
VCS	$\Delta x_{k+1}^{VCS} = \frac{v_{k+1}^2(t) - v_k^2(t)}{2b_{k+1}} + SM$	The minimum safe distance between trains can be minimized depending on the difference of current velocity between two adjacent trains, the braking rate of the following train and the safety margin (SM)

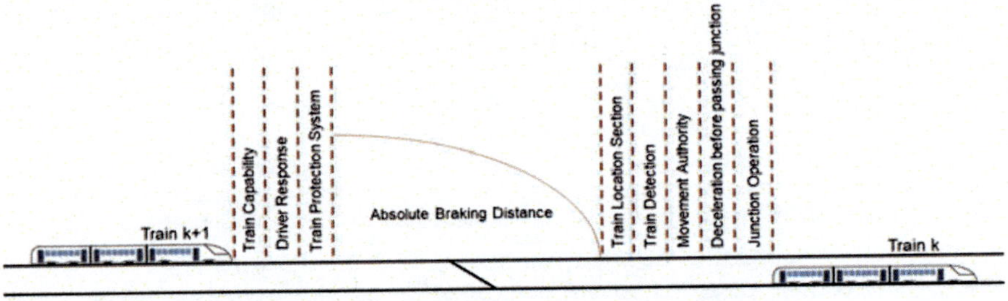

Fig. 2 Minimum separation distance for a diverging junction. Modified from McNaughton (2011)

2.2 *Controlling the Following Train's Movement*

To control the movement of trains under VCS, many approaches have been adapted as base-models, including discrete event models (Ming et al. 2014), discrete-time models (Yang et al. 2010), and car following models (Li and Gao 2009; Li et al. 2010; Ye and Li 2013). In particular, we can consider the car-following model as a relevant and appropriate model for simulating train movement. This is because the movement of trains following each other under VCS is essentially similar to car movement on a highway, in that the car's optimal velocity depends on the space between it and the car in front. Whilst many approaches have been introduced for controlling the following train under VCS, the general concept, based on previous studies, is to control the following train's movement by forcing it to decelerate or accelerate depending on the distance separating it from the train in front. In the case that separation distance between trains is less than the minimum safe distance, the following train will be forced to decelerate. If the space between trains is longer than the safe distance, the following train will be triggered to speed up in order to shorten the distance away from the leading train. The concept is to control the actual distance between trains close to the minimum safe distance. If actual distance is higher than the minimum safe distance, this will limit the benefit of the system (Ke-Ping and Li-Jia 2009; Li et al. 2010). The examples below show the previous proposed approaches and the results for the train's movement under VCS (Tables 2 and 3).

As can be seen, various approaches to controlling train movement under VCS have been proposed and most of them have been simulated for the simple plain lines. However, in real train operations, there are many constraints such as speed capability, restricted speed limits along the line and at junctions, the position of junctions, the number of trains built into each convoy, etc. which might limit the benefits of VCS. Thus, it is important to consider whether VCS increases line capacity under more realistic conditions of train operation. In this paper, approaches for controlling the following train under VCS are proposed with the objectives of controlling trains more safely and stably whilst obtaining high line capacity.

3 Train Movement Under the Virtual Coupling System

As mentioned above, the following train reacts to the train in front by adjusting the space separating the two trains. For increasing line capacity, the concept is to keep the space between trains close to the safe separation distance but not shorter than the safe distance. Thus, the safe separation distance can be considered as the key factor for controlling train's movement under VCS.

Table 2 Summary of previous proposed approaches for controlling trains under VCS

Year	Authors	Proposed approaches	Comment
2009	Ke-Ping, & Li-Jia	$a_{k+1}(t+\Delta t) =$ $C_1[1-\exp(-C_2\tau)]\left(\frac{1}{\tau}\right)\left[v^{opt}(\Delta x_{k+1}(t)) - v_{k+1}(t)\right]$	The additional term, $C_1[1-\exp(-C_2\tau)]$ is added into the traditional optimal car following model for limiting the range of acceleration/deceleration
2010	Li et al.	$a_{k+1}(t+\Delta t) =$ $C_1[1-\exp(-C_2\tau)]\left(\frac{1}{\tau}\right)\left[v^{opt}(\Delta x_{k+1}(t)) - v_{k+1}(t)\right]$	The term $v^{opt}(\Delta x_{k+1}(t))$ is calibrated by adding the coefficient factor (coe). It can be calculated by $v^{opt}(\Delta x_{k+1}(t)) =$ $\frac{v_{max}}{2}\left\{\tanh\left[coe^*\Delta x_{k+1}(t) - \Delta x_{k+1}^{safe}\right] + \tanh\left[coe^*\Delta x_{k+1}^{safe}\right]\right\}$
2011	Li and Gao	$a_{k+1}(t+\Delta t) = \left(\frac{1}{\tau}\right)\left[v^{opt}(\Delta x_{k+1}(t)) - v_{k+1}(t)\right]$	The optimal velocity term is classified based on the signal where: Green $v^{opt}(\Delta x_{k+1}(t)) = v_{max}$ Yellow $v^{opt}(\Delta x_{k+1}(t)) = (v_y - v_r)\left\{\tanh\left[\Delta x_{k+1}(t)\right]\right\} + v_r$ Red $v^{opt}(\Delta x_{k+1}(t)) = \frac{v_r}{2}\{\tanh\left[\Delta x_{k+1}(t) - \Delta x_{k+1}^{safe}\right] + \tanh \Delta x_{k+1}^{safe}$
2013	Ye and Li	$a_{k+1}(t+\Delta t) = \left(\frac{1}{\tau}\right)\left[v^{opt}(\Delta x_{k+1}(t)) - v_{k+1}(t)\right]$	The term optimal velocity $v^{opt}(\Delta x_{k+1}(t))$ is calculated by $v_{CR} = \frac{t_{co} - \sqrt{t_{co}^2 - 4\left(\frac{1}{2a_{k+1}} + \frac{1}{2b_{k+1}}\right)L}}{\frac{1}{a_{k+1}} + \frac{1}{b_{k+1}}}$
2015	Ye et al.	$a_{k+1}(t+\Delta t) = \left(\frac{1}{\tau}\right)\left[v^{opt}(\Delta x_{k+1}(t)) - v_{k+1}(t)\right]$	$v^{opt}(\Delta X_{k+1}(t)) =$ $\frac{v_{max}}{2}\left\{\tanh\left[\Delta x_{k+1}(t) - \Delta x_{k+1}^{safe}\right] + \tanh\left[\Delta x_{k+1}^{safe}\right]\right\}$ where $\Delta x_{k+1}(t) = \left[x_k(t) - x_{k+1}(t)\right] + x_{er}$ x_{er} is an error term

Table 3 Results obtained from the previous proposed approaches

Year	Authors	Purpose	Results	Comment
2009	Ke-Ping, & Li-Jia	Limits the range of acceleration and deceleration	Train's acceleration and deceleration are in the limited range and the separation distance between trains is higher than the safe stopping distance	The actual space between trains is higher than the safe distance which might limit the benefit of VCS. The line capacity is not significantly increased
2010	Li et al.	Limits the range of acceleration, and improves the train movement stability	The separation distance between trains is higher than the safety distance and the acceleration/deceleration rate is in the limited range	The actual space between trains is still higher than the minimum safe distance which might limit the benefit of VCS
2011	Li and Gao	Increases capacity under fixed block signaling system (FBS)	The trains can proceed safely without collision between trains. The actual gap between successive trains is longer than the safe distance (length of physical block)	The space between trains under VCS can be shorter than the length of block
2013	Ye and Li	Minimizes the gap between trains and control train arrives at the destination on-time	The optimal velocity curve can be created based on the proposed model. This will control the trains to arrive at the next station on-time. It can also be confirmed that the actual gap between trains is not less than the minimum safe distance confirming that the trains will run safely	The separation distance between trains is sometimes lower than the minimum safe distance which may increase the risk of collision
2015	Ye et al.	Improves safety by avoiding unsafe movement and Improves stability of the train under uncertain conditions	The proposed model can be used for controlling train movement under uncertain conditions	The separation distance between trains is sometime lower than the minimum safe distance and the movement of the train is not stable enough

3.1 Modified Safe Separation Distance

Assume that two trains have been proceeding along the same line. When the leading train (k) applies its brakes, the following train (k + 1) starts braking as well to maintain the safe space and avoid collision. The braking safe distance will be decided by the current position of the rear of the leading train (k) and the front of the following train. It must not be less than the full braking distance of the following train. Due to the delay caused by the detection, operation, and communication system, the safety margin (SM) must be added. The absolute braking distance of the following train ($x_{k+1}^{abs}(t)$) can be calculated by (Ning 1998)

$$x_{k+1}^{abs}(t) = \frac{v_{k+1}(t)^2}{2b_{k+1}} + SM \tag{1}$$

Under the VCS, we can assume that the probability that the leading train stops dead is very small, i.e. it can be assumed that the leading train will not suddenly stop at the point that it begins to applying its brakes, but will continue to move forwards at a decelerating rate.

The movement behaviour of both trains can be described as shown in Fig. 3. The difference of travelling distance between two trains after braking is;

$$\Delta x = \left(\frac{v_{k+1}^2}{2b_{k+1}} + SM\right) - \left(\frac{v_k^2}{2b_k}\right) \tag{2}$$

Under VCS, it is assumed that the braking rate of all trains proceeding in the same convoy is equal ($b_k = b_{k+1} = \ldots = b_{k+n}$). Then, the required braking distance for the following train ($\Delta x_{k+1}^{rel}(t)$) under VCS is (Ning 1998);

$$\Delta x_{k+1}^{rel}(t) = \left(\frac{v_{k+1}^2(t) - v_k^2(t)}{2b_{k+1}} + SM\right) \tag{3}$$

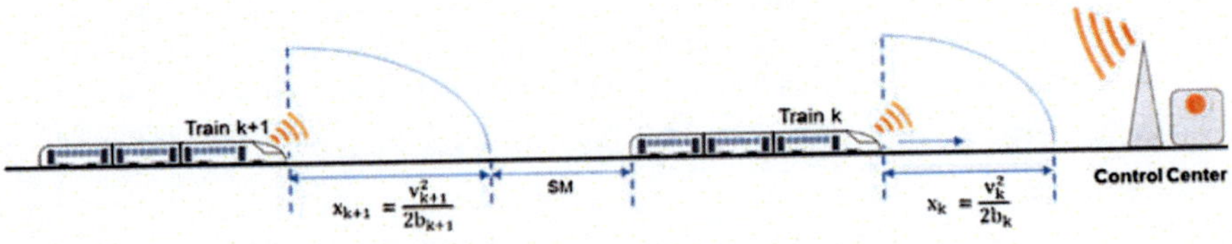

Fig. 3 Trains' movement after applying brake

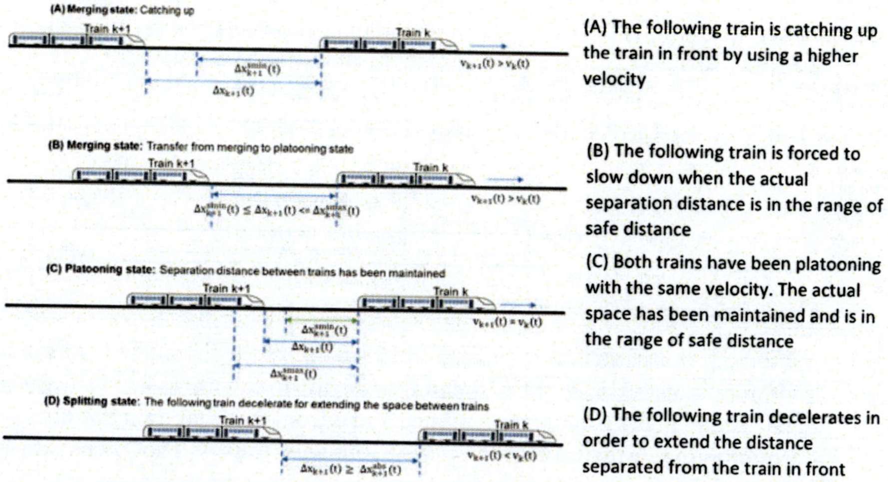

Fig. 4 Moving states under the VCS

3.2 *The Following Train's Movement Under the Virtual Coupling System*

There are 3 moving states under the VCS: merging, platooning, and splitting states. The movement of the following train in each state is described as Fig. 4.

To control a group of trains under VCS, the minimum safe distance as shown in Eq. (3) is considered as the reference point from which the following train must compute its optimal acceleration and velocity in order to adjust the space separated from the front train. However, unsafe movement might occur as the impact of system delay time means that the following train cannot decelerate instantly when the actual space is shorter than the minimum safe distance. Thus, to control trains more safely, the equation for calculating the minimum safe distance should be improved.

3.2.1 Modified Minimum Safe Distance ($\Delta x_{k+1}^{smin}(t)$)

Ones the trains have been operating under the VCS, it is important to ensure that a train is following each other by safe distance in that the gap separated from its front train (Δx_k (t)) must not be shorter than the minimum safe distance ($\llbracket \Delta x \rrbracket_k^{min}$) . According to the previous studies, the actual separation distance is sometime shorter than the minimum safe distance resulting an unsafe situation in that the collision between trains might occur when the leading train instantly brake (Xuan et al. 2013; Henke et al. 2008; and Ye et al. 2015).

- **Transferring from merging to platooning state**

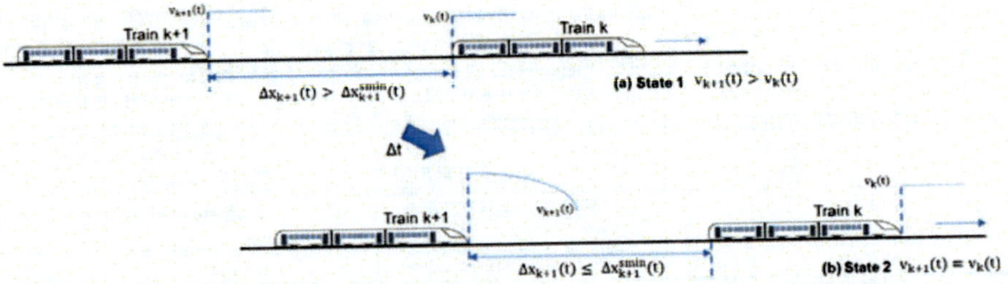

Fig. 5 Following train's movement in merging state

Before transferring to the platooning state, the following train's velocity is forced to be decelerated when the distance separated from the front train becomes less than the minimum safe distance. However, the following train will not decelerate instantly after receiving the data from the front train due to the communication delay time (Δt). As a result, the following train has moved with its constant velocity before slowing down (Fig. 5).

The difference of travelling distance in state 1 where the following train's velocity is higher than the leading train's velocity (Fig. 5a) is

$$\Delta x_{k+1}^{m1}(t) = ((v_{k+1}(t))(\Delta t) - ((v_k(t))(\Delta t) = ((v_{k+1}(t) - v_k(t))(\Delta t) \tag{4}$$

when the following train decelerates. The difference of distance in state 2 (Fig. 5b) is

$$\Delta x_{k+1}^{m2}(t) = \left[((v_{k+1}(t))(\Delta t) - 0.5b_{k+1}(\Delta t)^2\right] - ((v_k(t))(\Delta t) \tag{5}$$

According to the approach's constraint shown in the Eq. (22), the deceleration rate of the following train $b_{k+1} = \min\left[b_{k+1}^{max}, \frac{(v_{k+1}(t)-v_k(t))}{\Delta t}\right]$. Then,

$$\Delta x_{k+1}^{m2}(t) = \left[((v_{k+1}(t))(\Delta t) - 0.5\left[\frac{v_{k+1}(t) - v_k(t)}{\Delta t}\right](\Delta t)^2\right] - ((v_k(t))(\Delta t) \tag{6}$$

Thus, the difference of travelling distance of two successive trains in this situation is

$$\Delta x_{k+1}^{mer}(t) = \Delta x_{k+1}^{m1}(t) + \Delta x_{k+1}^{m2}(t) = 1.5\left[(v_{k+1}(t) - v_k(t))\Delta t\right] \tag{7}$$

This means that in the worst case, the actual separation distance between trains before transferring to the platooning state is less than the minimum safe distance by $\Delta x_{k+1}^{mer}(t)$. To avoid this unsafe situation, the term $\Delta x_{k+1}^{mer}(t)$ is added into the minimum safe distance in Eq. (3). Thus, the minimum safe distance between trains is modified and can be computed by

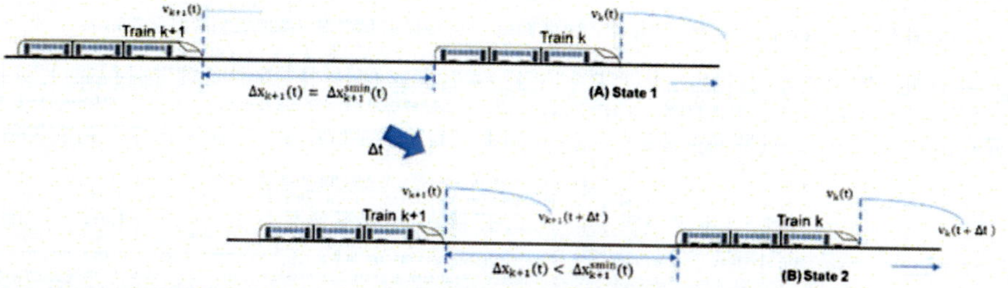

Fig. 6 Train movement in platooning state when the leading train decelerates

$$\Delta x_{k+1}^{smin}(t) = \left(\frac{v_{k+1}^2(t) - v_k^2(t)}{2b_{k+1}} + SM \right) + \Delta x_{k+1}^{mer}(t) \tag{8}$$

- **The leading train decelerates while it is in the platooning state**

Due to the impact from communication delay time (Δt) similarly to the unsafe movement explained above, the following train cannot decelerate instantly even when the actual space is shorter than the safe distance. In other words, the following train cannot decelerate at precisely the same time as the leading train. Consequently, unsafe movement might occur if the leading train slows down while it has been platooned. Assuming that the separation distance when two trains have platooned is equal to the minimum safe distance, the actual space between trains will be shorter than the minimum safe distance if the front train decelerates. To avoid this situation, the following train has to decelerate earlier in order to ensure that the distance is not shorter than the critical distance (minimum safe distance). The movement of the trains in this situation is shown in Fig. 6. It is seen that the following train has proceeded with its current velocity under the time period Δt while the leading train decelerates. The difference of traveling distance of both successive trains in this state is

$$\Delta x_{k+1}^{pl}(t) = (v_{k+1}(t)(\Delta t)) - \left[((v_k(t))(\Delta t) - \frac{1}{2} b_k(t)(\Delta t)^2 \right] \tag{9}$$

The data sent by the leading train is the current velocity and position. Thus, the actual braking rate is unknown. The worst case resulting in the shortest separation distance will occur when the leading train decelerates by the maximum braking rate (b_k^{max}). Before decelerating, both trains are platooned with the same velocity $v_{k+1}(t) = v_k(t)$. Thus, the distance difference is

$$\Delta x_{k+1}^{pl}(t) = \left[\frac{1}{2} b_k^{max} (\Delta t)^2 \right] \tag{10}$$

After state1 (Fig. 6a), the velocity of the following train is higher than the leading train's velocity. The movement behaviour of both trains can be seen in Fig. 6b. The

Table 4 Movement conditions

Conditions	Distance difference	AND	Vel. difference	$a_{k+1}(t)$
1	$\Delta x_{k+1}(t) > \Delta x^{smax}_{k+1}(t)$		$v_k(t) > v_{k+1}(t)$	Accelerate
2	$\Delta x_{k+1}(t) > \Delta x^{smax}_{k+1}(t)$		$v_k(t) = v_{k+1}(t)$	Accelerate
3	$\Delta x_{k+1}(t) > \Delta x^{smax}_{k+1}(t)$		$v_k(t) < v_{k+1}(t)$	Constant $v_{k+1}(t)$
4	$\Delta x_{k+1}(t) < \Delta x^{smin}_{k+1}(t)$		$v_k(t) > v_{k+1}(t)$	Constant $v_{k+1}(t)$
5	$\Delta x_{k+1}(t) < \Delta x^{smin}_{k+1}(t)$		$v_k(t) = v_{k+1}(t)$	Decelerate
6	$\Delta x_{k+1}(t) < \Delta x^{smin}_{k+1}(t)$		$v_k(t) < v_{k+1}(t)$	Decelerate
7	$\Delta x^{smin}_{k+1}(t) \le \Delta x_{k+1}(t) \le \Delta x^{smax}_{k+1}(t)$		$v_k(t) > v_{k+1}(t)$	Constant $v_{k+1}(t)$
8	$\Delta x^{smin}_{k+1}(t) \le \Delta x_{k+1}(t) \le \Delta x^{smax}_{k+1}(t)$		$v_k(t) = v_{k+1}(t)$	Constant $v_{k+1}(t)$
9	$\Delta x^{smin}_{k+1}(t) \le \Delta x_{k+1}(t) \le \Delta x^{smax}_{k+1}(t)$		$v_k(t) < v_{k+1}(t)$	Constant $v_{k+1}(t)$

problem is that we cannot predict the final velocity of the leading train (the velocity that the leading train reduces to). In the worst case, it is assumed that the leading train decelerates to a stop (The final velocity = 0). Thus, the difference of travelling distance in state 2 ($\Delta x^{p2}_{k+1}(t)$) is

$$\Delta x^{p2}_{k+1}(t) = \left[((v_{k+1}(t))(T) - \frac{1}{2}b_{k+1}(T)^2\right] - \left[((v_k(t))(t) - \frac{1}{2}b_k(T)^2\right] \tag{11}$$

As stated in the approach's constraints (Table 4), the acceleration and braking rate of all trains in the same convoy must be the same ($b_{k+1} = b_k$). Thus,

$$\Delta x^{p2}_{k+1}(t) = \left[(v_{k+1}(t) - (v_k(t))\right]T_k \tag{12}$$

where T_k refers to the time used for stopping,$T_k = \frac{v_k(t)}{b_k(t)}$. The following train will be forced to decelerate when it receives the deceleration data from the leading train. Due to the communication and operation time Δt, the difference of velocity of both successive trains is $v_k(t) = v_{k+1}(t) - b_k\Delta t$. Before the leading train decelerates, both trains have platooned with the same velocity $v_{k+1}(t) = v_k(t)$. Thus,

$$\Delta x^{p2}_{k+1}(t) = \left[(v_{k+1}(t) - (v_{k+1}(t) - b_k\Delta t)\right]\frac{v_k(t)}{b_k} \tag{13}$$

$$\Delta x^{p2}_{k+1}(t) = v_k(t)\Delta t \tag{14}$$

The adjusted distance for providing the unsafe situation is

$$\Delta x^{plaD}_{k+1}(t) = \Delta x^{p1}_{k+1}(t) + \Delta x^{p2}_{k+1}(t) \tag{15}$$

Thus, the minimum separation distance should be adjusted and can be calculated by

$$\Delta x_{k+1}^{smin}(t) = \left(\frac{v_{k+1}^2(t) - v_k^2(t)}{2b_{k+1}^{max}} + SM\right) + \Delta x_{k+1}^{plaD}(t) \tag{16}$$

In order to avoid such unsafe movement, the minimum safe distance ($\Delta x_{k+1}^{smin}(t)$) used in many previous studies needs to be improved. According to the 2 unsafe situations in both the transferring and platooning states explained above, the distance difference computed by Eqs. (8) and (16) is compared for considering the possible worst case. It is seen that the distance calculated from Eq. (16) is higher, resulting in a higher risk. This means that the second situation is more critical than another case. Thus, the minimum safe distance shown in Eq. (16) is used as the modified equation for calculating the minimum safe distance.

3.2.2 Modified Maximum Safe Distance ($\Delta x_{k+1}^{smax}(t)$)

For controlling the capacity and stabilizing the movement of the following train, the concept of maximum safe distance ($\Delta x_{k+1}^{smax}(t)$) is introduced. This refers to the upper bound of the range of safe distance provided in order to control the separation distance in the platooning state (capacity control) and stabilise the movement of the following train. The maximum safe distance can be considered through the movement situation below.

- **Leading train accelerates while it is in the platooning state**

According to the train movement under VCS shown in Fig. 4b, when the following train decelerates and the final velocity of the following train is equal to the leading train's velocity, both trains are transferred to the platooning state (Fig. 4c). After transferring to the platooning state, the actual space between trains will be maintained. As a result, the required minimum safe distance is reduced due to the decreasing in the current velocity of the following train proceeding in the platooning state.

According to the previous studies, the following train will be forced to accelerate again if the actual separation distance is still longer than the minimum safe distance resulting in unsmooth movement that the following train's velocity has adapted. In this study, after transferring to the platooning state where the velocity of both trains is the same, the idea is to maintain the separation distance between trains based on 2 conditions below:

- The following train is not forced to decelerate if the distance between trains is longer than the minimum safe distance.
- And the following train is not forced to accelerate if the distance between trains is lower than the maximum safe distance.

Therefore, the trains will be transferred to the platooning state when the actual distance between trains is in the safe distance range, the safe space which is longer than the minimum safe distance but still shorter than the maximum safe distance. The question is what is the optimal maximum safe distance? When the trains have platooned with constant separation distance, the following train should be forced to accelerate only in the case that the separation distance between trains is extended. The separation distance will be extended in the case that the leading train accelerates while it is in the platooning state. The idea is to maintain train velocity if the distance away from the front train has not been changed and to force the following train to accelerate instantly after the distance is extended. Assume that the leading train accelerates by its maximum acceleration rate (a_k^{max}). Under communication time (Δt), the actual separation distance will be increased by

$$\Delta x_{k+1}^{ext}(t) = \left[((v_k(t))\Delta t) + \frac{1}{2}a_k^{max}(\Delta t)^2\right] - [((v_{k+1}(t)\Delta t)] \tag{17}$$

but before the leading train accelerates $v_k(t) = v_{k+1}(t)$. The difference in traveling distance after the leading train accelerates under communication time, Δt is

$$\Delta x_{k+1}^{ext}(t) = \frac{1}{2}a_k^{max}(\Delta t)^2 \tag{18}$$

To force the following train to accelerate instantly when the leading train accelerates, the following train should accelerate when the actual separation distance is longer than the actual separation distance between trains in the platooning state ($\Delta x_{k+1}^{platoon}(t)$) plus the different distance due to the acceleration of the leading train ($\Delta x_{k+1}^{ext}(t)$). Thus, the upper bound of safe distance (maximum safe distance, $\Delta x_{k+1}^{smax}(t)$) should be

$$\Delta x_{k+1}^{smax}(t) = \Delta x_{k+1}^{platoon}(t) + \Delta x_{k+1}^{ext}(t) \tag{19}$$

However, the term $\Delta x_{k+1}^{platoon}(t)$) is unknown. As seen in Fig. 4c, after transferring from merging to the platooning state, the following train will start decelerating if the space between trains is equal or less than the minimum safe distance required in the merging state. Assuming that the following train decelerates when the actual gap is equal to the minimum safe distance. Thus, the actual gap after decelerating is less than the minimum safe distance in the merging state by $\frac{1}{2}b_{k+1}^{max}(\Delta t)^2$. It is seen that the actual separation distance between trains in the platooning state is slightly less from the minimum distance required in the merging state as the term $\Delta x_{k+1}^{tran}(t)$ is too small. To make the approach simpler, the minimum safe distance in the merging state could be assumed to be the actual separation distance in the platooning state $\Delta x_{k+1}^{platoon}(t)$. Thus, the maximum safe distance can be estimated to be approximately equal to the minimum safe distance required in the merging state (Fig. 7).

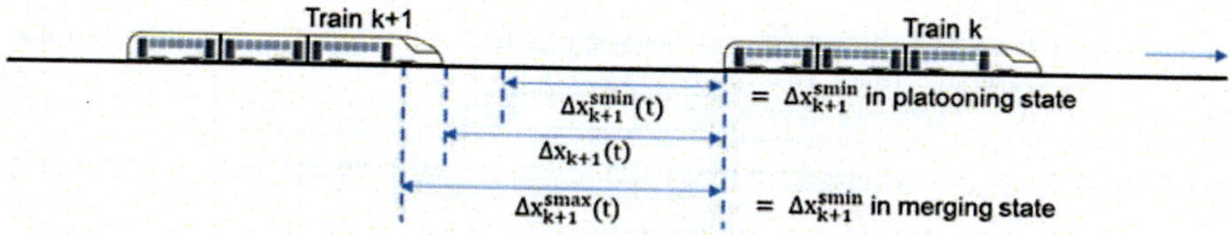

Fig. 7 Minimum and maximum safe separation distance

To stabilise the following train movement, the optimal gap between minimum and maximum safe distance is optimized. The maximum velocity (v_{k+1}^{max}) with which following train catches up the front train results in the highest maximum safe distance. Modified from Eq. (16), possible minimum safe distance in the splitting state $\Delta x_{k+1}^{plaM}(t)$ can be calculated by

$$\Delta x_{k+1}^{plaM}(t) = \left(\frac{\left(v_{k+1}^{max}\right)^2 - v_k^2(t)}{2b_{k+1}^{max}} + SM \right) + \Delta x_{k+1}^{plaD}(t) \tag{20}$$

Thus, the upper bound of the range of safe distance (maximum safe distance) can be computed by

$$\Delta x_{k+1}^{smax}(t) = \Delta x_{k+1}^{platM}(t) + \Delta x_{k+1}^{ext}(t) \tag{21}$$

To split the train out from the convoy, the following train has to decelerate and proceed with a lower velocity for extending the separation distance between trains. The control will be switched from VCS to MBS when the actual separation distance between trains is longer than the minimum safe distance required under MBS (full braking distance).

3.2.3 Optimal Acceleration and Deceleration Rates for Building the Convoy

As explained above, the range of safe distance can be set. The next step is to control the movement of the following train. The purposes are to control the distance between them to be close to but not shorter than the minimum safe distance. To achieve the purposes whilst controlling the following train's movement more stably, the conditions below are applied based on the assumption that the following train has been proceeding with a higher velocity for adjusting the space separated from its front train. In this thesis, the safe separation distance is set as the range of safe distance ranging between $\Delta x_{k+1}^{smin}(t)$ and $\Delta x_{k+1}^{smax}(t)$ as can be calculated by Eqs. (16) and (21) respectively. Table 4 shows the movement conditions that the following train reacts to the train in front. Before merging into the convoy, the actual space between successive trains is longer than the maximum safe distance ($\Delta x_{k+1}(t) > \Delta x_{k+1}^{smax}(t)$)

but the current velocity of the following train can be higher than, lower than, or equal to the leading train's velocity.

Approach's Constraint : One problem in the previous approaches is that the acceleration and deceleration rates calculated from these approaches are sometimes higher than the maximum acceleration capability of the trains. In this thesis, the acceleration and deceleration rates are limited and can be calculated by using the equations below. These are applied for limiting the acceleration and braking rate.

$$a_{k+1}(t) = \min\left[a_{k+1}^{max}, \frac{(v_{k+1}(t) - v_k(t))}{\Delta t}\right] \tag{22}$$

$$b_{k+1}(t) = \min\left[b_{k+1}^{max}, \frac{(v_{k+1}(t) - v_k(t))}{\Delta t}\right] \tag{23}$$

The following trains might need more than one time to adjust their velocity for transferring themselves to the platooning state. For example, assuming that the braking rate of the following train (b_{k+1}^{max}) is 0.5 m/s^2 and $\Delta t = 10$ s. If the following train is proceeding at 98 m/s and is catching up the leading train which is travelling at 90 m/s, then to maintain the actual gap between trains within the range of safe distance the following train will decelerate from 98 to 93 m/s and then decelerate from 93 to 90 m/s.

4 Simulation and Results

In this section the proposed approaches introduced in the methodology are applied in simulation test cases. Trains are controlled based on the virtual coupling system under normal conditions (i.e. no impact from weather and track elevation). The train movements are simulated based on the proposed approaches using MATLAB (R2018b). We assume that there are 2 trains operating along the same line with the operational parameters shown in Tables 5 and 6; Fig. 8.

The distance and velocity profiles of both leading and following trains is shown in Fig. 9. It is seen that the following train departed 3 min after the leading train departed. The velocity difference in the merging and splitting state is 10 and 5 m/s. respectively. The optimal distance that the following train should slow down in is approximately 17 km. before reaching the diverging junction. The following train accelerates to the operating velocity at 60 m/s. while the leading train has proceeded at 50 m/s. Due to the velocity difference between the two trains, the gap between them has been closed.

The trains have been platooned and then must split before approaching the diverging junction placed 2 km. before the next station. Figure 10 show the velocity and distance profile of trains in merging and splitting states.

Table 5 Operational parameters applied in the simulation

Parameters			Comment/references
(1) Restricted velocity along the line	70	m/s	
(2) Restricted velocity at junction	30	m/s	
(3) Operating velocity	60	m/s	
(4) Time step	10	s	Communication delay time
(5) Safety margin along the line	2.4	km	McNaughton (2011), Table 6
(6) Buffer at junction	300	m	Connor (2014)
(7) Max. acceleration rate	0.5	m/s^2	Hunyadi (2011)
(8) Max. braking rate	0.5	m/s^2	Hunyadi (2011)
(9) Junction operation time	12	s	McNaughton (2011)
(10) Junction's location (Diverging junction)	2	km	2 km. before station B
(11) Train length	100	m	HS2 (2011)
(12) Dispatching headway	180	s	

Table 6 Safety margin. Modified from McNaughton (2011)

Elements	Headway (s)	Commend
(1) Train capability	1	ETCS response time
	3	Brake operation time
(2) Driver response	6	Time of driver response to the displayed data to operate the control
(3) Train position system	5	Position detection tolerance
(4) Train location section	16	Distance between position detection devices
(5) Driver assist response	3	System operation time (Time to input data to display at the user interface)
Total	34	SM = 34 s × 70 m/s ≈ 2400 m

Fig. 8 Example

The movement of the following train is based on the proposed approaches. The train has moved with a higher velocity in order to merge itself into the convoy. It then decelerates to a lower velocity when it needs to split out from the convoy. Figure 11 shows the acceleration rates of two trains, it is seen that the following train decelerates at time 910 [when the space between away from the leading train is less than the maximum safe distance (Fig. 12 (Right))].

That will force the leading train to accelerate to the operating velocity at 60 m/s. Due to the increased velocity of the leading train, the following train is also forced

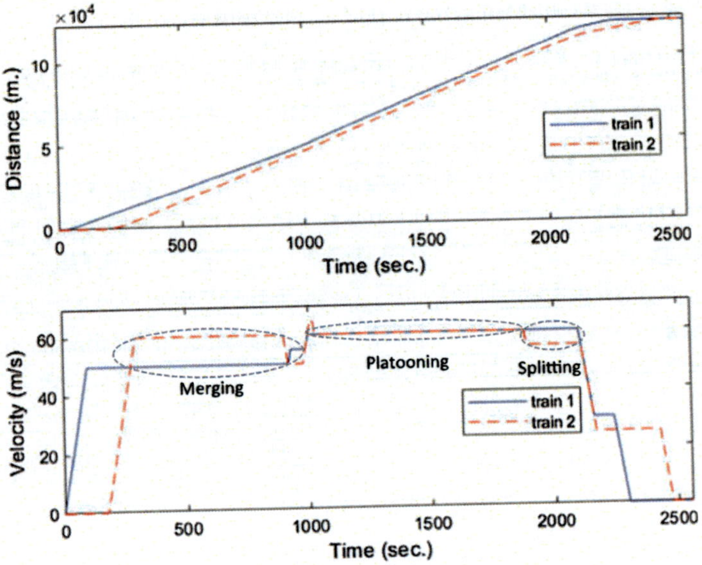

Fig. 9 Distance and velocity profile of trains proceeding under the VCS

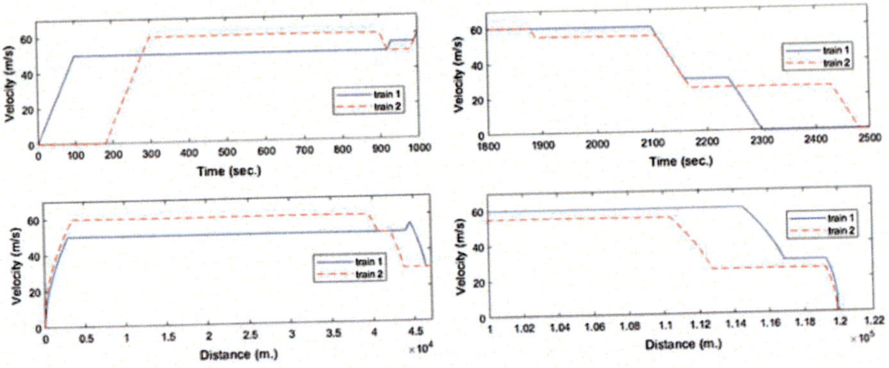

Fig. 10 Velocity and distance profile in merging (Left) and splitting state (Right)

to accelerate as shown in the dotted circle in Fig. 11 (Right). The following train's velocity has been adapted to be equal to the velocity of the leading train in order to maintain its velocity and space in the platooning state. Referring the acceleration profile, it is seen that the acceleration and braking rate of the trains is not higher than their capability.

Before transferring to the platooning state, the actual separation distance between trains has been higher than the maximum safe distance. The actual space has become closer to the range of safe distance. As seen in Fig. 12, the actual space between trains has been maintained when it is in the range of safe distance. It can be confirmed that

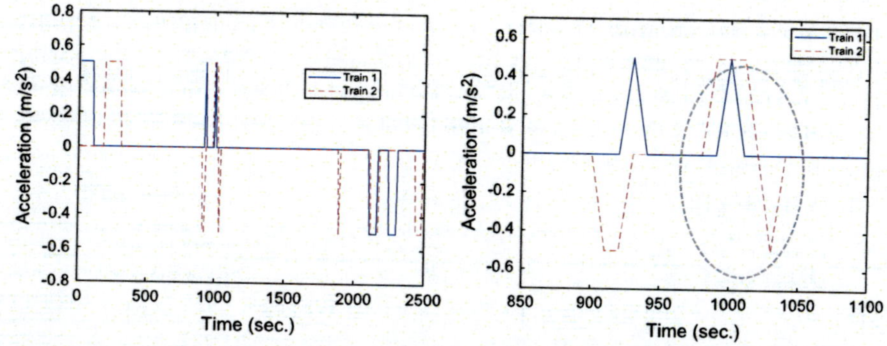

Fig. 11 Acceleration profile (Left) and acceleration rate before transferring to the platooning state (Right)

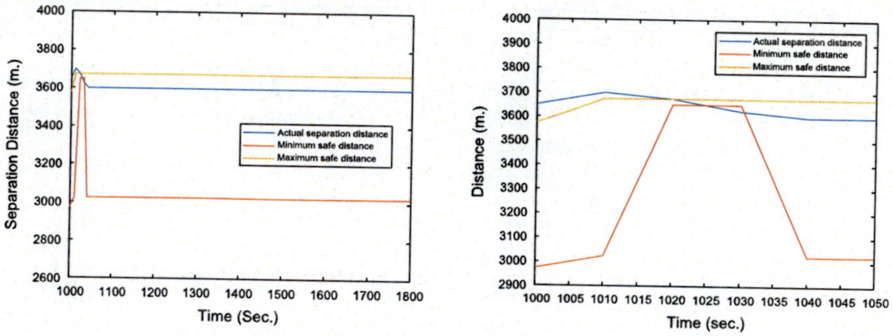

Fig. 12 Separation distance (Left) and separation distance before transferring to the platooning state (Right)

trains have been proceeding safely, in that the actual separation distance between them is not shorter than the minimum safe distance. In addition, the capacity can be controlled. When they are in the platooning state, the following train's velocity has been constant and maintained at a constant rate resulting in stable movement.

5 Conclusion and Recommendation

According to the simulation results shown in Part 4, it can be concluded that the trains have proceeded safely in that the actual separation distance between trains is not lower than the critical safe distance (minimum safe distance). The trains have been driven smoothly. The following train is catching up to the leading train by its constant velocity before decelerating and platooning with the front train with the same velocity. The capacity can be controlled by the range of safe distance. In the platooning state, the actual separation distance is in the range of safe distance. The space between trains is not more than the maximum safe distance resulting in the increasing in line capacity compared to the MBS system where the required space between them is longer than the maximum safe distance. In addition, the acceleration and deceleration rate of the train is realistic. This is because these rates are limited to be within acceleration and braking capabilities.

References

CDP (2016) The rail technical strategy capability delivery plan. RSSB

Flammini F, Marrone S, Nardone R, Petrillo A, Santini S, Vittorini V (2019) Towards railway virtual coupling. https://www.doi.org/10.1109/ESARS-ITEC.2018.860752

Harriss L (2016) Moving block signaling. Houses of Parliament: parliamentary office of science and technology

Henke C, Tichy M, Schneider T, Bocker J, Schafer W (2008) Organization and control of autonomous railway convoys. In: 9th international symposium on advance vehicle control (AVEC) 6–9 Oct 2008. Kobe, Japan

HS2 (2011) Operational concept study. Technical note: HS2 capacity and reliability. SYSTRA

Hunyadi B (2011) Capacity evaluation for ERTMS (European rail traffic management system) Level 2 operation on HS2. Bombardier

Ke-Ping L, Li-Jia G (2009) Simulating train movement in railway traffic using a car-following model. Chinese Phys B. 18

Li K, Gao Z (2009) An improved equation model for the train movement. Simul Modell Pract Theory 15:1156–1162

Li PK, Gao YZ (2011) An improved car-following model for railway traffic. J Adv Transp 47:475–482

Li-Xing Y, Feng L, Zi-You G, Ke-Ping L (2010) Discrete-time movement model of a group of trains on a rail line with stochastic disturbance. Chin Phys B 19

McNaughton A (2011) Signaling headways and maximum operational capacity on high speed two London to West Midlands Route

Ming XX, Ping KL, Xing LY (2014) Discrete event model-based simulation for train movement on a single-line railway. Chin Phys B Chin Phys Soc (IOP Publishing Ltd.) 23(8):080205

Mitchell I (2016) ERTMS Level 4, train convoys or virtual coupling. International technical committee report on topic 39. Institute of Railway Signal Engineers, IRSE news, Issue 219

Ning B (1998) Absolute braking and relative distance braking; train operation control modes in moving block systems. Computer in railway; transactions on the built environment. 34:991–1001

Rothery RW (1992) Car following models. In: Trac flow theory. Transportation research board. Special, report 165

Takeuchi H, Goodman JC, Sone S (2003) Moving block signaling dynamics: performance measures and re-starting queued electric trains. IEEE Proc-Electr Power Appl 150(4):483–492

Xuan et al (2013) Modelling and simulation of high-speed passenger train movements in the rail line

Yang L, Li F, Gao Z, Li K (2010) Discrete-time movement model of a group of trains on a rail line with stochastic disturbance. Chin Phys B 19(10):100510

Ye J, Li K, Jiang X (2015) Stability analysis of train movement with uncertain factors. Mathematical problems in engineering volume 2015. Hindawi Publishing Corporation, Article ID 230616

Zhao Y, Orlik P, Parsons K, Kataoka K, Nagy KT (2016) Improvement of train transportation performance by convoy signalling. Retrieved from http://dx.doi.org/10.1142/S0129183116500777

An Optimal Multi-objective Train Speed Profile for Mass Transit Systems Using a Genetic Algorithm-Based Technique

Chaiyut Sumpavakup and Phumin Kirawanich

Abstract This paper presents movement planning of a mass transit system between two stations with the use of Genetic Algorithm (GA) technique to minimize total energy consumption and total energy loss during the journey with appropriate weighting factors. The train movement is based on a sequence of four modes of operation, i.e., accelerating, constant speed or cruising, coasting, and braking modes. The train speed profile is genetically optimized by controlling the acceleration, the deceleration, and the location of coasting point. In this study, the investigation was carried out with a mass transit section between two station platforms with the service distance of 2 km, the variation of track gradient, and the maximum speed of 80 km/h. The results demonstrated that when compared with the use of GA-based single-objective functions, solving such a problem by using a GA-based multi-objective function can reduce the overall energy consumption (0.14% max) and total energy loss (3.53% max) while still being able to maintain the desired operation speed performance.

Keywords Train speed profile · Energy consumption · Optimization · Genetic algorithms

1 Introduction

In railway operation, the train speed profile is of great importance for such performance designs as an estimation of journey time, capacity analysis, train scheduling, route planning, an evaluation of energy consumption, etc. To accurately control the train movement through key behaviors, e.g., alternating powering, coasting, and braking, it, therefore, requires a precise and reasonable speed profile which is not

C. Sumpavakup (✉)
King Mongkut's University of Technology, North Bangkok, Bangkok, Thailand
e-mail: c.sumpavakup@gmail.com

P. Kirawanich
Mahidol University, Nakhon Pathom, Thailand
e-mail: phumin.kir@mahidol.ac.th

W. Weerawat et al. (eds.), *Urban Rail Transit*, Lecture Notes in Mobility,
https://doi.org/10.1007/978-981-15-5979-2_16

easily achieved by using analytical approaches (Jong and Chang 2005). One of state-of-the-art approaches is to use of a train performance simulator to determine an optimal speed profile (Goodman 1997).

Several train energy optimization techniques have been practically considered without considering the variations track slope profile (Ishikawa 1968; Kokotovic and Singh 1972; Strobel and Horn 1973; Aradiand and B´ecsi 2013). A number of experiments have shown that the slopes have significant effects on the energy consumption (Aradiand and B´ecsi 2013). Besides, without taking into account the slope profile for such an energy optimization, the lack of considering the energy loss in the feeder line could result in an underestimation of energy consumption.

This paper aims at optimizing both tractive energies consumed by the train and the energy loss in the train feeder line, taking into account the track gradient influence. The objective function as a constraint parameter includes maximum tractive effort, speed limit, journey time, acceleration and deceleration. The paper is composed of five sections. Section 2 illustrates basics of a single-train simulation consisting of the train movement calculation, speed control strategy, and power consumption. An optimal multi-objective train speed profile using Genetic Algorithms is described in Sect. 3. Section 4 gives simulation results and discussion. Lastly, the conclusion is drawn in Sect. 5.

2 Single Train Simulation Technique

The Single Train Simulation (STS) consists of two main components, i.e., train movement and performance simulation (TPS) and DC power flow simulation (DC-PFS). Such information as vehicle characteristics, line data, and operation data are the requirements of TPS. The outputs of TPS in terms of the running time, power consumption, and operation diagram interact with the network data such as substations, location, and source equivalent resistance, etc. through DC-PFS. The diagram in Fig. 1 describes the relationship of TPS and DC-PFS.

2.1 *Train Movement and Performance Calculation*

The train movement equation is expressed in (1) according to the free body diagram describing the forces acting on the traction vehicle running uphill as shown in Fig. 2.

$$F = F_T - R = M_{eff}\alpha \tag{1}$$

$$R = F_r + F_g + F_c \tag{2}$$

$$F_r = A + Bv + Cv^2 \tag{3}$$

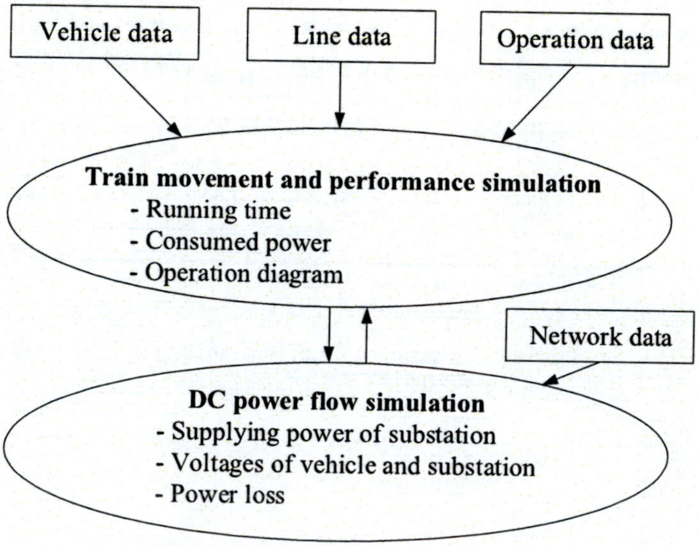

Fig. 1 Relationship of TPS and DC-PFS

Fig. 2 Free-body diagram of the train on the inclined plane

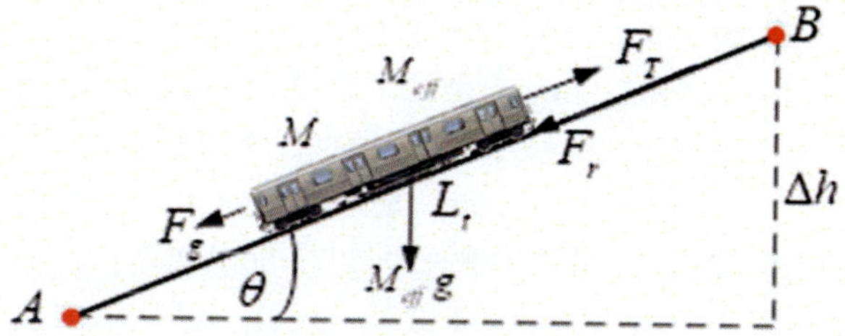

$$F_g = M_{eff} g \sin\theta \tag{4}$$

$$M_{eff} = M_t(1 + \lambda_w) + M_l \tag{5}$$

where the parameters used in this simulation are described in Table 1.

The power consumed by a train (P_{tr}) is constituted by the tractive power (P_{ta}), which depends on the tractive effort (F_T) and train speed (v), and the auxiliary power (P_{aux}) as given by the following expressions:

$$P_{tr} = P_{ta} + P_{aux} \tag{6}$$

$$P_{ta} = \begin{cases} \frac{F_T}{\eta} \times v; & \text{if } F_T \geq 0 \\ \eta F_T \times v; & \text{if } F_T < 0 \end{cases} \tag{7}$$

Table 1 Parameters

Parameter	Description	Unit
α	Acceleration	m/s^2
F_T	Tractive effort	N
R	Overall resistance force	N
F_r	Running resistance force	N
F_g	Gradient resistance force	N
F_c	Curve resistance force	N
v	Vehicle speed	Km/h
A	Constant (Rochard and Schmid 2000)	kN
B		kNh/km
C		kNh^2/km^2
θ	Slope angle	degree
M_t	Tare mass	kg
λ_w	Rotary allowance	–
M_l	Freight/passenger load	kg
P_{tr}	Power consumed by a train	W
P_{ta}	Tractive power	W
P_{aux}	Auxiliary power	W
η	Efficiency of the power conversion (electrical–mechanical)	–

where η denotes the efficiency of the power conversion from the electrical input power to the mechanical output power at the axles.

2.2 DC Power Flow Calculation

In this paper, a computer-based simulation for train movement integrated with the power supply interface is carried out. The power flow algorithm needs the location and consumed power output data. The location data are required for the creation of the conductance matrix differentiated at each calculation step. Consumed power data are used to determine the magnitude and the sign of current source for modelling electric vehicles.

The iterative calculation of the power network is based on nodal analysis consisting of two power supply substations and a railway vehicle as illustrated in Fig. 3. This can be modeled by a Norton equivalent circuit detailed in Fig. 4 where x is the train position, L is the distance between two substations, I_{tss} is substation short-circuit current, I_{tr} is the current drawn by the train, R_{tss} is substation short-circuit resistance, R_f and R_r are railway feeder and running rail resistances, respectively. Equation (8) expresses Nodal equations for equivalent circuit shown in Fig. 4.

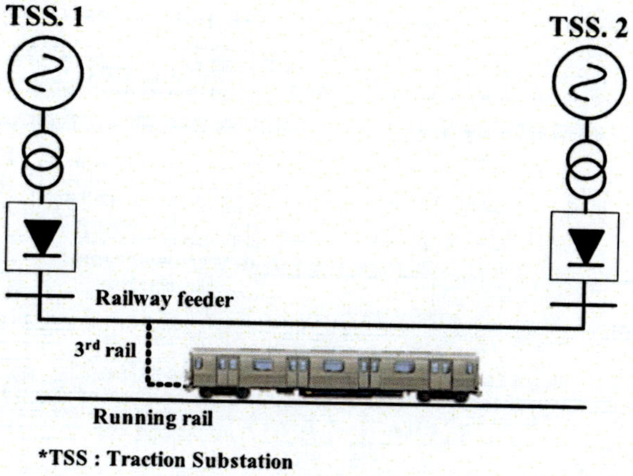

Fig. 3 DC railway power network used for the simulation

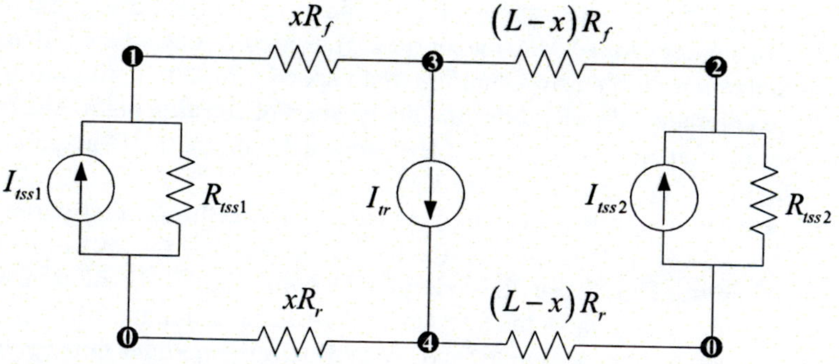

Fig. 4 Norton equivalent circuit

$$\begin{bmatrix} \frac{1}{xR_f}+\frac{1}{R_{tss1}} & 0 & -\frac{1}{xR_f} & 0 \\ 0 & \frac{1}{(L-x)R_f}+\frac{1}{R_{tss2}} & -\frac{1}{(L-x)R_f} & 0 \\ -\frac{1}{xR_f} & -\frac{1}{(L-x)R_f} & \frac{1}{xR_f}+\frac{1}{(L-x)R_f} & 0 \\ 0 & 0 & 0 & \frac{1}{xR_r}+\frac{1}{(L-x)R_r} \end{bmatrix} \begin{bmatrix} V_1 \\ V_2 \\ V_3 \\ V_4 \end{bmatrix} = \begin{bmatrix} I_{tss1} \\ I_{tss2} \\ -I_{tr} \\ I_{tr} \end{bmatrix} \tag{8}$$

In fact, since I_{tr} is not constant due to the fluctuations of train power loads and the feeder voltage, modelling I_{tr} with (9) takes into account the relationship between I_{tr}, P_{tr} and the vehicle voltage.

$$I_{tr} = \frac{P_{tr}}{V_3 - V_4} \tag{9}$$

Generally, the power flow calculation in the power system has an aim to determine the voltage, angle, real power and reactive power at any snapshot (Lee et al.

2010). However, the movement of vehicles is essential in the calculations because the power consumption is changed as the train moves along the railway route depending on the operation mode with periodic acceleration, constant speed, coasting, and braking processes. Therefore, the power flow calculation considering the movement of vehicle updates the admittance matrix at each calculation step. Each power flow calculation gets the initial current vector from the initial voltage vector by using (9) and renews the voltage vector by using (8). The iteration process stops when the error is less than the pre-determined value.

3 Optimal Multi-objective Train Speed Profile Using a Genetic Algorithm-Based Technique

This section describes a movement planning of a mass transit system between two station platforms where the goal is to minimize the total energy consumption and the total energy loss during the journey. The case presented here is a multi-objective optimization problem in which the train speed profile is optimized by controlling the acceleration, deceleration, and the location of coasting point. The overall energy consumption and energy loss of the proposed journey therefore can be minimized with the use of Genetic Algorithms.

A genetic algorithm technique is simple, powerful, general-purpose, derivative free, stochastic global optimization method inspired by the laws of natural selection and genetics. Thus, it uses the concept of Darwin's theory of evolution, which is based on the rule of survival of the fittest (Vas 1999). This technique initializes a population of individuals (chromosomes) and probabilistically modifies the population by three genetic operators: reproduction (selection), crossover and mutation, in such a way that the population number is maintained with the fittest survivals.

The genetic algorithm involves three basic steps, i.e., Random initialization of population, Evaluation of fitness of individuals in the population, and New population generation. Given a solution space, individuals (possible-solutions) are randomly created. An individual is characterized by a fixed-length binary bit string which is called a chromosome; the individuals are evaluated by means of fitness function. The new generation is then formed by applying the three genetic operators. First a selection (reproduction) operator is applied where evaluated individuals are selected with probability proportional to their fitness values, the selected individuals are reproduced (copied one or more times to obtain a mating pool) such that the population number is maintained. Secondly individuals in the mating pool are paired randomly to obtain couples (parents) and the crossover operator is applied. The crossover operator is then followed by a mutation operator; the mutation rate is normally kept low so that good chromosomes are preserved. Mathematically the mutation ensures that, given any population, the entire search space is connected (Vas 1999). The GAs process can be described by the diagram as shown in Fig. 5.

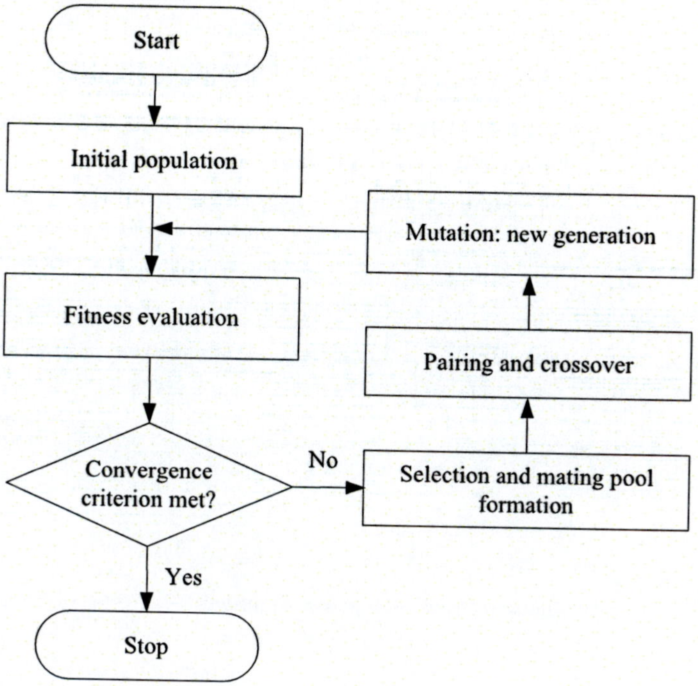

Fig. 5 Process of a genetic algorithm

As mention previously, a GA technique has been applied to determine optimal values of three parameters: (i) acceleration, (ii) deceleration, and (iii) coasting point, such that the tractive energy consumption and energy loss of the proposed journey are minimized. The objective function to be minimized is defined as

$$\phi = \omega_E \times E + \omega_L \times L \tag{10}$$

where the minimization is subject to $\alpha_{acc}^{\min} \leq \alpha_{acc} \leq \alpha_{acc}^{\max}$, $\alpha_{dec}^{\min} \leq \alpha_{dec} \leq \alpha_{dec}^{\max}$, $L_{coast}^{\min} \leq L_{coast} \leq L_{coast}^{\max}$, and time-distance services. E is the total energy consumption defined in (11), L is the total energy loss given in (12), ω_E and ω_L are weighting factors for the total energy consumption and the total energy loss, respectively, and $P_{loss} = \sum P_{tss} - P_{tr}$.

$$E = \sum_{i=1}^{N} P_i \times \Delta t_i \tag{11}$$

$$L = \sum_{i=1}^{N} P_{loss,i} \times \Delta t_i \tag{12}$$

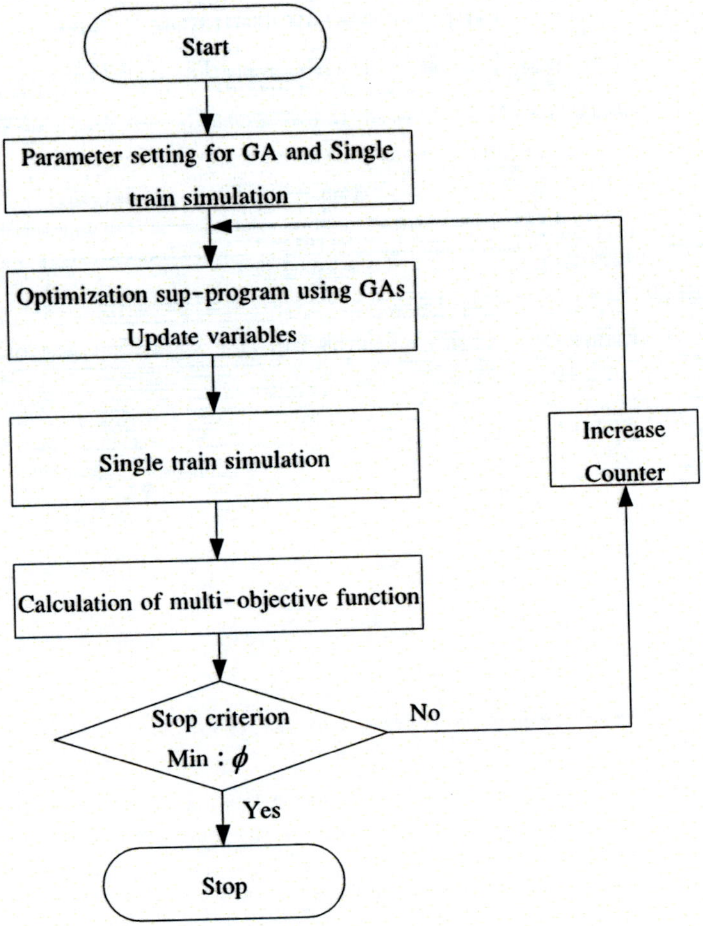

Fig. 6 Flow chart of using the GA to optimize multi-objective function

The procedure to perform the optimal multi-objective function using GAs process is described in Fig. 6, with the parameters as follows: populations = 50, generation = 100, crossover = 0.8, and TolFun = 1e−6.

4 Simulation Results and Discussion

The system under study in this paper is a simple mass transit section between two-passenger stopping stations with the service distance of 2 km as shown in Fig. 7. The application was implemented in MATLAB and run on the Intel Core i5 2.53 GHz

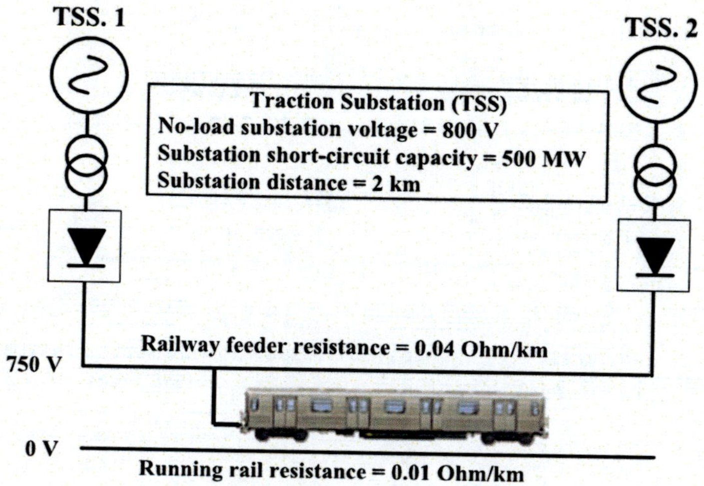

Fig. 7 Test System

with 4.0 GB of RAM. The system conditions for simulation are listed in Table 2. In this work the simulation is carried out under three cases of minimization as follows.

Case 1: Single-objective function with the minimization of total energy consumption.
Case 2: Single-objective function with the minimization of total energy loss.
Case 3: Multi-objective function with the minimization of total energy consumption and total energy loss.

Table 3 shows the parameters with deviation determined through the GA process for all 3 cases of operation with energy minimization where the averaged values of control variables were chosen. As a result, Figs. 10 and 11 demonstrate the train power consumption and the train speed, respectively, as a function of the journey distance between stations.

The Simulation results suggest that under the variation of track gradient, the GA application on the multi-objective function (Case 3) can perform the minimization evidently both during acceleration and deceleration while that of case 1 and case 2 can accommodate only the minimization during acceleration or deceleration, respectively. Besides, the algorithm of multi-objective function with the GA technique in case 3 can slightly improve the total energy consumption E and the total energy loss L while being able to maintain the speed performance as compared to others.

The simulations were carried over for case 3 to study the key parameters at traction substation TSS1, at the train, and at the rail. Figure 12 shows the voltage and current at TSS1, while the train and rail voltages are demonstrated in Fig. 13. Finally, the total power loss is shown in Fig. 14 as a function of the journey distance. It is seen that the track gradient obviously contributes to the train traction performance especially during the cruising period with constant speed. However, the train with

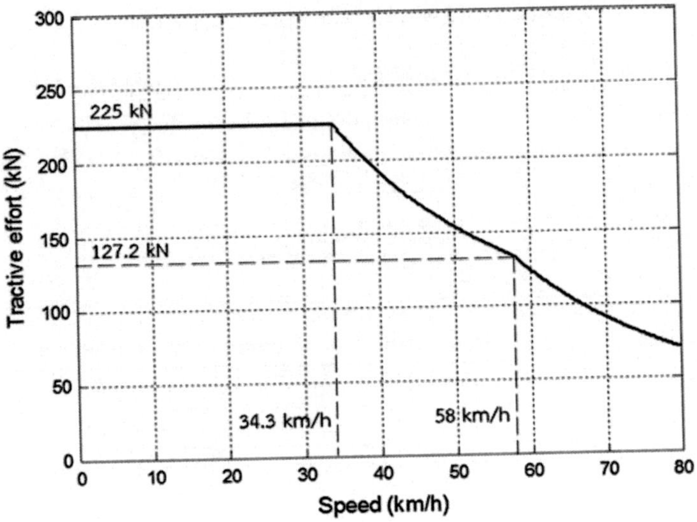

Fig. 8 Tractive effort diagram

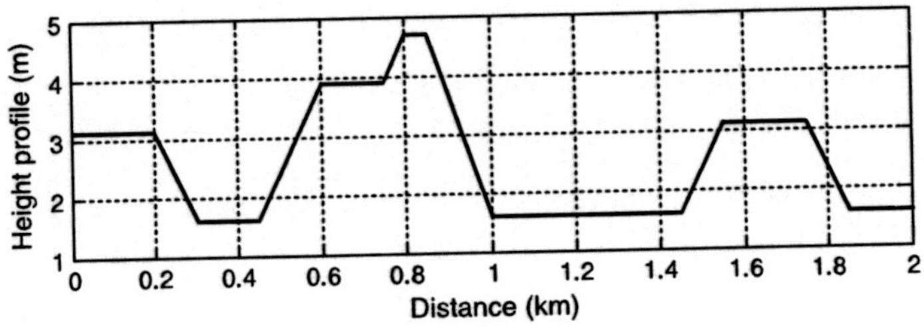

Fig. 9 Gradient profile

the GA control is capable of maintaining the operation speed performance while drawing less power with low energy loss.

5 Conclusion

This paper presents the use of a Genetic Algorithms technique in a movement planning of a mass transit system between two stations aiming to minimize the total energy consumption and the total energy loss during the journey. The train movement in this work is based on a sequence of four modes of operation which are accelerating mode, constant speed or cruising mode, coasting mode, and braking mode. In this study, it can be demonstrated that the train speed profile can be optimized by controlling the acceleration, deceleration, and the location of coasting point. The system under test

Table 2 System conditions for simulation

System condition	Information	
Voltage conditions	Nominal voltage	750 V
Weight conditions	Tare weight	153 ton
	Max. payload	85 ton
Movement features	Max. speed	80 km/h
	Max. acceleration	0.87 m/s^2
	Max. deceleration	1.00 m/s^2
Efficiencies	Gear, motor, inverter	98, 88, 98%
Max. auxiliaries	constant load	270 kW
Train resistance	$A = 4025$, $B = 118.67$, $C = 0.871$ [Eq. (3)]	
Tractive effort	The train characteristics in terms of the relationship between the tractive effort and the speed as given in Fig. 8	
Gradient profile	The gradient profile of the infrastructure as a function of the distance as given in Fig. 9	

Table 3 Comparison of simulation results obtained from each case

Control variables	Genetic algorithm (GA)			
	Min	Average	Max	SD
Case I : $\omega_E = 1, \omega_L = 0$				
α_{acc} (m/s^2)	0.7977	0.8468	0.8696	0.0232
α_{dec} (m/s^2)	−0.9179	−0.9739	−0.9997	0.0284
L_{coast} (km)	0.3416	0.6142	0.6609	0.1265
E (kWh)	12.0066	12.0306	12.0434	0.0111
L (kWh)	0.1557	0.1645	0.1668	0.0038
ϕ (kWh)	12.0066	12.0306	12.0434	0.0111
Case II : $\omega_E = 0, \omega_L = 1$				
α_{acc} (m/s^2)	0.8659	0.8675	0.8695	0.0012
α_{dec} (m/s^2)	−0.8371	−0.8908	−0.9894	0.0575
L_{coast} (km)	0.3318	0.3867	0.6183	0.1251
E (kWh)	12.0137	12.0442	12.0504	0.0137
L (kWh)	0.1551	0.1593	0.1635	0.0033
ϕ (kWh)	0.1551	0.1593	0.1635	0.0033
Case III : $\omega_E = 1, \omega_L = 1$				
α_{acc} (m/s^2)	0.8173	0.8687	0.8693	0.0170
α_{dec} (m/s^2)	−0.8765	−0.8959	−0.9989	0.0415
L_{coast} (km)	0.3596	0.3817	0.6392	0.1220
E (kWh)	12.0108	12.0277	12.0493	0.0110
L (kWh)	0.1565	0.1587	0.1663	0.0035
ϕ (kWh)	12.1684	12.1864	12.2156	0.0142

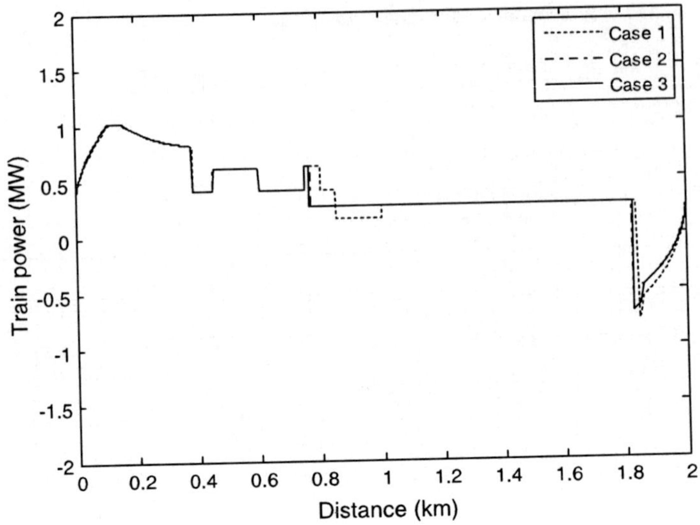

Fig. 10 Train power

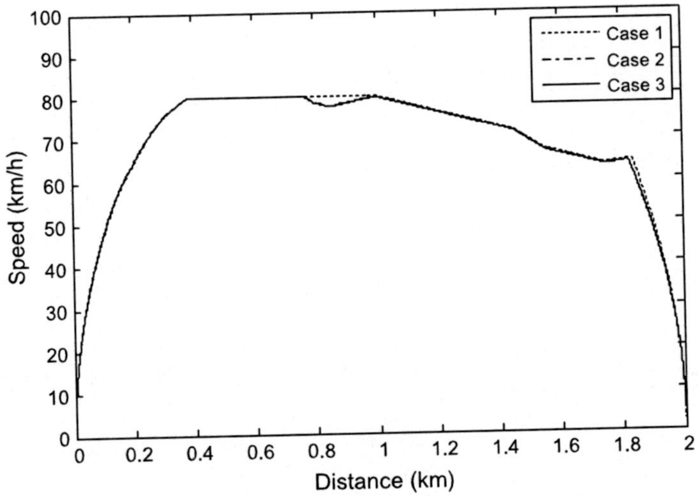

Fig. 11 Speed-distance trajectory

is a simple mass transit section consisting of two stations with the service distance of 2 km, track gradient variation, and maximum speed limit of 80 km/h. The results showed that when compared with the application of GA-based single-objective function, the use of a GA-based technique with multi-objective functions can reduce the overall energy consumption (0.14% max) and the total energy loss (3.53% max) without reducing the desired operational performance.

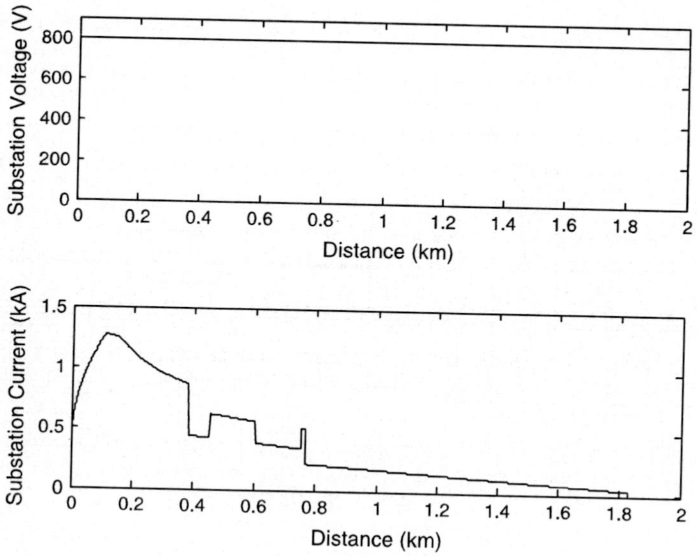

Fig. 12 Voltage and current of TSS 1 for Case 3

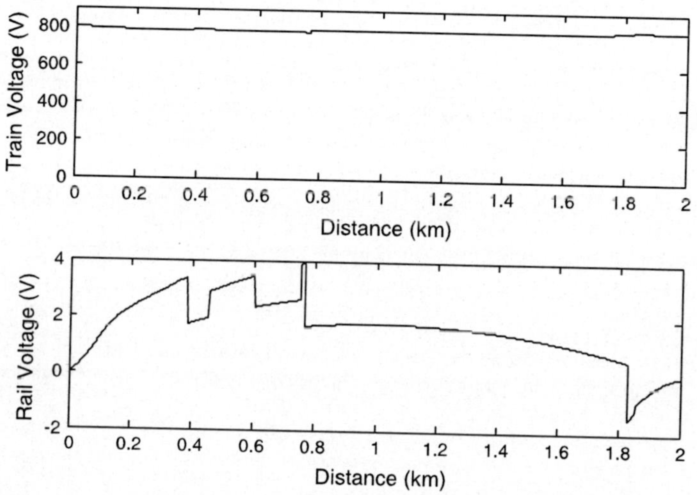

Fig. 13 Train and rail voltage for Case 3

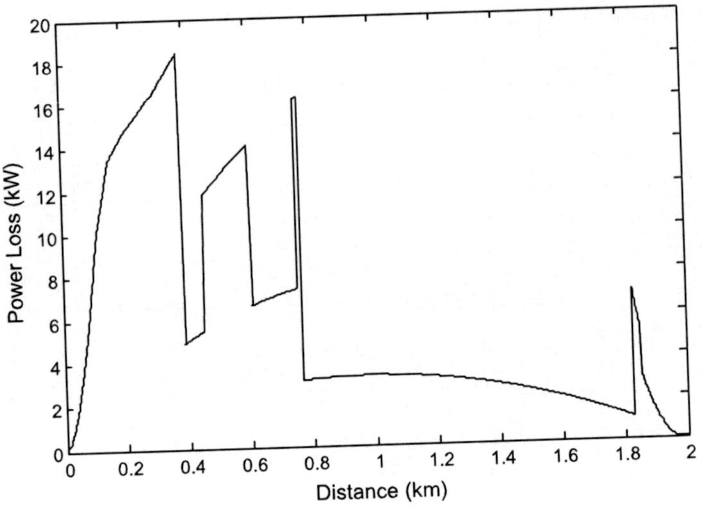

Fig. 14 Power loss for Case 3

References

Aradiand S, B´ecsi T (2013) A predictive optimization method for energy-optimal speed profile generation for trains. In: Proceedings of 14th IEEE international symposium on computational intelligence and informatics, Hungary

Goodman CJ (1997) Train performance and simulation. Fourth Vocation School on Electric Traction Systems, IEE Power Division

Ishikawa K (1968) Application of optimization theory for bounded state variable problems to the operation of trains. Bull JSME Nagoya Univ 11(47):857–865

Jong JC, Chang S (2005) Algorithms for generating train speed profiles. J Eastern Asia Soc Transp Stud 6:356–371

Kokotovic P, Singh G (1972) Minimum-energy control of a traction motor. IEEE Trans Autom Control 17(1):92–95

Lee HM, Jeon EJ, Jeong SC (2010) A study on calculation of DC railway loadflow with energy storage system. In: International conference on control, automation and systems, Korea

Rochard BP, Schmid F (2000) A review of methods to measure and calculate train resistances. Proc Inst Mech Eng Part F: J Rail Rapid Transit 214(4):185–199

Strobel H, Horn P (1973) On energy-optimum control of train movement with phase constraints. Electr Inf Energy Tech J 6:304–308

Vas P (1999) Genetic algorithms," . Artificial intelligence based electrical machines and drives: Application of fuzzy, neural, fuzzy-neural, and genetic algorithms based techniques. Oxford University Press, New York, pp 263–277

Design of Optimal Train Speed Profile for PMSM Railway Traction System Using Dynamic Programming with MTPA Control Method

Klara Dwi Kristianingtyas and Masafumi Miyatake

Abstract Permanent magnet synchronous motor (PMSM) as a railway traction system has begun to be introduced in some commuter lines. In comparison with the common induction motor (IM), PMSM has different loss characteristics. Namely, during coasting, the production of iron loss still exists due to the rotating permanent magnet which increases energy consumption. Conventionally, no traction energy is consumed during coasting and thus, coasting is often exploited to configure optimal train speed profile. However, this feature might not be applied in PMSM. In this paper, the optimal train speed profile for PMSM is configured with dynamic programming (DP). To include the effect of coasting, the train energy consumption is modeled using the Maximum Torque Per Ampere (MTPA) control method. The result shows that the energy is consumed during coasting due to d-axis current flowing to the motor to suppress the back-emf. Moreover, the regenerative braking can achieve an efficiency of more than 88%. In addition, the optimal train speed profile for PMSM utilizes less coasting, and during braking, maximum deceleration is utilized more compared to the conventional model.

Keywords Energy-efficient driving · PMSM · Dynamic programming · MTPA · Train speed profile

1 Introduction

In recent years, with the increasing concerns about environmental issues, the railway industries focus to improve the utilization of energy in order to reduce energy consumption. The reduction of energy consumption is not only beneficial for the environment but it also improves the economics of railway operators by reducing the operational costs.

In response to that, there have been many attempts to improve the efficiency of the traction system. For instance, in Japan, Tokyo Metro, incorporation with Toshiba, has

K. D. Kristianingtyas (✉) · M. Miyatake
Sophia University, 7-1, Kioicho, Chiyoda-ku, Tokyo 102-8554, Japan
e-mail: klara_dks@hotmail.com

W. Weerawat et al. (eds.), *Urban Rail Transit*, Lecture Notes in Mobility,
https://doi.org/10.1007/978-981-15-5979-2_17

started developing the PMSM traction system as the solution for energy-saving vehicles (Kawai and Tasaka 2010). Toshiba claims that PMSM has three major benefits, which are: energy savings, low noise emissions, and less maintenance. In its application to the Ginza Line 01 series, the running tests showed a total of 20% energy reduction in comparison with the IM traction system (Kawai and Tasaka 2010). The arrival of PMSM to the industry, with even higher efficiency than IM, shows some potentials in more energy efficient traction systems.

Optimal technical operations also have the potential of creating better energy utilization. The energy efficient driving is one of the solutions to reduce train energy consumption from technical operations. Driving the PMSM with optimal train speed profile is inevitable to exploit the benefits of the traction system and to achieve maximum energy reduction. Therefore, the objective of this study is to find the optimal train speed profile for PMSM. The optimal train speed profile is obtained by using DP to solve the minimum energy consumption problem. In order to include the characteristics of PMSM, the energy consumption model based on the Maximum Torque Per Ampere (MTPA) control method is proposed in this paper. MTPA control method produces the desired torque based on the motor's load while minimizing the current magnitude.

First, the application of dynamic programming to solve the optimal speed profile problem based on the previous study are explained. In addition, the different loss characteristics between IM and PMSM are reported based on previous research. Second, the mathematical model for PMSM and the application of the MTPA control method in DP are elaborated. Finally, the simulation condition and results of the proposed model are provided in comparison with the conventional model.

2 Literature Review

Reducing energy consumption while maintaining the quality of the service draws attention from the railway industry. Previous studies have found that energy reduction is feasible through appropriate driving strategies. Train speed profile optimization or train trajectory optimization may provide energy reduction. There are various methods used in the speed profile optimization. Using the Pontryagin maximum principle, it has been reported that the optimal driving strategy should consist of four driving modes which are powering, speed-holding or cruising, coasting, and braking as defined in Table 1 (Howlett et al. 1994). Figure 1 shows the common optimal train

Table 1 Optimal driving modes

Powering	Maximum tractive acceleration
Speed-holding	Tractive acceleration = resistive acceleration
Coasting	Tractive acceleration = 0
Braking	Maximum braking

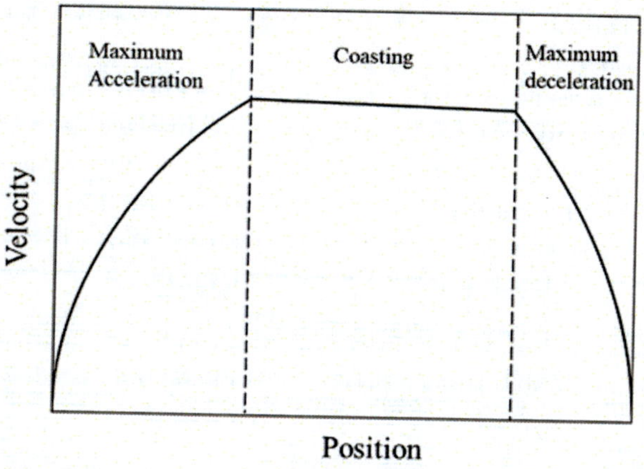

Fig. 1 Typical optimal train speed profile

speed profile consists of maximum acceleration, coasting, and maximum deceleration. In general, coasting is mainly exploited to obtain lower energy consumption, however, this driving mode prolongs the running time. Thus, to keep the punctuality of the schedule, sometimes coasting is mixed with cruising.

2.1 *Application of DP in Finding Optimal Train Speed Profile*

DP is one of the methods used in train speed profile optimization. In DP, the search for the optimum speed at each position is done through the n-stage decision process. The application of DP in train speed profile optimization is reported by Ko et al. (2004). It has been shown that DP can handle the non-linear characteristics in train speed profile optimization problems such as speed limit, track gradient, and track curves. Following these features, Oba and Miyatake (2018) were able to implement DP in train speed profile optimization under a fixed-block signaling system. In addition, DP was implemented in train trajectory optimization under a moving-block signaling system (Ichikawa and Miyatake 2019). In comparison with genetic algorithm (GA) and ant colony optimization (ACO), DP is shown to be the best method in this application as the other methods failed to converge onto a good solution (Lu et al. 2013). Furthermore, DP has been implemented to the driver advisory systems on an Android environment (Ghaviha et al. 2017).

In DP calculation, the conventional method using constant efficiency is commonly used in the speed profile optimization problem to calculate energy consumption. In 2017, Gaviha used a new model which considers the dynamic loss of the train in the optimization problem to estimate the energy consumption. The results show that the accuracy of the energy calculation is improved and the train speed profile is more energy efficient compared to the conventional model (Ghaviha et al. 2017).

2.2 *Loss Characteristics Between PMSM and IM*

One of the major benefits of PMSM is that PMSM has high efficiency of up to 97% (Matsuoka 2007). In comparison with IM, although energy consumption of IM is higher, PMSM and IM have different loss characteristics and sometimes it is neglected. Kondo et al. (2005) conducted a study on energy consumption for railway traction motors between IM and PMSM. The result shows that the total energy loss in traction motor IM is higher than in PMSM, yet, stator iron loss during coasting only exists in PMSM. Conventionally, no loss is consumed during coasting. The stator iron loss during coasting in PMSM is due to the currents that are fed to the network in order to suppress the back-emf produced by the rotating permanent magnet.

2.3 *Summary*

Based on the above literature reviews, if the conventional model with constant efficiency is used to obtain the optimal train speed profile for the PMSM traction system, the coasting loss is not evaluated. To consider the iron loss during coasting, another model needs to be used to calculate the train energy consumption. In this study, the estimation of energy consumption is obtained from the current and voltage commands based on the control method of PMSM in addition to the conventional model. Then, the minimum train energy consumption problem is solved using DP in the MATLAB environment to obtain optimal train speed profiles for PMSM.

3 Methodology: Simulation Model

As described in the previous section, the new model to estimate the energy consumption of PMSM is required to evaluate the loss characteristics of PMSM. In this section, the mathematical model of PMSM is constructed to model the MTPA control method. The energy consumption calculation based on MTPA is explained. Then, the application of DP to obtain an optimal speed profile is described. Finally, the application of the conventional model with constant efficiency and the new model with MTPA to DP is described at the end of this section along with the algorithm of DP in solving the train optimization problem.

3.1 Mathematical Model of PMSM

The PMSM is modeled in a d-q axis reference frame fixed to the rotor synchronous side. The d-axis is aligned with the rotor flux linkage. The mathematical model is developed under the following assumptions.

- Magnetic saturation is neglected.
- The total flux of the permanent magnet is assumed constant and has a linear demagnetization curve.
- Back-EMF is assumed to be sinusoidal.
- Core losses such as hysteresis and eddy currents are negligible.

Using the above assumptions, the stator voltage equations are described in Eqs. (1) and (2) (Yang et al. 2012).

$$v_d = R_s i_d - \omega_e L_q i_q \tag{1}$$

$$v_q = R_s i_q + \omega_e (L_d i_d + \psi_m) \tag{2}$$

where $v_d(v_q)$ are the voltages in d(q) axis, $i_d(i_q)$ is the current in the d(q) axis, ψ_m is the flux linkage by the permanent magnet, $L_d(L_q)$ is the inductance in the d(q) axis, R_s is the winding resistance, and ω_e is the electrical angular frequency.

The generated electromagnetic torque T_e by PMSM is calculated using Eq. (3) where P is the number of pole pairs (Yang et al. 2012).

$$T_e = \frac{3}{2} \cdot P \cdot i_q \{\psi_m + (L_d - L_q) i_d\} \tag{3}$$

3.2 Modeling Control Method of PMSM Based on MTPA

In this paper, the control method is modeled under the assumption of electrical steady-state operation. Hence, no error is assumed between the controller's reference signal and the real output.

The motor is operated under two operating zones, which are constant torque (CT) region and field weakening (FW) region. In the CT region, the distribution of the currents is designed to produce maximum torque with the minimum current. The voltage increases almost linearly with the increase of the angular speed. When the voltage reaches the maximum value, the motor is operating under the FW region to protect the device from overvoltage in the windings. In the FW region, the distribution of i_d and i_q is obtained under the voltage and current limits.

The MTPA control in the CT region is defined in Eq. (4) (Yang et al. 2012).

$$\text{Max } |i| = \sqrt{i_d^2 + i_q^2} \tag{4}$$

$$s.t \quad T_e = \frac{3}{2} \cdot P \cdot i_q \{\psi_m + (L_d - L_q) i_d\}$$

The relationship between i_d and i_q can be obtained by solving the problem in Eq. (4) using the *Lagrangian multiplier method*. Hence, the relationship between generated electromagnetic torque and q-axis current only can be described in Eq. (5) (Yang et al. 2012).

$$T_e = \frac{3}{4} P i_q \left(\psi_m + 2\sqrt{\psi_m^2 + 4(L_d - L_q)^2 i_q^2} \right) \tag{5}$$

If the torque is known, Eq. (5) is used to find i_q. Then, i_d is obtained using Eq. (3).

When the maximum voltage has been reached, the motor enters the FW region. In this region, the motor is operated under the constraints in Eqs. (6) and (7).

$$i_d^2 + i_q^2 = I_{\max}^2 \tag{6}$$

$$v_d^2 + v_q^2 = V_{\max}^2 \tag{7}$$

The definition of voltages in Eqs. (1) and (2) are substituted to Eq. (7). By neglecting the voltage drop in the winding resistance since FW regions occur in the high-speed range, Eq. (8) can be obtained.

$$(-\omega_e L_q i_q)^2 + (\omega_e (L_d i_d + \psi_m))^2 = V_{\max}^2 \tag{8}$$

If the maximum current and maximum voltage are given, Eq. (6) can be substituted in Eq. (8) to obtain i_d. Then, i_q is found using Eq. (6). The generated electromagnetic torque in this speed range is obtained by Eq. (3) with the corresponding currents.

3.3 *Train Energy Consumption Calculation Based on MTPA*

Using the train's load, namely torque and rotational speed, the d-q axis current is calculated using Eqs. (3) and (5). The torque is calculated from the tractive/braking effort and rotational speed is calculated from the train's speed. If the train is accelerating, the q-axis current is the real and positive solution. Conversely, during braking, the q-axis current is real and negative. The d-axis current is always real and negative values for all modes to utilize the reluctance torque and to make a field weakening effect in the high-speed area.

The voltages are calculated using Eqs. (1) and (2) with the obtained currents. If the terminal voltage exceeds the voltage limit, the new i_d and i_q are computed by solving Eqs. (6) and (8), respectively. Then, the new voltages are recomputed.

$$E_{in} = \int_{0}^{T} \frac{3}{2}(v_d \cdot i_d + v_q \cdot i_q)dt \tag{9}$$

Once the currents and voltages are obtained, the energy consumption can be calculated using Eq. (9) where T is the total running time. Figure 2 shows the flowchart of the calculation for energy consumption based on the MTPA control strategy from the train's load.

3.4 Train Speed Profile Optimization Problem with DP

In this paper, DP based on Bellman's principle of optimality is used to solve the optimal train speed profile problem.

The principle of optimality states that 'An optimal policy has the property that whatever the initial state and initial decision are, the remaining decisions must constitute an optimal policy with regard to the state from the first decision' (Bellman and Kalaba 1965).

Although DP has a high computational cost, it is powerful in this problem because inequality constraints such as speed limit can be dealt with. In other words, train running conditions can be easily handled.

The optimal train speed profile is obtained by finding the minimum energy consumption for the trip. It is done by solving the train's equation of motion with the minimum energy. The mathematical model for the optimization problem is as follows (Ko et al. 2004).

If the train's front position and speed are assumed as $x(t)$ [m] and $v(t)$ [m/s], the train's equations of motion are expressed by differential equations in Eqs. (10) and (11).

$$\frac{dx(t)}{dt} = v(t) \tag{10}$$

$$\frac{dv(t)}{dt} = F(u(t), v(t)) - R(x(t), v(t)) \tag{11}$$

Here, t [s] is time, $F(u(t), v(t))$ [m/s^2] is the tractive effort per unit mass and $R(x(t), v(t))$ [m/s^2] is the total resistance per unit mass. The control input or notch command u is defined as the state variable in which it corresponds to the train's acceleration, coasting, and deceleration mode as defined in Table 2.

To be applied to DP, Eq. (11) is linearized using the Taylor expansion technique and discretized using the trapezoidal rule of the approximation of integral and expressed in Eq. (12).

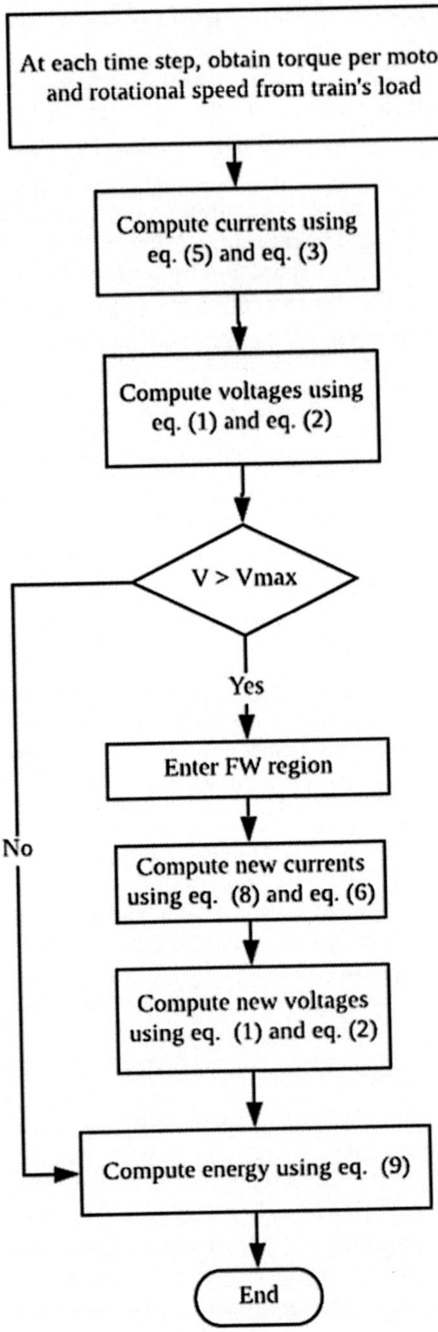

Fig. 2 Flowchart of calculation for train energy consumption based on the MTPA control strategy

Table 2 Control input and the corresponding traction mode

Control input/notch command u	Traction mode
1	Maximum acceleration
$0 < u < 1$	Accelerating
0	Coasting
$-1 < u < 0$	Decelerating
-1	Maximum deceleration

$$X_{k+1} = \left[I - \frac{A}{2}\Delta t\right]^{-1}\left\{\left[I + \frac{A}{2}\Delta t\right]X_k + BF_0\Delta t\right\} \quad (12)$$

where

$X_k = \begin{bmatrix} x_k \\ v_k \end{bmatrix}$, $A = \begin{bmatrix} 0 & 1 \\ \alpha & \beta \end{bmatrix}$, $B = \begin{bmatrix} 0 \\ 1 \end{bmatrix}$, $I = \begin{bmatrix} 1 & 0 \\ 0 & 1 \end{bmatrix}$,

$F_0(u, x_0, v_0) = F(u, v_0) - F_v \cdot v_0 - R(x_0, v_0) + R_x \cdot x_0 + R_v \cdot v_0$,

$\alpha = -R_x$, $\beta = -(F_v - R_v)$,

$\Delta t = \frac{v_k}{v_k - v_{k+1}}(t_{k+1} - t_k)$.

The initial and final condition of the problem is set in Eqs. (13) and (14) where L [m] is the total running distance and T [s] is the total running time.

$$x(0) = 0,\ v(0) = 0 \quad (13)$$

$$x(T) = L,\ v(T) = 0 \quad (14)$$

When applying DP, the terminal boundary needs to be transformed into a penalty function in order to satisfy the terminal constraints. The penalty function is shown in Eq. (15).

$$\phi(x_N, v_N) = c_x|x_N - L| + c_v|v_N - 0| \quad (15)$$

x_N, v_N, and L are position in the final state, velocity in the final state, and the total running distance, respectively. c_x and c_v are arbitrary constants. The value of the constants has a great effect on the solution accuracy. Namely, less error can be achieved with greater constants. However, the constants are also proportional to the energy consumption resulting in bigger energy consumption if bigger constants are implemented. Thus, the constants and the energy consumption are in the trade-off relationship. The appropriate value of the penalty constants needs to be set to achieve good solution accuracy with minimum energy consumption.

The speed limit constraints $v_{\max}(x, t)$ are expressed in Eq. (16).

$$v(x, t) - v_{\max}(x, t) \leq 0 \quad (16)$$

Finally, the train minimum energy consumption operation problem is formulated in Eq. (17).

$$\min_{\{u_k\}_{k=1}^{N}} \{J + \phi(x_N, v_N)\} \text{ subject to (11)} - \text{(16)} \tag{17}$$

In here, J is train energy consumption. This paper used two models to estimate energy consumption which are the conventional model and the new model.

3.4.1 Conventional Model

In the conventional model, the train energy consumption J is calculated based on Eq. (18). In this way, the energy consumption is calculated from the output of mechanical energy using constant efficiency. Following the definition of notch command in Table 2, the notch command is proportional with tractive effort F_a and braking effort F_b for u $\geq$ 0 and u < 0, respectively.

$$\begin{aligned}\frac{\mathrm{d}J(t)}{\mathrm{d}t} &= \begin{cases} M \cdot \frac{1}{\eta_a} \cdot F_a(u, v) \cdot v & u \geq 0 \\ M \cdot \eta_b \cdot F_b(u, v) \cdot v & u < 0 \end{cases} \\ &= p_c(t)\end{aligned} \tag{18}$$

η_a and η_b are the motor and generator efficiency, respectively, which are regarded as constant. M [kg] is the total train's mass. Moreover, because the regenerative braking is assumed to be applied at all times, when $u < 0$, the energy is generated, instead of consuming. The feeder circuit is not considered.

Using this energy consumption model, when a train is coasting, the model calculates zero energy consumption because no tractive/braking effort is produced. Therefore, no energy is consumed by the train when the control command is zero.

3.4.2 New Model

In this study, a new model to evaluate train energy consumption based on the MTPA control method is proposed. In the new model, the energy consumption is calculated from the currents and voltages commands obtained from the MTPA control method. The state equation to total energy consumption $J(t)$ [J] of the train and input/output power $p(t)$ [W] is described in Eq. (19). The currents and voltages are obtained as described in Fig. 2.

$$\begin{aligned}\frac{\mathrm{d}J(t)}{\mathrm{d}t} &= \frac{3}{2}\left(v_d(t) \cdot i_d(t) + v_q(t) \cdot i_q(t)\right) \\ &= p_n(t)\end{aligned} \tag{19}$$

In an account of the notch command, the q-axis current is proportional to the notch command while the d-axis current is not affected. In this way, when the train coasts, d-axis current is still supplied to the motor, thus the power accounts for supplying d-axis current contributes to the total energy consumption.

Figure 3 shows the algorithm to solve the minimization problem based on DP as explained below. Figure 4 illustrates the forward search to construct the optimal speed profile.

1. State space of position-speed or x-v plane is configured into the grid and partitioned along time axis t. x-v plane at time point t_k is called the kth phase plane. The lattice points are $X_k = (x_i, v_j)$. Nth phase plane is the arrival time T.
2. The penalty is assigned to every lattice in the Nth phase plane by using Eq. (15). The values are called the evaluation scores.
3. The backward calculation is begun.
4. $k = \mathrm{N} - 1$ is set.
5. At the kth phase plane, Eq. (12) is solved in every lattice point with respect to all control inputs u. The energy consumption J is calculated to obtain the local evaluation function. The local evaluation function is recorded as the evaluation score of the corresponding lattice points.
6. If X_{k+1} does not exist in the lattice points in $(k + 1)$th plane, the evaluation score is calculated using bilinear interpolation.
7. The optimal control input u_{opt} is determined and recorded by evaluating the local evaluation function.
8. $k = k - 1$ is set. If $k = 0$, go to (9), else go to (5).
9. The forward search is begun.
10. $k = 1$ is set. $X_1 = (0, 0)$ is set.
11. Using the recorded optimal control input u_{opt}, X_{k+1} in $(k + 1)$th plane is found using Eq. (12).
12. If the lattice point of X_k does not exist in the phase plane, bilinear interpolation is used to acquire the optimal control input.
13. The energy consumption J is integrated and recorded.
14. $k = k + 1$ is set. If $k = N$, the search is terminated, else go to (11).

4 Results: Case Study

In this study, the optimal train speed profile for PMSM is configured using the conventional model and new model by carrying out case studies in the MATLAB environment.

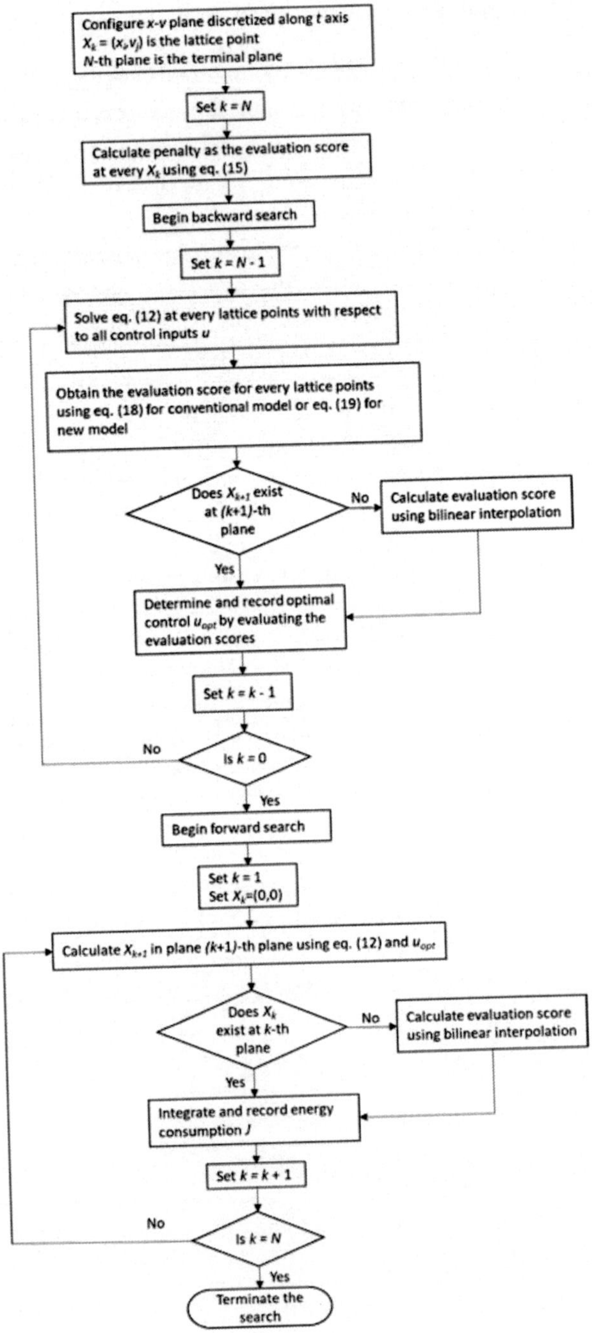

Fig. 3 The algorithm to solve the optimal train speed profile using dynamic programming

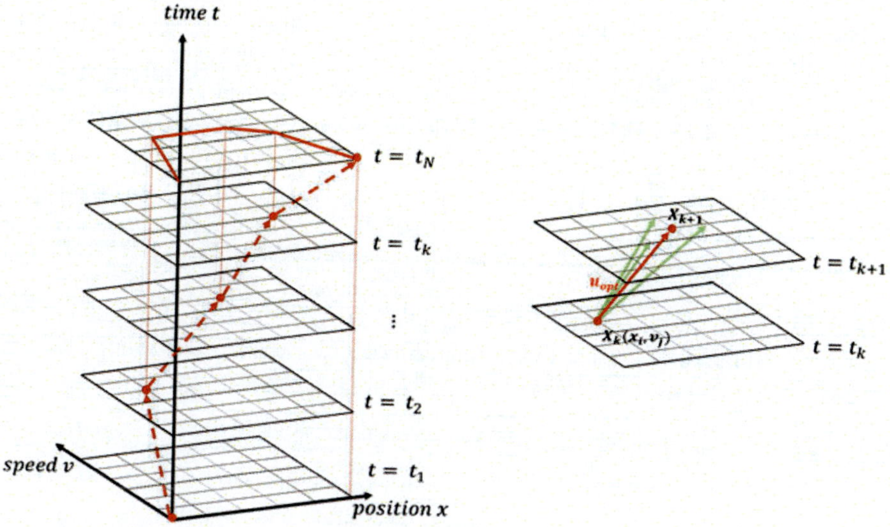

Fig. 4 Generation of train speed profile using optimal control command on grid points

4.1 Simulation Conditions

The vehicle and motor parameters are defined in Table 3. The motor parameters are based on the study by Kondo and Kondo (2003). The tractive and braking effort with its proportionality to the notch command are presented in Fig. 5. The notch command u is selected from −1, −0.9, −0.5, −0.1, 0, 0.1, 0.5, 0.9 and 1. The current commands obtained from the parameters of the motor based on MTPA are given in Fig. 6. The maximum voltage during accelerating and braking of 1550 V and 1650 V, respectively, are assumed.

The tractive resistance is defined as Eq. (20). The starting resistance is 30 N/ton, which is the force from resistance per train weight in ton.

$$R = (14.2974 + 0.1949 \cdot v)M + 0.8095 \cdot v^2 \text{ [N]} \tag{20}$$

Two cases were simulated with MATLAB. Case 1 has a track length of 2000 m and a running time of 120 s. In case 2, the track's length is 5000 m and the running time is 250 s. Case 2 is designed to have a longer track. The gradient of the track is assumed to be zero, thus, no gradient resistance was considered.

4.2 Evaluation of New Model

Before the optimal train speed profile is configured using the new model, the model is evaluated by calculating the energy consumption of a test speed profile of 2000 m. The

Table 3 Vehicle and motors specifications

Parameters	Symbol	Unit	Value
Mass	M	(ton)	353
Length	L	(m)	144
Diameter of wheel	D	(mm)	820
Gear ratio	Gr	–	7.07
Maximum speed	v_{max}	(km/h)	120
Maximum acceleration	–	(km/h/s)	3.3
Maximum deceleration	–	(km/h/s)	3.5
Number of motors[a]	n	–	16
Number of pole-pair[a]	P	–	4
Armature flux linkage[a]	ψ_m	(Wb)	0.837
Armature winding resistance[a]	R_s	(Ω)	0.1
d-axis inductance[a]	L_d	(mH)	4.923
q-axis inductance[a]	L_q	(mH)	20.68
Gear efficiency[a]	η_{gear}	–	0.98
Motor efficiency[b]	η_a	–	0.97
Generator efficiency[b]	η_b	–	0.88

[a]Used only in the new model, [b]used only in the conventional model

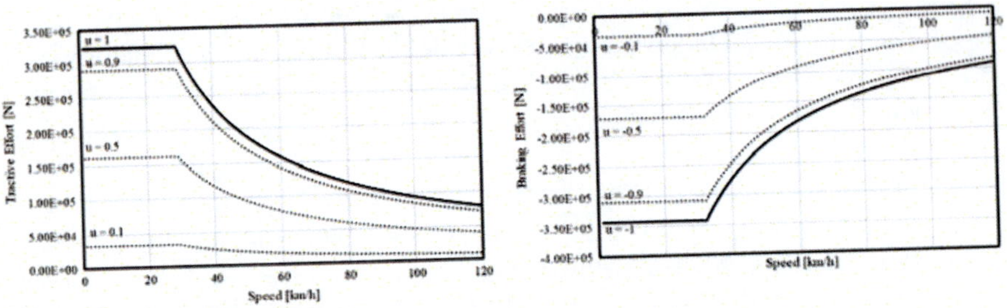

Fig. 5 The tractive and braking performance proportional to notch command

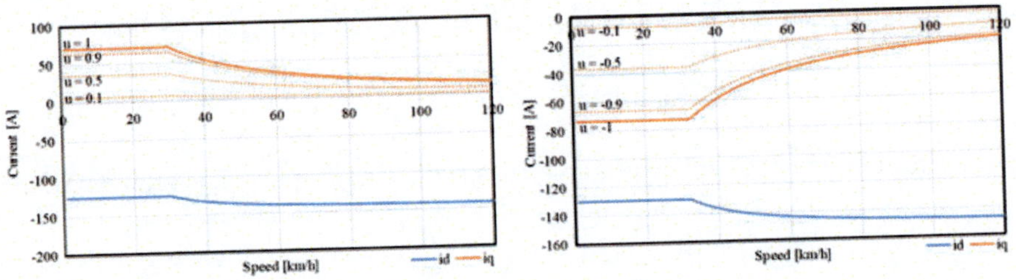

Fig. 6 d-q axis current commands and its proportionality to notch command

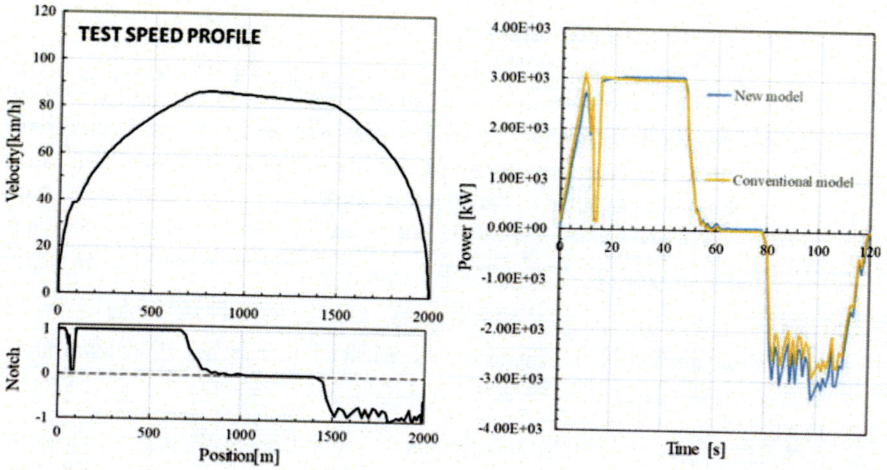

Fig. 7 Test speed profile and the power consumption calculated by the conventional model and new model

Table 4 Optimization results

		Terminal error		Energy consumption J (kWh)
		Position x(T) (m)	Velocity v(T) (m/s)	
Case 1	Conventional model	−0.06	0.00	13.7
	New model	0.09	0.00	10.7
Case 2	Conventional model	−0.07	0.00	28.4
	New model	−0.05	0.00	27.8

calculated energy by the conventional model and the new model is compared in Fig. 7. The result shows that during acceleration both models consume power almost the same. A slight difference can be noticed during coasting at around 60–80 s in which the new model calculates some power consumption while the power consumption in the conventional model is zero. The significance difference appears in the braking mode. The new model has more regenerative power than the conventional model which suggests that the regenerative efficiency might be higher than 88%, which is higher than the regenerative (or generator) efficiency that was set for the conventional model.

4.3 Simulation Results

The optimization results are presented in Table 4. The optimal train speed profile results are presented in Fig. 8. Based on the definition in Table 2, the driving modes are classified and the percentage of each driving mode is compared in Fig. 9.

In case 1, the train speed profiles generated using the new model and conventional model are almost identical. Either model uses maximum acceleration with

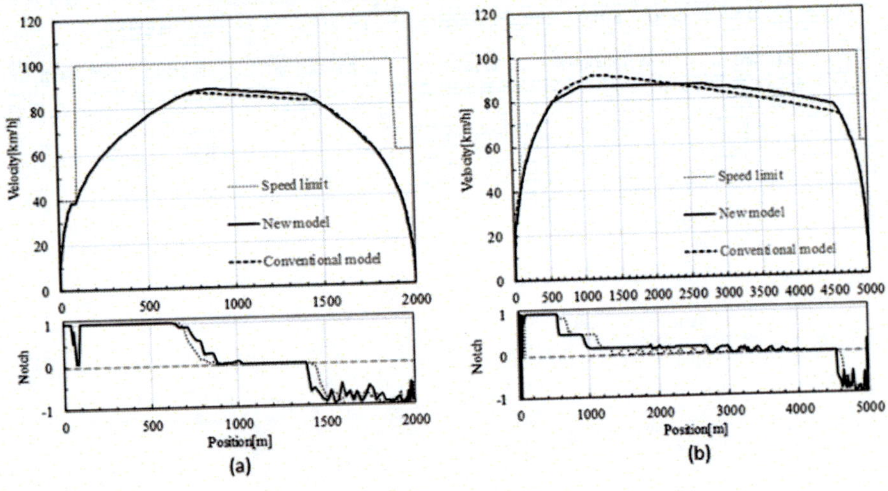

Fig. 8 The optimal train speed profile or **a** case 1 and **b** case 2

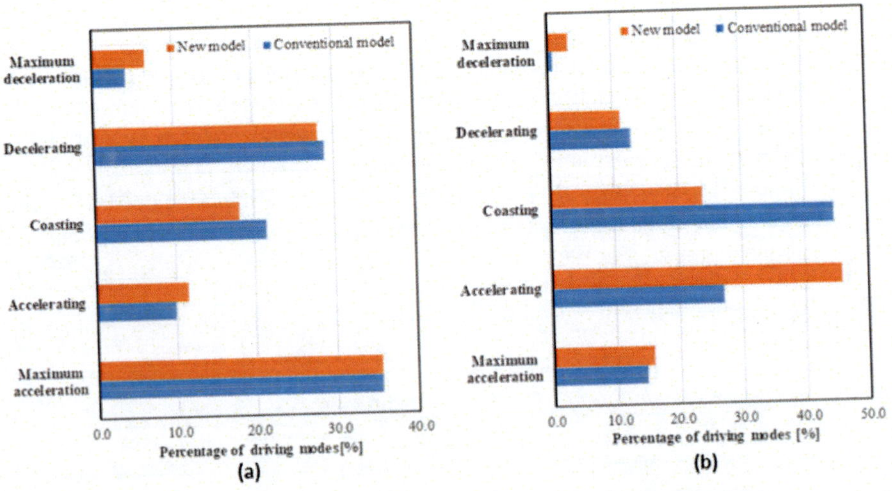

Fig. 9 The comparison of driving modes between the new model and conventional model in **a** case 1 and **b** case 2. The percentage of driving modes were calculated by the length of time for each mode

Table 5 Energy consumption during coasting and the regenerative energy

		Energy consumption (coasting) (kWh)	Regenerative energy (kWh)
Case 1	Conventional model	0.1	22.6
	New model	0.4	27.1
Case 2	Conventional model	3.3	17.7
	New model	3.3	22.5

the same proportion. The slight difference is observed in the coasting mode. The conventional model has a longer coasting period and therefore, shorter accelerating mode. However, during braking, the new model uses more maximum deceleration mode than the conventional model.

In case 2, the train runs on a long track with no gradient. However, unlike case 1, the train speed profile between the new model and the conventional model shows a major difference. For the conventional model, the typical optimal train speed profile consists of maximum acceleration-coasting-maximum deceleration are still present. Yet, the new model utilizes cruising which does not present in the conventional model.

In case 2, the new model begins with accelerating using maximum acceleration. After that, the train reduces its acceleration rate and starts holding the speed until it slowly changes to coasting at point 2800 m with some fluctuating notch. As a result, the new model coasts less than the conventional model. After coasting, the train starts to brake. Similar to the case 1, the new model uses more maximum deceleration mode to brake than the conventional model.

Table 5 shows the energy consumption during coasting and regenerative energy between the conventional model and the new model. The coasting energy in the conventional model, especially in case 2, comes from the impure coasting period. In case 2, the conventional model experiences the decrease in speed from 1000 to 3000 m but the notch command is not constantly zero.

In general, the energy consumption of the new model is generally less than the conventional model. This result corresponds to the finding in the previous section which suggests that the regenerative efficiency is higher than 88%. Utilizing more braking mode using maximum deceleration rate increases the regenerated energy, which reduces the total energy consumption.

5 Conclusions and Recommendations

In this paper, the optimal train speed profile for PMSM is obtained with the new model based on the MTPA control method and DP calculation. The new model improved the accuracy of train energy consumption estimation. As a result, the new model suggests that the regenerative efficiency (or the generator efficiency) of PMSM is higher than 88%. Moreover, the energy is consumed during coasting due to the d-axis current flowing to suppress the generated back-emf. In comparison with the conventional model which uses constants efficiency, the new model suggests that shortening coasting times leads to energy reduction. In a longer track, the mixture between cruising and coasting is used in the optimal train speed profile for PMSM. In addition to that, utilizing a maximum deceleration rate during braking is beneficial.

6 Limitations and Next Steps

In this study, DP is used to deal with the optimal speed profile problem. Even though it can handle most of the train running conditions such as speed limit, the terminal constraints can only be handled using the terminal penalty. Using the appropriate penalty constants improves the accuracy of the calculation but tuning the penalty can only be done through trial and error which is time consuming. Special tuning of penalty constants is necessary for future works. Moreover, the proposed model only considers motor dynamics. The results might be improved by including the model of converters.

References

Bellman R, Kalaba RE (1965) Dynamic programming and modern control theory, vol 81. Academic Press, New York

Ghaviha N, Bohlin M, Holmberg C, Dahlquist E, Skoglund R, Jonasson D (2017) A driver advisory system with dynamic losses for passenger electric multiple units. Transp Res Part C Emerg Technol 85:111–130

Howlett PG, Milroy IP, Pudney PJ (1994) Energy-efficient train control. Control Eng Pract 2(2):193–200

Ichikawa S, Miyatake M (2019) Energy efficient train trajectory in the railway system with moving block signaling scheme. IEEJ J Ind Appl 8(4):586–591

Kawai H, Tasaka Y (2010) Evaluation of the energy-saving performance of the PMSM drive system. In: The 2010 international power electronics conference-ECCE ASIA, June 2010. IEEE, pp 1605–1608

Ko H, Koseki T, Miyatake M (2004) Application of dynamic programming to the optimization of the running profile of a train. WIT Trans Built Environ 74

Kondo K, Kondo M (2003) Design of the parameters of permanent magnet synchronous motors for railway vehicle traction. In: The fifth international conference on power electronics and drive systems, 2003. PEDS 2003, vol 2, Nov 2003. IEEE, pp 992–997

Kondo M, Kawamura J, Terauchi N (2005) Energy consumption calculation of permanent magnet synchronous motor for railway vehicle traction using equivalent circuit. IEEJ Trans Ind Appl 125:313–320

Lu S, Hillmansen S, Ho TK, Roberts C (2013) Single-train trajectory optimization. IEEE Trans Intell Transp Syst 14(2):743–750

Matsuoka K (2007) Development trend of the permanent magnet synchronous motor for railway traction. IEEJ Trans Electr Electron Eng 2(2):154–161

Miyatake M, Ko H (2010) Optimization of train speed profile for minimum energy consumption. IEEJ Trans Electr Electron Eng 5(3):263–269

Oba N, Miyatake M (2018) Design of energy-efficient train speed profiles considering fixed-block signalling system. Electr Eng Jpn 205:26–35

Yang N, Luo G, Liu W, Wang K (2012) Interior permanent magnet synchronous motor control for electric vehicle using look-up table. In: Proceedings of the 7th international power electronics and motion control conference, vol 2, June 2012. IEEE, pp 1015–1019